ATOMIC AND MOLECULAR DATA AND THEIR APPLICATIONS

Related Titles from AIP Conference Proceedings

534 Nucleation and Atmospheric Aerosols 2000: 15th International Conference
Edited by Barbara N. Hale and Markku Kulmala, August 2000, 1-56396-958-0

525 Multiphoton Processes: ICOMP VIII: 8th International Conference
Edited by Louis F. DiMauro, Richard R. Freeman, and Kenneth Kulander, June 2000, 1-56396-946-7

506 X-Ray and Inner Shell Processes: 18th International Conference
Edited by R. W. Dunford, D. S. Gemmell, E. P. Kanter, B. Krässig, S. H. Southworth, and L. Young, February 2000, 1-56396-713-8

477 Atomic Physics 16: Sixteenth International Conference on Atomic Physics
Edited by William E. Baylis and Gordon W. F. Drake, May 1999, 1-56396-752-9

467 Spectral Line Shapes: Volume 10, 14th ICSLS
Edited by Roger M. Herman, March 1999, 1-56396-754-5

454 Resonance Ionization Spectroscopy: Ninth International Symposium
Edited by J. C. Vickerman, I. Lyon, N. P. Lockyer, and J. E. Parks, December 1998, 1-56396-810-X

434 Atomic and Molecular Data and Their Applications: ICAMDATA — First International Conference
Edited by Peter J. Mohr and Wolfgang L. Wiese, June 1998, 1-56396-751-0

ATOMIC AND MOLECULAR DATA AND THEIR APPLICATIONS

ICAMDATA—Second International Conference

Oxford, England 26–30 March 2000

EDITORS

Keith A. Berrington

Sheffield Hallam University, England

Kenneth L. Bell

The Queen's University of Belfast, Northern Ireland

AMERICAN INSTITUTE OF PHYSICS

Melville, New York, 2000

AIP CONFERENCE PROCEEDINGS ■ VOLUME 543

Editors:

Keith A. Berrington
School of Science and Mathematics
Sheffield Hallam University
Sheffield S1 1WB
ENGLAND

E-mail: k.berrington@shu.ac.uk

Kenneth L. Bell
School of Mathematics and Physics
The Queen's University of Belfast
Belfast BT7 1NN
NORTHERN IRELAND

E-mail: kl.bell@qub.ac.uk

L.C. Catalog Card No. 00-109204
ISBN 1-56396-971-8
ISSN 0094-243X
Printed in the United States of America

CONTENTS

*Italicized name indicates the author who presented the paper.

ATOMIC DATA PRODUCTION

PANEL DISCUSSIONS

*Italicized name indicates the author who presented the paper.

Preface

These proceedings contain the written versions of the invited talks presented at the second International Conference on Atomic and Molecular Data and Their Applications (ICAMDATA), which was held in Keble College, Oxford, England from March 26 through March 30, 2000. This is the second in the series of conferences, the first being held in Gaithersburg, Maryland, USA in 1997.

Eighty-four registered participants from 18 countries attended the conference. Twenty-four invited talks and about 57 contributed poster papers were presented. Representatives from six atomic and molecular data centres from around the world gave short presentations with regard to access to and content of their databases.

In broad terms, the invited presentations addressed the major areas of importance to atomic and molecular activities. In the opening talk, the topic of fundamental constants was presented. Two sessions were devoted to data needs of the main user communities, including the plasma physics (discharge, laser and fusion) communities, the atmospheric and astrophysics communities, the medical radiotherapy community and the lighting industries. Secondly, several talks reviewed the production of data in attempting to meet these needs. Both theoretical and experimental approaches were reviewed and covered a broad spectrum of topics, including electron impact with atoms and molecules, atomic structure and transition probabilities, heavy particle collisions, quantum chemistry and thermochemical data. Most presentations focussed not only on the means of production of data but also on providing some idea of the accuracy of the data produced. The third aspect of the conference examined closely the various databases around the world. It was clearly recognized that the volume of data both needed and produced required systematic and careful co-ordination, accessability and, preferably, a guide to the accuracy of the data extracted from any given database. The need for worldwide co-ordination came clearly from the discussions.

This conference series grew out of a Workshop convened at the Harvard Smithsonian Center for Astrophysics on June 14 and 15, 1996. The main objectives of the conferences, arising out of that meeting, were to bring users, producers, and database developers closer together and, in addition, to serve as a focal point for getting messages to the general community on data needs as well as on the general status of data production activities. These proceedings are a part of the attempt to meet this latter objective. Internationally recognized experts have contributed to this volume, presenting current information on data needs, production and assembly to the whole scientific community.

It is a pleasure to acknowledge the competent assistance of Tara Spencer in the preparation for the conference. The International Program Committee greatly assisted in the selection of invited speakers and other organizational matters. Finally, we gratefully acknowledge the financial support of the organisations listed below as sponsors.

Keith Berrington
Sheffield Hallam University

Kenneth Bell
Queens University of Belfast

SPONSORS

The Queen's University of Belfast:

Theoretical & Computational Physics
Research Division
(School of Mathematics & Physics)

International Union of Pure & Applied Physics

National Institute of
Standards and Technology:

Physics Laboratory
Atomic Physics Division
Standard Reference Data Program

LOCAL COMMITTEE

Keith Berrington	Sheffield Hallam University (Co-Chair)
Kenneth Bell	The Queen's University of Belfast (Co-Chair)
Keith Burnett	University of Oxford
Nigel Mason	University College, London
Hugh Summers	Strathclyde University
Bill Graham	The Queen's University of Belfast
Jim Lang	Rutherford Appleton Laboratory
Derrick Crothers	The Queen's University of Belfast

INTERNATIONAL PROGRAM COMMITTEE

A Dalgarno	USA
R Janev	Austria
D Schultz	USA
K Berrington	UK
L Brown	USA
M Capielli	Italy
G Lister	USA
H Mason	UK
K Niemax	Germany
L Presnyakov	Russia
E Roueff	France
H Tawara	Japan
W Wiese	USA

INTERNATIONAL ADVISORY BOARD

N Bardsley	USA
P Burke	UK
R Crompton	Australia
J L Delcroix	France
G Dunn	USA
F Gianturco	Italy
R Hulse	USA
J M Li	PR China
I Martinson	Sweden
C Mendoza	Venezuala
J Rumble	USA
E Salzborn	Germany
M Seaton	UK
I Sobelman	Russia
K Takayangi	Japan

Laboratory Data Needs and Applications

CODATA Recommended Values of the Fundamental Constants*

Peter J. Mohr and Barry N. Taylor

National Institute of Standards and Technology
Gaithersburg, MD 20899-8401

Abstract. A review is given of the latest Committee on Data for Science and Technology (CODATA) adjustment of the values of the fundamental constants. The new set of constants, referred to as the 1998 values, replaces the values recommended for international use by CODATA in 1986.

The values of the constants, and particularly the Rydberg constant, are of relevance to the calculation of precise atomic spectra. The standard uncertainty (estimated standard deviation) of the new recommended value of the Rydberg constant, which is based on precision frequency metrology and a detailed analysis of the theory, is approximately 1/160 times the uncertainty of the 1986 value.

The new set of recommended values as well as a searchable bibliographic database that gives citations to the relevant literature is available on the World Wide Web at physics.nist.gov/constants and physics.nist.gov/constantsbib, respectively.

INTRODUCTION

A new constants adjustment, based on data available through 31 December 1998, determines "best" values of the fundamental physical constants, such as the fine structure constant, Rydberg constant, Avogadro constant, Planck constant, electron mass, muon mass, and many others. The best values are those which provide the most consistency between the results of the critical experiments and the corresponding predictions of theory, particularly condensed matter theory and quantum electrodynamics (QED).

*) Authors' email addresses: mohr@nist.gov; barry.taylor@nist.gov.

CP543, *Atomic and Molecular Data and Their Applications,* edited by K. A. Berrington and K. L. Bell
2000 American Institute of Physics 1-56396-971-8

PUBLICATIONS AND WEB SITE

The details and results of the least-squares adjustment are published as an article entitled **CODATA Recommended Values of the Fundamental Physical Constants: 1998**, by Peter J. Mohr and Barry N. Taylor, in the *Journal of Physical and Chemical Reference Data* **28**, 1713 (1999) and in the *Reviews of Modern Physics* **72**, 351 (2000). The 1998 CODATA recommended values of the constants are available on the World Wide Web at physics.nist.gov/constants.

CODATA

Established in 1966, the Committee on Data for Science and Technology (CODATA) is an interdisciplinary scientific committee of the International Council for Science (ICSU), which works to improve the quality, reliability, management, and accessibility of data of importance to all fields of science and technology.

The CODATA Task Group on Fundamental Constants was established in 1969 to periodically provide the scientific and technological communities with a self-consistent set of internationally recommended values of the basic constants and conversion factors of physics and chemistry based on all relevant data available at a given point in time.

At the time of completion of the 1998 adjustment, the Task Group, which sanctions the data selection and adjustment methodology, consisted of the following members:

F. Cabiati, Istituto Elettrotecnico Nazionale "Galileo Ferraris," Italy;

E. R. Cohen, Science Center, Rockwell International (retired), USA;

T. Endo, Electrotechnical Laboratory, Japan;

R. Liu, National Institute of Metrology (retired), China (Peoples Republic of);

B. A. Mamyrin, A. F. Ioffe Physical-Technical Institute, Russian Federation;

P. J. Mohr, National Institute of Standards and Technology, USA;

F. Nez, Laboratoire Kastler-Brossel, France;

B. W. Petley, National Physical Laboratory, United Kingdom;

T. J. Quinn, Bureau International des Poids et Mesures;

B. N. Taylor, National Institute of Standards and Technology, USA;

V. S. Tuninsky, D. I. Mendeleyev All-Russian Research Institute for Metrology, Russian Federation;

W. Wöger, Physikalisch-Technische Bundesanstalt, Germany;

B. M. Wood, National Research Council, Canada.

BACKGROUND

The first comprehensive adjustment of the values of the constants was published by Raymond T. Birge in 1929. Major adjustments done in the past half century, which includes those sponsored by the National Research Council/National Academy of Sciences, USA (NRC/NAS) and by CODATA, are the following:

Bearden and Watts (1951) [1–3]

NRC/NAS: DuMond and Cohen (1951) [4–6];

Bearden and Thomsen (1957) [7];

NRC/NAS: Cohen and DuMond (1965) [8];

Taylor, Parker, and Langenberg (1969) [9];

CODATA: Cohen and Taylor (1973) [10];

CODATA: Cohen and Taylor (1987) [11].

Thus, the 1998 adjustment is the latest in a series of efforts that began some 70 years ago.

METHOD OF LEAST SQUARES

The method of least squares, as it is applied to the adjustment of the values of the fundamental constants, can be summarized as follows. Relevant results of measurements, or in some cases calculations, are termed *observational data* (or *input data*) and denoted by $q_1,\ q_2, \ldots, q_N$. The subset of the fundamental constants that are adjusted are viewed as *unknowns* (or *variables*) and denoted by $z_1,\ z_2, \ldots, z_M$; $(M \leq N)$. The observational data and the unknowns are connected through theoretical expressions called *observational equations* and expressed as the relations

$$q_i \doteq f_i(z) \equiv f_i(z_1, z_2, \ldots, z_M) \ . \tag{1}$$

Here, the symbol $\doteq$ is used, because in general, the observational equations are overdetermined and do not express equalities.

The observational equations are non-linear, but we take advantage of the fact that the constants are already known precisely and only need small corrections by making a linear approximation about starting values s for z:

$$q_i \doteq f_i(s) + \sum_{j=1}^{M} \frac{\partial f_i(s)}{\partial s_j}(z_j - s_j) + \cdots \tag{2}$$

or

$$y_i \doteq \sum_{j=1}^{M} A_{ij} x_j + \cdots \ , \tag{3}$$

where $y_i = q_i - f_i(s)$, $x_j = z_j - s_j$, and $A_{ij} = \dfrac{\partial f_i(s)}{\partial s_j}$.

Equation (3) can be written compactly in terms of matrices as

$$Y \doteq AX \tag{4}$$

with the uncertainties (estimated standard deviations) of the observational data expressed in terms of the *covariance matrix*:

$$V = \mathrm{cov}(Y) \ . \tag{5}$$

The least-squares adjustment is based on the solution $\hat{X}$ for X that minimizes

$$(Y - AX)^{\top} V^{-1} (Y - AX) \ . \tag{6}$$

Note that if the observations Y are uncorrelated, this reduces to the simple weighted sum of differences given by

$$(Y - AX)^{\top} V^{-1} (Y - AX) = \sum_{i=1}^{N} \frac{(Y - AX)_i^2}{(\delta Y)_i^2} \ , \tag{7}$$

where $(\delta Y)_i^2 = V_{ii}$ is the square of the standard uncertainty associated with the input datum q_i.

This system has the solution [12]

$$\hat{X} = (A^{\top} V^{-1} A)^{-1} A^{\top} V^{-1} Y \tag{8}$$

$$\mathrm{cov}(\hat{X}) = (A^{\top} V^{-1} A)^{-1} \ , \tag{9}$$

and so the constants and their covariance matrix are given by

$$\hat{Z} = S + \hat{X} \tag{10}$$

$$\mathrm{cov}(\hat{Z}) = \mathrm{cov}(\hat{X}) \ , \tag{11}$$

and, as a corollary, the best estimates of the measured quantities are given by

$$\hat{Y} = A\hat{X} \tag{12}$$

$$\hat{q}_i = f_i(s) + \hat{y}_i \ . \tag{13}$$

VARIABLES IN THE LSA

The adjusted constants, or variables z_j, of the 1998 least-squares adjustment are given in Table 1. The last five entries termed "corrections" are correction terms that are added to the theoretical expressions for the corresponding quantities (hydrogen and deuterium energy levels, anomalous magnetic moment of the electron and muon, and muonium hyperfine splitting) to take into account the uncertainty of the theory. There are a total of 25 correction terms for the hydrogen and deuterium energy levels. In the table, the term "Si crystal X" refers to a specific lattice spacing for any of a number of different Si crystals used in various experiments that provide information on the constants.

Obviously, the list of constants in Table 1 is not complete. However, it is a mutually independent subset of constants from which the other constants are calculated.

TABLE 1. Variables of the 1998 least-squares adjustment.

Variable name	Symbol
electron relative atomic mass	$A_r(\mathrm{e})$
proton relative atomic mass	$A_r(\mathrm{p})$
neutron relative atomic mass	$A_r(\mathrm{n})$
deuteron relative atomic mass	$A_r(\mathrm{d})$
helion relative atomic mass	$A_r(\mathrm{h})$
alpha particle relative atomic mass	$A_r(\alpha)$
electron-muon mass ratio	m_e/m_μ
fine-structure constant	α
Planck constant	h
Rydberg constant	R_∞
proton rms charge radius	R_p
deuteron rms charge radius	R_d
molar gas constant	R
Newtonian constant of gravitation	G
electron-proton magnetic moment ratio	$\mu_\mathrm{e}/\mu_\mathrm{p}$
deuteron-electron magnetic moment ratio	$\mu_\mathrm{d}/\mu_\mathrm{e}$
electron to shielded proton magnetic moment ratio	$\mu_\mathrm{e}/\mu'_\mathrm{p}$
shielded helion to shielded proton magnetic moment ratio	$\mu'_\mathrm{h}/\mu'_\mathrm{p}$
neutron to shielded proton magnetic moment ratio	$\mu_\mathrm{n}/\mu'_\mathrm{p}$
{220} lattice spacing of Si crystal X	$d_{220}(\mathrm{X})$
Cu x unit	$\mathrm{xu}(\mathrm{Cu\,K}\alpha_1)$
Mo x unit	$\mathrm{xu}(\mathrm{Mo\,K}\alpha_1)$
Angstrom star	Å*
corrections to hydrogen level theory	$\delta_\mathrm{H}(n\mathrm{L}_j)$
corrections to deuterium level theory	$\delta_\mathrm{D}(n\mathrm{L}_j)$
correction to $a_\mathrm{e}(\mathrm{th})$	δ_e
correction to $a_\mu(\mathrm{th})$	δ_μ
correction to $\Delta\nu_\mathrm{Mu}(\mathrm{th})$	δ_Mu

The choice of constants for this subset is not unique, but as long as they form an independent set, the final values of all the constants will be the same.

For example, to calculate the elementary charge e we take advantage of its relation to the fine structure constant α and the Planck constant h which do appear in Table 1:

$$\alpha = \frac{e^2}{4\pi\epsilon_0 \hbar c} \quad \Rightarrow \quad e = \sqrt{2\epsilon_0 hc\alpha} \; . \tag{14}$$

Similarly, for the electron mass we have

$$R_\infty = \frac{\alpha^2 m_e c}{2h} \quad \Rightarrow \quad m_e = \frac{2hR_\infty}{\alpha^2 c} \; , \tag{15}$$

and for the Avogadro constant

$$A_r(e) = \frac{m_e}{m_u} = \frac{N_A m_e}{10^{-3}\ \text{kg/mol}} \; , \tag{16}$$

$$\Rightarrow \quad N_A = A_r(e)\frac{1\ \text{kg/mol}}{10^3 m_e} \tag{17}$$

$$= A_r(e)\,\frac{1\ \text{kg/mol}}{2000}\,\frac{\alpha^2 c}{hR_\infty} \; , \tag{18}$$

where m_u is $1/12$ times the mass of a ^{12}C atom. More than 300 constants obtained either directly through the least-squares adjustment or calculated this way are included in the 1998 CODATA set of recommended values.

OBSERVATIONAL EQUATIONS

Some of the observational equations, represented by Eq. (1), employed in the 1998 least-squares adjustment are given in Table 2. In each observational equation, the symbol on the left-hand side represents one of the 93 measured, or in some cases calculated, input data of the adjustment. The right-hand side expresses that quantity as a function of the constants that are taken as variables of the adjustment. (This distinction explains the appearance of equations such as $\delta_e \doteq \delta_e$, which expresses the relation between an input datum and the corresponding adjusted constant, even though their values will not be equal in general.)

For example, in the first equation in Table 2, the quantity $A_r(^1H)$ represents the measured value of the relative atomic mass of the hydrogen atom, *i.e.*, the ratio of its mass to m_u. The right-hand side is the sum of the relative atomic masses of the proton and electron, both of which are variables of the adjustment, minus the binding energy equivalent mass in the same units, which is relatively small and sufficiently well known to be taken as exact.

In the remaining sections, the observational equations associated with the anomalous magnetic moment of the electron (fourth and fifth) and those associated with transition frequencies in hydrogen (the last three) are discussed in more detail.

TABLE 2. Some observational equations used in the 1998 adjustment.

$$A_{\rm r}(^1{\rm H}) \doteq A_{\rm r}({\rm p}) + A_{\rm r}({\rm e}) - E_{\rm b}(^1{\rm H})/m_{\rm u}c^2$$

$$\frac{\lambda_{\rm meas}}{d_{220}({\rm ILL})} \doteq \frac{\alpha^2 A_{\rm r}({\rm e})}{R_\infty d_{220}({\rm ILL})} \frac{A_{\rm r}({\rm n}) + A_{\rm r}({\rm p})}{[A_{\rm r}({\rm n}) + A_{\rm r}({\rm p})]^2 - A_{\rm r}^2({\rm d})}$$

$$\frac{m_{\rm e}}{m(^{12}{\rm C}^{6+})} \doteq \frac{6A_{\rm r}({\rm e})}{12 - 6A_{\rm r}({\rm e}) + E_{\rm b}(^{12}{\rm C})/m_{\rm u}c^2}$$

$$a_{\rm e} \doteq a_{\rm e}(\alpha, \delta_{\rm e})$$

$$\delta_{\rm e} \doteq \delta_{\rm e}$$

$$\frac{\mu_{\rm e^-}({\rm H})}{\mu_{\rm p}({\rm H})} \doteq \frac{g_{\rm e^-}({\rm H})}{g_{\rm e^-}} \left(\frac{g_{\rm p}({\rm H})}{g_{\rm p}}\right)^{-1} \frac{\mu_{\rm e^-}}{\mu_{\rm p}}$$

$$\nu(f_{\rm p}) \doteq \nu\left(f_{\rm p}; R_\infty, \alpha, \frac{m_{\rm e}}{m_\mu}, \frac{\mu_{\rm e^-}}{\mu_{\rm p}}, \delta_{\rm e}, \delta_\mu, \delta_{\rm Mu}\right)$$

$$\frac{\mu_{\mu^+}}{\mu_{\rm p}} \doteq -\frac{1 + a_\mu(\alpha, \delta_\mu)}{1 + a_{\rm e}(\alpha, \delta_{\rm e})} \frac{m_{\rm e}}{m_\mu} \frac{\mu_{\rm e^-}}{\mu_{\rm p}}$$

$$\Delta\nu_{\rm Mu} \doteq \Delta\nu_{\rm Mu}\left(R_\infty, \alpha, \frac{m_{\rm e}}{m_\mu}, \delta_\mu, \delta_{\rm Mu}\right)$$

$$\delta_{\rm Mu} \doteq \delta_{\rm Mu}$$

$$K_{\rm J} \doteq \left(\frac{8\alpha}{\mu_0 ch}\right)^{1/2}$$

$$R_{\rm K} \doteq \frac{\mu_0 c}{2\alpha}$$

$$K_{\rm J}^2 R_{\rm K} \doteq \frac{4}{h}$$

$$\frac{h}{m_{\rm n} d_{220}({\rm W04})} \doteq \frac{cA_{\rm r}({\rm e})\alpha^2}{2R_\infty A_{\rm r}({\rm n}) d_{220}({\rm W04})}$$

$$\frac{d_{220}({\rm X}) - d_{220}({\rm Y})}{d_{220}({\rm Y})} \doteq \frac{d_{220}({\rm X}) - d_{220}({\rm Y})}{d_{220}({\rm Y})}$$

$$d_{220}({\rm X}) \doteq d_{220}({\rm X})$$

$$\frac{\lambda({\rm W\,K}\alpha_1)}{d_{220}({\rm N})} \doteq \frac{0.209\,010\,0\ \text{Å}^*}{d_{220}({\rm N})}$$

$$\nu_{\rm H}(n_1{\rm L}_{1j_1} - n_2{\rm L}_{2j_2}) \doteq \Big[E_{\rm H}\left(n_2{\rm L}_{2j_2}; R_\infty, \alpha, A_{\rm r}({\rm e}), A_{\rm r}({\rm p}), R_{\rm p}, \delta_{\rm H}(n_2{\rm L}_{2j_2})\right)$$
$$- E_{\rm H}\left(n_1{\rm L}_{1j_1}; R_\infty, \alpha, A_{\rm r}({\rm e}), A_{\rm r}({\rm p}), R_{\rm p}, \delta_{\rm H}(n_1{\rm L}_{1j_1})\right)\Big]\Big/h$$

$$\delta_{\rm H}(n{\rm L}_j) \doteq \delta_{\rm H}(n{\rm L}_j)$$

$$R_{\rm p} \doteq R_{\rm p}$$

ELECTRON MAGNETIC MOMENT ANOMALY

It is well known that the magnetic moment of the electron is not equal to the value predicted by the Dirac equation $g_e(\text{Dirac}) = -2$, and the deviation from that value is given in terms of the electron magnetic moment anomaly a_e by

$$g_e = -2(1 + a_e) .$$

The anomaly has been measured for the electron and positron at the University of Washington with the results [13]

$$\begin{aligned} a_{e^-}(\text{exp}) &= 1\,159\,652\,188.4(4.3) \times 10^{-12} \\ a_{e^+}(\text{exp}) &= 1\,159\,652\,187.9(4.3) \times 10^{-12} . \end{aligned} \tag{19}$$

The average of these values (assuming as we do that the value of the anomaly is CPT invariant) yields the input datum corresponding to a_e on the left-hand side of the fourth equation in Table 2. The theoretical expression on the right-hand side of that equation is

$$a_e(\alpha, \delta_e) = C_e^{(2)}\left(\frac{\alpha}{\pi}\right) + C_e^{(4)}\left(\frac{\alpha}{\pi}\right)^2 + C_e^{(6)}\left(\frac{\alpha}{\pi}\right)^3 \tag{20}$$

$$+ C_e^{(8)}\left(\frac{\alpha}{\pi}\right)^4 + a_e(\text{had}) + a_e(\text{weak}) \tag{21}$$

$$+ \delta_e , \tag{22}$$

which gives the anomaly as a function of the variables α and δ_e. The coefficients $C_e^{(n)}$, the strong interaction correction $a_e(\text{had})$, and the weak interaction correction $a_e(\text{weak})$ are calculated from theory. The largest uncertainty in the theory arises from numerical integration uncertainty in the massive calculation of $C_e^{(8)}$ [14]. The total uncertainty of the theory is estimated to be 1.1×10^{-12}, so the input datum corresponding to δ_e is 0 with that uncertainty. Hence, the observational equation for δ_e is

$$0.0(1.1) \times 10^{-12} \doteq \delta_e . \tag{23}$$

If δ_e is the kth item of input data, then $q_k = 0$ and $V_{kk} = (1.1 \times 10^{-12})^2$.

The electron anomalous magnetic moment data provide the most influential information on the value of the fine-structure constant α. When considered alone they yield

$$\alpha^{-1} = 137.035\,999\,58(52) . \tag{24}$$

Of course, in the final adjustment, all sources of information on α contribute to the 1998 recommended value:

$$\alpha^{-1} = 137.035\,999\,76(50) . \tag{25}$$

This value has an uncertainty that is about 1/12 times the uncertainty of the 1986 recommended value, and differs from the earlier value by about 1.7 times the uncertainty of the earlier value. The adjusted value of δ_e is $0.1(1.1) \times 10^{-12}$, very nearly its input value in Eq. (23).

RYDBERG CONSTANT

The Rydberg constant is determined primarily by comparison of theory and experiment for energy levels in hydrogen and deuterium. For example, the observational equation corresponding to the the 1S – 2S transition frequency of hydrogen is given approximately by the expression

$$\nu_{\rm H}(1{\rm S}_{1/2} - 2{\rm S}_{1/2}) \doteq \frac{3}{4} R_\infty c \left[1 - \frac{m_{\rm e}}{m_{\rm p}} + \frac{11}{48}\alpha^2 - \frac{28}{9}\frac{\alpha^3}{\pi} \ln \alpha^{-2} - \frac{14}{9}\left(\frac{\alpha R_{\rm p}}{\lambda\!\!\!^{-}_{\rm C}}\right)^2 + \cdots \right], \qquad (26)$$

where the input datum for this transition is [15]

$$\nu_{\rm H}(1{\rm S}_{1/2} - 2{\rm S}_{1/2}) = 2\,466\,061\,413\,187.34(84)\ {\rm kHz}\ , \qquad (27)$$

which has a relative uncertainty of 3.4×10^{-13}. The expression on the right-hand side of Eq. (26) is approximate and only indicates the leading term of each of several contributions. In particular, the four terms beyond the "1" correspond to contributions from reduced mass, relativistic, radiative, and finite nuclear size effects, respectively.

The complete set of hydrogen and deuterium transition frequencies that are included in the 1998 adjustment are given in Table 3. The relevant theory is considered in more detail in the following.

Theory relevant to the Rydberg constant

Effects that have been considered to compile expressions for the energy levels in hydrogen and deuterium are

Dirac eigenvalue,
relativistic recoil,
nuclear polarization,
self energy,
vacuum polarization,
two-photon corrections,*
three-photon corrections,
finite nuclear size,*
nuclear size correction to self energy and vacuum polarization,
radiative-recoil corrections, and
nucleus self energy.

Not all of these effects make a significant contribution to the energy levels. However, the two that make the largest contributions to the uncertainty have been marked with an asterisk (*) and are considered separately below.

TABLE 3. Transition frequencies in hydrogen (H) and deuterium (D) used in the 1998 adjustment.

Transition frequency	Lab[a]-year
$\nu_{\rm H}(1{\rm S}_{1/2} - 2{\rm S}_{1/2}) = 2\,466\,061\,413\,187.34(84)$ kHz	MPQ-97 [15]
$\nu_{\rm H}(2{\rm S}_{1/2} - 8{\rm S}_{1/2}) = 770\,649\,350\,012.1(8.6)$ kHz	LK/LP-97 [16]
$\nu_{\rm H}(2{\rm S}_{1/2} - 8{\rm D}_{3/2}) = 770\,649\,504\,450.0(8.3)$ kHz	LK/LP-97 [16]
$\nu_{\rm H}(2{\rm S}_{1/2} - 8{\rm D}_{5/2}) = 770\,649\,561\,584.2(6.4)$ kHz	LK/LP-97 [16]
$\nu_{\rm H}(2{\rm S}_{1/2} - 12{\rm D}_{3/2}) = 799\,191\,710\,472.7(9.4)$ kHz	LK/LP-99 [17]
$\nu_{\rm H}(2{\rm S}_{1/2} - 12{\rm D}_{5/2}) = 799\,191\,727\,403.7(7.0)$ kHz	LK/LP-99 [17]
$\nu_{\rm H}(2{\rm S}_{1/2} - 4{\rm S}_{1/2}) - \frac{1}{4}\nu_{\rm H}(1{\rm S}_{1/2} - 2{\rm S}_{1/2}) = 4\,797\,338(10)$ kHz	MPQ-95 [18]
$\nu_{\rm H}(2{\rm S}_{1/2} - 4{\rm D}_{5/2}) - \frac{1}{4}\nu_{\rm H}(1{\rm S}_{1/2} - 2{\rm S}_{1/2}) = 6\,490\,144(24)$ kHz	MPQ-95 [18]
$\nu_{\rm H}(2{\rm S}_{1/2} - 6{\rm S}_{1/2}) - \frac{1}{4}\nu_{\rm H}(1{\rm S}_{1/2} - 3{\rm S}_{1/2}) = 4\,197\,604(21)$ kHz	LKB-96 [19]
$\nu_{\rm H}(2{\rm S}_{1/2} - 6{\rm D}_{5/2}) - \frac{1}{4}\nu_{\rm H}(1{\rm S}_{1/2} - 3{\rm S}_{1/2}) = 4\,699\,099(10)$ kHz	LKB-96 [19]
$\nu_{\rm H}(2{\rm S}_{1/2} - 4{\rm P}_{1/2}) - \frac{1}{4}\nu_{\rm H}(1{\rm S}_{1/2} - 2{\rm S}_{1/2}) = 4\,664\,269(15)$ kHz	Yale-95 [20]
$\nu_{\rm H}(2{\rm S}_{1/2} - 4{\rm P}_{3/2}) - \frac{1}{4}\nu_{\rm H}(1{\rm S}_{1/2} - 2{\rm S}_{1/2}) = 6\,035\,373(10)$ kHz	Yale-95 [20]
$\nu_{\rm H}(2{\rm S}_{1/2} - 2{\rm P}_{3/2}) = 9\,911\,200(12)$ kHz	Harvard-94 [21]
$\nu_{\rm H}(2{\rm P}_{1/2} - 2{\rm S}_{1/2}) = 1\,057\,845.0(9.0)$ kHz	Harvard-86 [22]
$\nu_{\rm H}(2{\rm P}_{1/2} - 2{\rm S}_{1/2}) = 1\,057\,862(20)$ kHz	Sussex-79 [23]
$\nu_{\rm D}(2{\rm S}_{1/2} - 8{\rm S}_{1/2}) = 770\,859\,041\,245.7(6.9)$ kHz	LK/LP-97 [16]
$\nu_{\rm D}(2{\rm S}_{1/2} - 8{\rm D}_{3/2}) = 770\,859\,195\,701.8(6.3)$ kHz	LK/LP-97 [16]
$\nu_{\rm D}(2{\rm S}_{1/2} - 8{\rm D}_{5/2}) = 770\,859\,252\,849.5(5.9)$ kHz	LK/LP-97 [16]
$\nu_{\rm D}(2{\rm S}_{1/2} - 12{\rm D}_{3/2}) = 799\,409\,168\,038.0(8.6)$ kHz	LK/LP-99 [17]
$\nu_{\rm D}(2{\rm S}_{1/2} - 12{\rm D}_{5/2}) = 799\,409\,184\,966.8(6.8)$ kHz	LK/LP-99 [17]
$\nu_{\rm D}(2{\rm S}_{1/2} - 4{\rm S}_{1/2}) - \frac{1}{4}\nu_{\rm D}(1{\rm S}_{1/2} - 2{\rm S}_{1/2}) = 4\,801\,693(20)$ kHz	MPQ-95 [18]
$\nu_{\rm D}(2{\rm S}_{1/2} - 4{\rm D}_{5/2}) - \frac{1}{4}\nu_{\rm D}(1{\rm S}_{1/2} - 2{\rm S}_{1/2}) = 6\,494\,841(41)$ kHz	MPQ-95 [18]
$\nu_{\rm D}(1{\rm S}_{1/2} - 2{\rm S}_{1/2}) - \nu_{\rm H}(1{\rm S}_{1/2} - 2{\rm S}_{1/2}) = 670\,994\,334.64(15)$ kHz	MPQ-98 [24]

[a] MPQ: Max-Planck-Institut für Quantenoptik, Garching.
LKB: Laboratoire Kastler-Brossel, Paris.
LPTF: Laboratoire Primaire du Temps et des Fréquences, Paris.

Two-Photon Corrections

The contribution to hydrogen and deuterium energy levels from radiative corrections involving two virtual photons can be written as

$$E^{(4)} = \left(\frac{\alpha}{\pi}\right)^2 \frac{(Z\alpha)^4}{n^3} m_e c^2 F^{(4)}(Z\alpha) \,, \tag{28}$$

where the leading terms in a power series expansion of $F^{(4)}(Z\alpha)$ are given by

$$F^{(4)}(Z\alpha) = B_{40} + B_{50}\,(Z\alpha) + \cdots \,. \tag{29}$$

The first term in this series has been known for decades, but the second has been calculated only relatively recently. They are given by

$$\begin{aligned} B_{40} &= \left[2\pi^2 \ln 2 - \tfrac{49}{108}\pi^2 - \tfrac{6131}{1296} - 3\zeta(3)\right]\delta_{l0} \\ &\quad + \left[\tfrac{1}{2}\pi^2 \ln 2 - \tfrac{1}{12}\pi^2 - \tfrac{197}{144} - \tfrac{3}{4}\zeta(3)\right]\frac{1}{\kappa(2l+1)} \end{aligned} \tag{30}$$

and [25–28]

$$B_{50} = -21.5561(31)\delta_{l0} \,. \tag{31}$$

Subsequent terms in this series are only partially known, and the associated uncertainty is one of the largest in the theory of the energy levels. However, it is expected that as for the first two terms, the dominant contribution of the higher-order terms will be proportional to $1/n^3$. Hence, the theoretically estimated input data for the correction terms $\delta_X(nL_j)$, $X = \mathrm{H}, \mathrm{D}$, are highly correlated, especially for S states, with some correlation coefficients reaching 0.999.

Finite Nuclear Size

The effect of the finite size of the proton or deuteron on the H or D energy levels is

$$E_{NS} = \frac{2}{3}\frac{(Z\alpha)^2}{n^3} m_e c^2 \left(\frac{Z\alpha R_N}{\lambda\!\!\!^{-}_C}\right)^2 \delta_{l0} \,, \tag{32}$$

where R_N is the root-mean-square (rms) charge radius of the nucleus and $\lambda\!\!\!^{-}_C$ is the Compton wavelength of the electron divided by 2π.

Potential input data for the charge radius of the proton, from the results of electron scattering experiments, are

$R_p = 0.847(8)$ fm [29]
$R_p = 0.862(12)$ fm [30]
$R_p = 0.805(11)$ fm [31].

In the 1998 adjustment, the initial input datum for R_{p} was the simple average of the first two of these values, which are based on considerably more data than the third. It is clear from Eq. (32) that the uncertainty in the contribution to the energy levels arising from uncertainty in R_{p} is proportional to $1/n^3$ as in the case of the two-photon corrections.

Uncertainty

It is evident that the types of uncertainty encountered in the previous two sections are of a different nature. The uncertainty arising from uncertainty in the radius of the proton originates from an uncertainty in the value of the input datum, whereas the uncertainty in the contribution of the two-photon radiative correction arises from the uncertainty of the theoretical expression for the energy level. However, the device of introducing the additive correction terms δ, such as $\delta_{\mathrm{H}}(n\mathrm{L}_j)$, shifts the source of the uncertainty from the theoretical expression to the uncertainty in the value of the input datum for the corresponding δ, which puts it on the same footing as the uncertainty in the proton radius. Since the covariances of the input data enter as the off-diagonal matrix elements of the covariance matrix V in Eq. (5), the strong correlations among the various δ's are taken into account by those matrix elements.

Result of LSA for the Rydberg Constant

The values of the Rydberg constant and the proton and deuteron radii resulting from two least-squares adjustments (LSAs) are shown in Table 4. In the first case (A), the data in Table 3, values for the proton and deuteron charge radii, as well as the rest of the input data of the 1998 least-squares adjustment are included. However, in this adjustment the value of the normalized residual

$$r_i = \frac{x_i - \hat{x}}{u(x_i)} \tag{33}$$

of each $\delta_{\mathrm{X}}(n\mathrm{S}_{1/2})$, $n = 1, 2, 3, 4, 6, 8$, is in the range $-1.410 < r_i < -1.406$. [In Eq. (33), $u(x_i)$ is the uncertainty of x_i.] This indicates a deviation between theory and experiment corresponding to about $126/n^3$ kHz for $n\mathrm{S}_{1/2}$ states. The most likely sources for this difference are a deviation of the value of the proton charge radius predicted by the spectroscopic data from the values deduced from scattering experiments, an uncalculated contribution to the energy levels from the two-photon QED correction that exceeds the estimated uncertainty for this term, or a combination of these.

In the second adjustment (B) shown in Table 4, no input datum was included for either the proton radius or deuteron radius. As a result, the value of the Rydberg constant shifted by more than one third of its uncertainty, even though its

TABLE 4. Two LSA results for R_∞, $R_{\rm p}$, and $R_{\rm d}$.

Case	$R_\infty/{\rm m}^{-1}$	$R_{\rm p}/{\rm fm}$	$R_{\rm d}/{\rm fm}$
A	10 973 731.568 521(81)	0.859(10)	2.1331(42)
B	10 973 731.568 549(83)	0.907(32)	2.153(14)

uncertainty increased only slightly. In this case, the normalized residuals satisfied $|r_i| < 0.0001$. Because this latter adjustment, with the proton and deuteron radii taken as free parameters, gives greater consistency between theory and experiment, it was taken as the final adjustment for the 1998 recommended values.

Hence, the 1998 recommended value for the Rydberg constant is

$$R_\infty = 10\,973\,731.568\,549(83)\ {\rm m}^{-1}\ , \tag{34}$$

where the relative uncertainty is 7.6×10^{-12}. This uncertainty is about 1/160 times the uncertainty of the 1986 recommended value. The 1998 recommended value for R_∞ differs from the 1986 recommended value by about 2.7 times the uncertainty of the latter value, due to an incorrect experimental result that strongly influenced the 1986 value.

Transition frequencies between levels with $n = 2$

As a further check on the consistency of the results of the least-squares adjustment and the transition frequency data in hydrogen, a reduced least-squares adjustment was carried out with input data for the proton and deuteron charge radii omitted and the measured data for the $2{\rm P}_{1/2} - 2{\rm S}_{1/2}$ and $2{\rm S}_{1/2} - 2{\rm P}_{3/2}$ transitions omitted as well. The best estimate for these transition frequencies from that least-squares adjustment is

$$\nu_{\rm H}(2{\rm P}_{1/2} - 2{\rm S}_{1/2}) = 1\,057\,844.9(3.2)\ {\rm kHz} \tag{35}$$
$$\nu_{\rm H}(2{\rm S}_{1/2} - 2{\rm P}_{3/2}) = 9\,911\,196.3(3.2)\ {\rm kHz}\ , \tag{36}$$

while the corresponding experimental values are

$$\nu_{\rm H}(2{\rm P}_{1/2} - 2{\rm S}_{1/2}) = 1\,057\,845.0(9.0)\ {\rm kHz}\ [22] \tag{37}$$
$$\nu_{\rm H}(2{\rm P}_{1/2} - 2{\rm S}_{1/2}) = 1\,057\,862(20)\ {\rm kHz}\ [23] \tag{38}$$
$$\nu_{\rm H}(2{\rm S}_{1/2} - 2{\rm P}_{3/2}) = 9\,911\,200(12)\ {\rm kHz}\ [21]\ . \tag{39}$$

The good agreement between the theory and experiment is apparent.

An indication of the effect of taking the covariances of the additive constants δ for the theory into account is the fact that the uncertainty of the Lamb shift $\nu_{\rm H}(2{\rm P}_{1/2} - 2{\rm S}_{1/2})$ obtained this way is less than one third the *a priori* theoretical uncertainty of the $2{\rm S}_{1/2}$ level itself.

REFERENCES

1. J. A. Bearden and H. M. Watts, Phys. Rev. **81**, 73 (1951).
2. J. A. Bearden and H. M. Watts, Phys. Rev. **81**, 160 (1951).
3. J. A. Bearden, M. D. Earle, J. M. Minkowski, and J. S. Thomsen, Phys. Rev. **93**, 629 (1954).
4. J. W. M. DuMond and E. R. Cohen, Phys. Rev. **82**, 555 (1951).
5. J. W. M. DuMond and E. R. Cohen, Rev. Mod. Phys. **25**, 691 (1953).
6. E. R. Cohen, J. W. M. DuMond, T. W. Layton, and J. S. Rollett, Rev. Mod. Phys. **27**, 363 (1955).
7. J. A. Bearden and J. S. Thomsen, Nuovo Cimento Suppl. **5**, 267 (1957).
8. E. R. Cohen and J. W. M. DuMond, Rev. Mod. Phys. **37**, 537 (1965).
9. B. N. Taylor, W. H. Parker, and D. N. Langenberg, Rev. Mod. Phys. **41**, 375 (1969).
10. E. R. Cohen and B. N. Taylor, J. Phys. Chem. Ref. Data **2**, 663 (1973).
11. E. R. Cohen and B. N. Taylor, Rev. Mod. Phys. **59**, 1121 (1987).
12. A. C. Aitken, Proc. R. Soc. Edinburgh **55**, 42 (1934).
13. R. S. Van Dyck, Jr., P. B. Schwinberg, and H. G. Dehmelt, Phys. Rev. Lett. **59**, 26 (1987).
14. V. W. Hughes and T. Kinoshita, Rev. Mod. Phys. **71**, S133 (1999).
15. T. Udem *et al.*, Phys. Rev. Lett. **79**, 2646 (1997).
16. B. de Beauvoir *et al.*, Phys. Rev. Lett. **78**, 440 (1997).
17. C. Schwob *et al.*, Phys. Rev. Lett. **82**, 4960 (1999).
18. M. Weitz *et al.*, Phys. Rev. A **52**, 2664 (1995).
19. S. Bourzeix *et al.*, Phys. Rev. Lett. **76**, 384 (1996).
20. D. J. Berkeland, E. A. Hinds, and M. G. Boshier, Phys. Rev. Lett. **75**, 2470 (1995).
21. E. W. Hagley and F. M. Pipkin, Phys. Rev. Lett. **72**, 1172 (1994).
22. S. R. Lundeen and F. M. Pipkin, Metrologia **22**, 9 (1986).
23. G. Newton, D. A. Andrews, and P. J. Unsworth, Philos. Trans. R. Soc. London, Ser. A **290**, 373 (1979).
24. A. Huber *et al.*, Phys. Rev. Lett. **80**, 468 (1998).
25. K. Pachucki, Phys. Rev. A **48**, 2609 (1993).
26. M. I. Eides, H. Grotch, and V. A. Shelyuto, Phys. Rev. A **55**, 2447 (1997).
27. M. I. Eides and V. A. Shelyuto, Phys. Rev. A **52**, 954 (1995).
28. K. Pachucki, Phys. Rev. Lett. **72**, 3154 (1994).
29. P. Mergell, U.-G. Meißner, and D. Dreschsel, Nucl. Phys. **A596**, 367 (1996).
30. G. G. Simon, C. Schmitt, F. Borkowski, and V. H. Walther, Nucl. Phys. **A333**, 381 (1980).
31. L. N. Hand, D. G. Miller, and R. Wilson, Rev. Mod. Phys. **35**, 335 (1963).

Data Needs for Diagnostics of Low Pressure Plasmas

Bill Graham

School of Mathematics and Physics
The Queen's University of Belfast
Belfast BT7 1NN
Northern Ireland

Abstract. The low pressure plasma processing environment is complex and presents many diagnostic challenges. Here the diagnostic techniques used for accurate and detailed measurement of the density and energy distributions of charged and neutral species are reviewed. Most of the techniques rely heavily on atomic and molecular data. The specific data needs of each diagnostic are outlined. It is shown that in total these data needs are vast and diverse and cannot all be met from specific measurements or calculations. The real need is for generic scaling rules for each of the significant atomic and molecular processes.

INTRODUCTION

Low pressure plasmas, already of pivotal importance in the lighting and microelectronics industry are finding an increasing number of applications in new technologies, such as electronic displays, micromachines and bioengineering, finding increasing applications in areas such as surface treatment and even penetrating more established industries such as textiles. An extensive review of the applications of low pressure plasmas has been provided by Bardsley in the proceedings of the first ICAMDATA [1].

In the past most of the advances in plasma processing have come from empirical engineering approaches i.e. systematically changing the operating parameters of the plasma to optimize a particular process. As more sophisticated processing is required, generally driven by the relentless quest for smaller feature size, faster processing, new materials and larger wafer size in the microelectronics industry, this approach has become inefficient. There is therefore an increasing awareness of the need to understand the physical and chemical processes which initiate and sustain both the plasma and the plasma-induced process. Other technological processes are benefiting and new applications are being identified as a result of these microelectronic industry motivated developments.

CP543, *Atomic and Molecular Data and Their Applications*, edited by K. A. Berrington and K. L. Bell

The growing number of plasma processes is paralleled by the development of different types of plasma sources. Detailed descriptions of processes and plasma sources can be found in a number of textbooks and references 2 to 5 provide some examples. Here the focus will be the diagnostic techniques used to measure the relevant properties of the plasma. The purpose of the diagnostics is to understand the plasma production and sustaining mechanisms and the plasma chemistry associated with particular processes. Since experiments cannot cover the complete gambit of sources and processes the other main role for diagnostics is to provide accurate and detailed information to benchmark computer-based simulations.

There are a vast number of diagnostic techniques and thus some focus is necessary and this paper concentrates on measurement of the densities and energy distributions of the plasma particles. An important diagnostic tool, the detailed electrical characterisation of the plasma and the associated circuit, has modest atomic and molecular data needs and will not be discussed. In addition there will be a focus on those diagnostics which have been applied to rf driven plasma systems. Most of these diagnostic techniques however are widely used in other low pressure systems such as dc discharges, the magnetically enhanced magnetrons and helicons and microwave systems, such as electron cyclotron resonance (ECR) sources.[3,4] .

THE STRUCTURE OF A TYPICAL LOW PRESSURE PLASMA

The rf driven capacitively coupled plasma source has been a workhorse of the semiconductor industry for over a decade. A generic source would consist of two parallel metal electrodes, of 100 to 1000 cm^2 area, mounted in a grounded, vacuum chamber and separated by a few centimetres. The chamber is evacuated before the working gas mixtures are introduced at a rate to sustain a pressure of about 10 Pa. A 13.56 MHz radio frequency voltage (a frequency allocated to the industry) is applied to one electrode (the powdered electrode) while the other electrode is grounded. The use of RF rather than DC both reduces the required breakdown potential and allows the use of insulating material on the powered electrode.

At 13.56 MHz the ions are too massive to move in the oscillating electric fields and so respond only to the time-averaged field. The electrons on the other hand, pick up energy from the oscillating electric field which they may then lose through collisions with neutral particles. This produces the plasma sustaining ionisation along with excitation of the neutral particles and dissociation of molecular constituents. Importantly the plasma constituent are not in thermal equilibrium with one another. The input power is efficiently transferred to electrons but not, in the binary collisions at these low pressures, to the ions. The electron temperature is therefore very much greater than the ion temperature which is usually the same or slightly higher than the gas temperature which is close to that of the surrounding walls (~ 300 K). This means that low melting point or fragile materials can be processed in the plasma. Often the electrons have a non-Maxwellian distribution which makes the measurement of their energy distribution particularly important.

Ions and electrons diffuse out of the plasma centre at different rates so the plasma self adjusts to a positive potential relative to any contact surface such that the rate of loss of both species is equal. This potential is called the plasma or space potential. As a result, near electrodes or other surfaces in the plasma there are less electrons than ions present. This region is called the sheath. Here there are electric fields which can accelerate and hence provide energy to any charged particles which diffuse to the sheath.

If the powered electrode is coupled to the power supply via a capacitor the differing mobility of electrons and ions causes this electrode to attain a constant negative self bias relative to the plasma. Ions respond to this time averaged self bias so positive ions diffusing to the sheath edge are accelerated to the electrode surface.

Thus low pressure plasma system has three distinct regions.

- The bulk, central region where the plasma is sustained through ionising collisions of energetic electrons with the neutral gas which also produce chemically active species in dissociating collisions with molecular gas.
- The sheath region to where ions, electrons and active species diffuse. On average the sheath fields accelerate the electrons into the plasma and the positive ions onto the surface.
- The surface region where ions and active species interact with the surface atoms. The impacting ions can carry sufficient energy to remove atoms from the surface in a process called sputtering or can merely enhance the surface atom mobility and thus their reactivity.

THE LOW PRESSURE PLASMA ENVIRONMENT

The capacitively coupled plasmas generally operate at

- gas pressures ranging from 1 to 70 Pa ,
- electron densities of 10^{14} to 5×10^{16} m^{-3}
- fractional ionisation ~10^{-5} to 10^{-6}
- fractional dissociation of molecular gases ~0.01.

Recently inductively coupled discharges have become more prevalent. Here the power is coupled into the plasma through time varying magnetic fields generated by an rf current in a coil. There normal operating regime is

- gas pressures 0.1 to 1 Pa
- electron densities of 10^{17} to 10^{18} m^{-3}
- fractional ionisation ~$2x10^{-2}$
- fractional dissociation of molecular gases~0.2.

The lower working pressures and high plasma densities give such systems the advantage of high etch rates with low ion collisionality in the sheath region so maintaining ion directionality which improves etch profiles.

Each plasma process requires a specific gas or gas mixture. Inert gases, are probably the most often used, with argon predominant. These are used for both sputtering applications and are added to other gases where their metastable states improve the ionisation. When depositing protective films there is often a need to promote chemistry either in the bulk plasma or at the surface itself. Different precursor gases are used depending on the atoms required. These include

H_2 for H,
O_2 , H_2O for O,
N_2 , NH_3 for N,
CH_4 , C_2H_2 for C
SiH_4 for Si.

The etching processes require complex chemistry often involving both etching and the deposition of passivating polymer layers. Some of the gases used are
Cl_2
F_2
halocarbons such as CF_4, C_2F_6, HBr, BCl_3 and SF_6.

The plasma environment therefore consists of a wide variety of particles:

- electrons
- positive ions (atomic and molecular)
- negative ions (atomic and molecular)
- excited atoms, molecules and ions
- dissociation products (atomic, molecular, neutral, positive, negative)
- process products (atomic, molecular, neutral, positive, negative)
- clusters and micron sized particle

The plasma environment presents many diagnostic challenges. These are primarily in the accurate measurement of the density and energy distributions of charged and neutral species, with the neutral species of interest being the reactive dissociation products of the precursor gases and the products of the process. In addition since the plasma environment is structured, spatially resolved measurements to millimeter precision are often required. There is currently interest in pulsed plasmas, plasma initiation and instabilities which requires time resolved measurements, often to a temporal resolution of a few nanoseconds.

DIAGNOSTIC TECHNIQUES AND THEIR DATA NEEDS

There follows a brief discussion of some diagnostic techniques with an illustrative list of atomic and molecular data requirements. These data are required for some or all of the plasma environments outlined above. The references are provided to guide the reader to a source which describes the technique and points to other papers using the technique.

Electrostatic probe [3,5,6]

The measurement of the current collected as a function of the voltage applied to a small metal probe inserted into the plasma can be interpreted to provide information on the ion and electron density, the electron energy distribution function and the plasma potential. In addition to an appropriate model for the sheath formed around the probe, a knowledge of collisions in the sheath and of surface effects influencing the current measurement are required, for example

- electron and ion scattering cross sections (0 to 50 eV)
- secondary electron emission coefficients for metals and insulators (0 to 50 eV)
- electron emission by metastables (thermal upwards)

Mass spectroscopy [5,7,8,9]

The identification of ionised and neutral species extracted or effusing from the plasma, often using a differentially pumped quadrupole mass filter with an ionising source at the entrance to facilitate neutral particle detection. This technique provides information on plasma produced species and so the species likely to be incident on a processing surface. Interpretation requires

- electron impact ionisation and dissociative ionisation branching ratios (from threshold energy to 100 eV)
- charge exchange cross sections (0.1 to 200 eV)

Attachment spectroscopy [10]

A recent mass spectroscopic alternative for detection of electronegative neutral species by attaching electrons in the entrance source and detecting the resultant negative ions. Data requirements are

- electron attachment cross sections (from threshold energy up)

Ion energy analysis [5,11,12]

The analysis of the energy of ions at the sampling surface, using an energy analyser in conjunction with a mass spectrometer. These measurements can not only provide information on the energy distribution of particles onto a surface but, if the analyser is mounted on the powered electrode, also measures the self bias and the ion collisionality in the sheath. Requires data on

- charge exchange cross sections (0.1 eV to 200 eV)

Plasma induced emission [5,7,13,14,15]

The analysis of light produced from the decay of species excited in the plasma usually by electron collisions. This technique identifies the constituents of the plasma and also can be used to determine the rovibrational temperatures of molecules which may be present. Data needs are

– spectroscopy of neutral and charged atoms, molecules and radicals.
– electron impact excitation cross sections (from threshold to 100 eV)
– quenching rates, particularly for molecular gases, radicals and process products.

These requirements are common to most photon-based diagnostic techniques.

Actinometry [5,16,17]

Well-documented photon emitting additives (often argon) are used to provide convenient calibration of optical detection systems and so allow absolute density determination from emission/fluorescence measurements. This technique requires

– electron impact excitation cross sections (threshold to 30 eV)
– metastable state cross sections (threshold to 30 eV)
– quenching rates, particularly for inert gas atomic emission lines in molecular gases, plasma generated radicals and process products
– lifetimes of excited states, particularly Ar, F, Cl atoms.

Absorption [5,18,19]

The measurement of particle densities by absorption spectroscopy at an appropriate wavelength. Often carried out in multiple pass or more recently in cavity ring down mode. Data requirements include

– photoabsorption cross sections
– dipole moments and oscillator strengths of molecules and dissociation products
– absorption cross sections for hot bands of IR spectra.

Laser induced fluorescence [7,20,21]

The plasma species of interest is excited by a spatially and spectrally well-defined laser beam. Interest is often in detection of the predominant ground state populations of a specific plasma constituent which cannot be excited to a higher electronic level by a single photon in which case multiphoton techniques are required. In addition to the same data requirements as emission spectroscopy also needed are

- multiphoton absorption cross sections
- photoionisation and photodissociation cross sections from excited states
- Frank Condon factors and oscillator strengths, particularly for plasma produced radicals and process products.

Thomson scattering [22]

The spectra of photons scattered from plasma electrons is Doppler broadened and so contains information about the velocity distributions of the electrons. The Raleigh scattering from a background gas is used for calibration of the detection system so that absolute electron densities can be obtained. The main data requirements relate to possible competing factors associated with the high power laser beams used as a scatter source e.g.

- photoionisation of metastable states
- photochemistry of molecular gases and plasma produced radicals.

Rydberg state stark spectroscopy [7, 23,24]

The Stark Effect, energy shifts in allowed transitions and the occurrence of transitions which are normally forbidden, can be used to determine sheath electric field strengths. The data required includes

- energy level shifts of Rydberg levels of argon, chlorine and fluorine.

Photodetachment of negative ions [25, 26,27]

A laser is used to photodetach negative ions in the plasma. The electrons produced have a much higher mobility than the negative ions and can be detected by the change in current collected on an electrostatic probe or the change in microwave absorption by the plasma. Absorption spectroscopy has also been used for negative ion detection. There is a need for data on

- Photodetachment cross sections for radical negative ions, particularly CF^-, CF_3^-, $C_2F_x^-$, etc. and for small clusters.

CONCLUSIONS

The data needs for low temperature plasma diagnostics are vast and diverse and they cannot all be met from specific measurements or calculations. The real need is for generic scaling rules. The low temperature plasma physics community needs fast, user-friendly access to existing and emerging atomic and molecular data.

ACKNOWLEDGEMENTS

Special thanks are due to Jean-Paul Booth, Gerrit Kroesen, Greg Hebner, Tom Morrow, Stanislav Moshkalyov and the postdocs and graduate students in the QUB Low Temperature Plasma Physics Group (Elm Costa i Bricha, Cormac Corr, Sergi Gomez, Charlie Mahony, Jill McFarland, Philip Steen and Catherine Thompson) for sharing their thoughts on the data needs for diagnostics.

REFERENCES

1. Bardsley, J.N., *Atomic and Molecular Data and Their Applications*, edited by P.J. Mohr and W.L. Wiese, The American Institute of Physics, New York (1998) pp. 333
2. Chapman, B., *Glow Discharge Processes*, A Wiley & Sons Inc., (1980)
3. Lieberman, M.A. and Lichtenberg, A.J., *Principles of Plasma Discharges and Materials Processing*, John Wiley & Sons, Inc. (1994)
4. Roth, J.R., *Industrial Plasma Engineering*, Inst. of Physics Publishing, Bristol and Philadelphia (1995)
5. Sugawara, M., *Plasma Etching*, Oxford University Press Inc., New York., (1998)
6. Godyak, V.A., Piejak R.B.and B.M. Alexandrovich, *Plasma Sources Sci. Technol.*, **1**, 36 (1992)
7. *Glow Discharge Spectroscopies*, edited by R.H. Marcus, Plenum Press, New York (1993)
8. Kojima,H., Toyoda,H. and Sugai,H, Appl. Phys, Lett. **55**, 7292 (1989)
9. Kae-Nune, P., Perrin, J., Gullion, J. and Jolly, J. , *Plasma Sources Sci. Technol.*, **4**, 250 (1995)
10. Stoffels, W.W., Stoffels, E., and Tachibana, K., *J. Appl. Phys.* **36** 4638-4643 (1997)
11. Coburn, J.W. and Kay, E., *J. Appl. Phys.* **50**, 4965 (1972)
12. Kuypers, A.D. and Hopman, H.J., *J. Appl. Phys.* **67**, 1229 (1990)
13. Coburn, J.W. and Chen, M., *J. Appl. Phys.* **51**, 3134 (1980)
14. Hancock, G. and Sucksmith, J.P., J. Vac. Sci. Technol. A **13**. 2945 (1995)
15. Donnelly, V.M., Flamm, D.L. and Bruce, R.H., *J. Appl. Phys.* **58**, 2135 (1985)
16. Coburn, J. and Chen, M., *J. Appl. Phys.* **51**, 3134 (1980)
17. Walkup, R.E., Saenger, K.L. and Selwyn, G.S., J.Chem. Phys. **84**, 2668 (1986)
18. Gousset, G., Panafieu, P., Touzeau, M. and Vaille, M. *Plasma Chem. and Plasma Process.* **7**, 409 (1987)
19. Booth, J.-P., Cunge, G., Neuilly, F. and Sadeghi, N. J. Phys. D, **31**, 1168 (1998)
20. Moore, C.A., Davis, G.P. and Gottscho, R.A., *Phys. Rev. Lett.* **52**, 538 (1984)
21. Preppernau, B.L., Dolson, D.A., Gottscho, R.A. and Miller, T.A., *Plasma Chem. and Plasma Process.* **9**, 157 (1989)
22. Hori, T., Kogano, M., Bowden, M.D., Uchino, K., and Muraoka, K., *J. Appl. Phys.*, **83**, 1909-1916 (1998)
23. Ganguly, B.N., Shoemaker, J.R., Preppernau, B.L., and Garscadden, A., *J. Appl. Phys.* **61**, 2778-2783 (1987)
24. Czarnetzki, U., Luggenholscher, D., and Döbele, H.F., *Plasma Sources Sci. Technol.* **8**, 230-248 (1999)
25. Taillet, J., Acad. Sci., Ser. B **269**, 52 (1969)
26. Bacal, M., and Hamilton, G.W., *Phys. Rev. Lett.*, **42**, 1538-1540 (1979)
27. Quandt, E., Döbele, H.F., and Graham, W.G., *Appl. Phys. Lett.*, **72**, 2394-2396 (1998)

Heavy-Ion Radiotherapy

Tatsuaki Kanai

Research Center of Charged Particle Therapy, National Institute of Radiological Sciences, 4-9-1 Anagawa, Inage-ku, Chiba-shi, CHIBA 263-8555 JAPAN

Abstract. Heavy-ion radiotherapy using high-energy carbon beams has been performed at the National Institute of Radiological Sciences, Japan. The physical frame works for heavy-ion radiotherapy are established using physical understandings of radiation physics. In order to increase the accuracy of heavy-ion radiotherapy, many physical problems should be solved. Unsolved problems, such as the depth dose distributions, range of heavy-ion in patients and heavy-ion dosimetry in the radiation therapy, are discussed.

INTRODUCTION

Since June of 1994, a clinical trial of heavy-ion radiotherapy has been performed at National Institute of Radiological Sciences using the HIMAC (Heavy Ion Medical Accelerator in Chiba) accelerator. During these 6 years, over 750 patients have been treated using carbon beams of 290, 350 and 400 MeV/u [1]. For these radiation therapies using high-energy carbon beams, highly precise irradiations are required, because of the high biological effectiveness for tumor and also for surrounding normal tissues. Thus, the physical characteristics of the beam should be examined in detail.

An accuracy of less than 1% is sometimes required in heavy-ion radiotherapy. Some physical quantities can not be estimated by theoretical calculations to the required accuracy. In this paper, the present status of physical data in heavy-ion radiotherapy is discussed, for example, the range of heavy-ions in a human body, the depth dose distribution of the carbon beams in water, the dosimetry of heavy-ion beams, and the W-values of air for the beams.

PRESENT STATUS OF HIMAC RADIOTHERAPY

HIMAC consists of PIG and two ECR ion sources, a RFQ linac, an Alvaretz linac, two synchrotrons and a beam delivery system [2]. Three treatment rooms are available for therapy, one irradiation room for biological experiments and three general experiment rooms. Table 1. lists the physical characteristics of the HIMAC facility. For radiotherapy, 290, 350 and 400 MeV/n carbon beams are used. The ranges of a beam in water are 16 cm, 22 cm and 28 cm, respectively.

CP543, *Atomic and Molecular Data and Their Applications,* edited by K. A. Berrington and K. L. Bell

In order to uniformly irradiate tumors in patients, accelerated narrow beams are broadened laterally to about 22 cm in diameter by using a scatterer and a pair of wobbler magnets. Various shapes of the irradiation fields are made by a multileaf collimator or patient collimators, which are made of brass. A sharp Bragg peak of the carbon beam is broadened in the depth direction using ridge filters according to the size of the tumor. Thus, the doses can be localized in the target region [1].

TABLE 1. Specification of HIMAC facility

Ion species	p to Kr	
Maximum Energy	800 MeV/u (for q/A = 1/2)	
Minimum Energy	100 MeV/u	
Beam Intensity	1.1×10^{10} pps(particles per second)	for proton
	1.2×10^{10} pps	for ^{4}He Ions
	2.0×10^{9} pps	for ^{12}C
	2.7×10^{7} pps	for ^{40}Ar
	1.1×10^{8} pps	for Kr
Beam duration	400 ms	
Repetition rate	1/2 to 1/3 Hz	
Beam emittance	10 πmm mrad (unnormalized)	
Momentum spread	± 0.2 %	
Irradiation Facility		
Treatment rooms	1 (Horizontal beam)	
	1 (Vertical beam)	
	1 (Horizontal and vertical beams)	
	4 Experimental rooms	
Beam Characteristics		
Field size	22 cm (Maximum diameter)	
Dose Uniformity	± 2 %	
Maximum Range	30 cm	

The results of heavy-ion radiotherapy are described on the home page of the National Institute of Radiological Sciences, Japan (http://www.nirs.go.jp/newinfo /press/: unfortunately it is written only in Japanese). According to the documents, the local control rate for cancer of the head and neck, lung, liver, prostate, uterine cervix were about 80, 60, 80, 100 and 50% two years after treatments, respectively. Also, the survival rate for the same cancer were about 45, 70, 55, 94 and 40%, three years after treatments, respectively. The therapy started with a relatively lower dose, and gradually increased to find the optimum dose for carbon beam. Also, patients eligible for the HIMAC clinical trial were limited to cancers not expected to be cured by conventional treatments, including surgery. Considering these situations, the obtained clinical result shows that heavy-ion radiotherapy is very effective, especially for advanced cancer [3].

Depth Dose Distributions in Water

High dose localization as well as very high biological effectiveness can be achieved simultaneously when using heavy-ions for radiotherapy. Because of these reasons, more accurate irradiations are required in heavy-ion radiotherapy. First of all, depth-dose distributions of high-energy carbon beams are discussed in this section. Heavy-ion beams show the very sharp Bragg peak in their depth-dose distributions. Fig. 1 shows the depth-dose distributions in water of carbon beams, with energies of 290, 350 and 400 MeV/n. The solid curves in the figure show the results of theoretical calculations using a code developed by Sihver et al.[4]. The closed triangles and circles in the figure show the experimental results. The experimental depth-dose distributions were shifted so that the measured Bragg-peak positions coincided with the calculated results. The heights of the relative-dose distributions are adjusted so as to coincide with dose at the surface region of the calculated curves. The calculation can reproduce the experimental result for a 290 MeV/n carbon beam. For higher energies, the resultant doses of the calculations are larger than the experimental results at deeper regions. The calculation may over-estimate the dose contributed by fragmented particles. A complete analysis of the comparison should be made for these depth dose distributions of high-energy heavy ions in water.

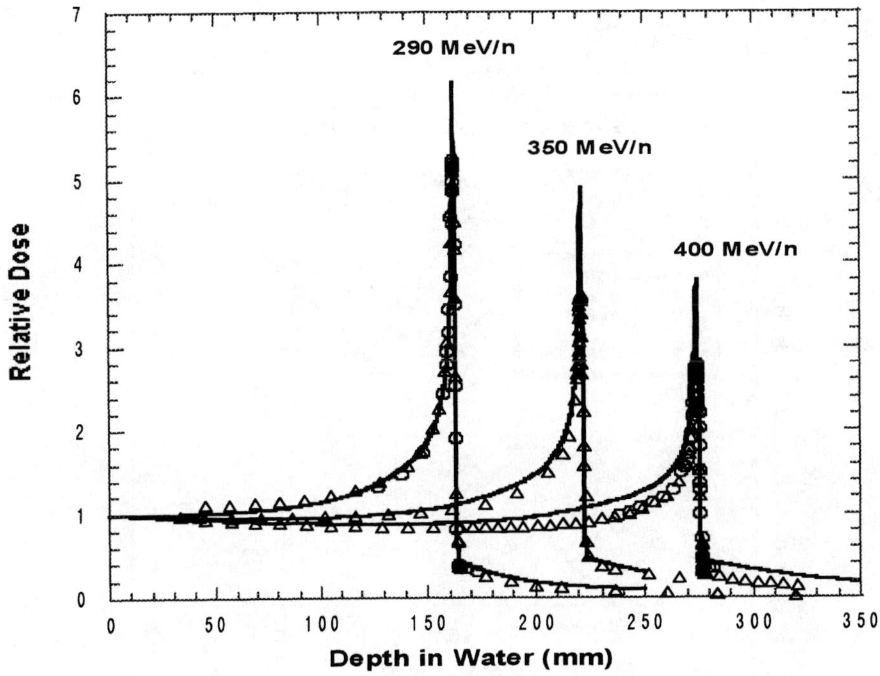

FIGURE 1. Depth-dose distributions of carbon beams in water.

Estimation of Heavy-Ion Ranges in Patients

In the heavy-ion radiation therapy, a medical doctor determines the target regions using various tomographic images of the patients. The medical doctor draws the target region on each slice of the CT (Computed Tomography) images with help of various information from MR (Magnetic Resonance) images as well as PET (Positron Emission Tomography) images. On the basis of these CT images, the ranges in patients or thickness of absorbers needed to localize the Bragg peak onto the desired positions in the patients should be calculated. CT images are images of absorption coefficient of X-rays in the patients. The absorption coefficient in the patient does not have a linear relationship with the electron density of the patient, which is roughly proportional to the stopping power of the heavy-ions. It is necessary to convert the CT values in the CT images to electron density or the water-equivalent length corresponding to the pixel[5]. Now, almost all facilities use an empirical relationship between the CT values and the electron densities, which are obtained by measurements. Figure 2 shows the experimental relationship of the CT number and the electron-density near the water used in HIMAC facility. The CT values and electron density for materials which have lower CT values than water were measured using solusions of C_2H_5OH of various density. Those values for materials having higher CT values were obtained by measuring those using K_2HPO_4.

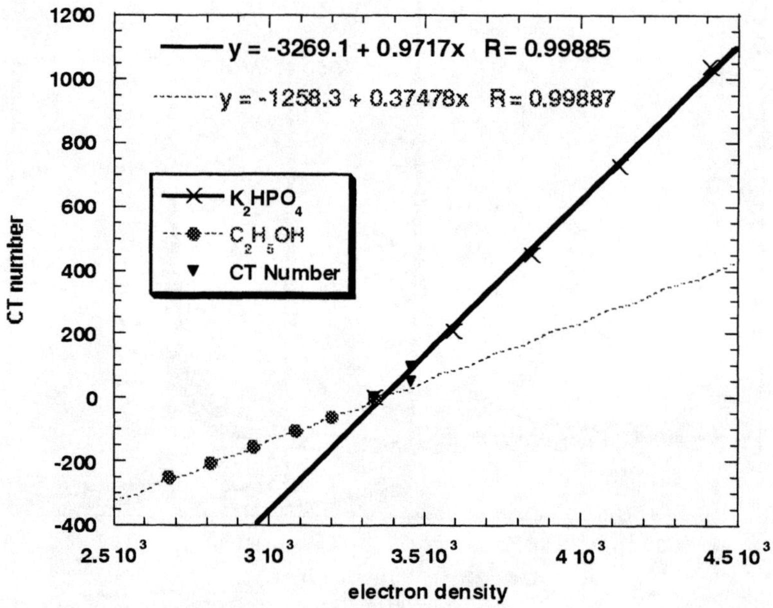

FIGURE 2. Empirical relationship between the CT number and the electron density which is experimentally determined using solutions of K_2HPO_4 and C_2H_5OH.

The electron density for patients can not be uniquely obtained from the CT numbers. For real tissues, the values of the electron density scatter around the mean values. The error was estimated about ±1% [6]. In order to reduce these errors in an estimation of the electron density of patients, several ideas have been proposed in the past. For example, the development of a dual energy X-ray CT, a high-energy X-ray CT using the Compton effect, and a charged particle CT have been proposed. None of the ideas completed.

In heavy-ion radiotherapy, positions of stopped heavy ions can be detected from the outside of the patient if the beam is a positron emitter, for example, ^{11}C, ^{10}C, ^{19}Ne, etc [7,8]. A primary ^{11}C beam or ^{11}C induced in-patient by therapy irradiation of a ^{12}C beam (autoactivation) can be used for this estimation of the irradiated region in the patient directly. At Gesellschaft fur Schwerionenforschung (GSI) in Germany, autoactivated ^{11}C after a single treatment is detected to check the irradiated region. At HIMAC, a spot scanning system using ^{11}C is now under construction for this purpose.

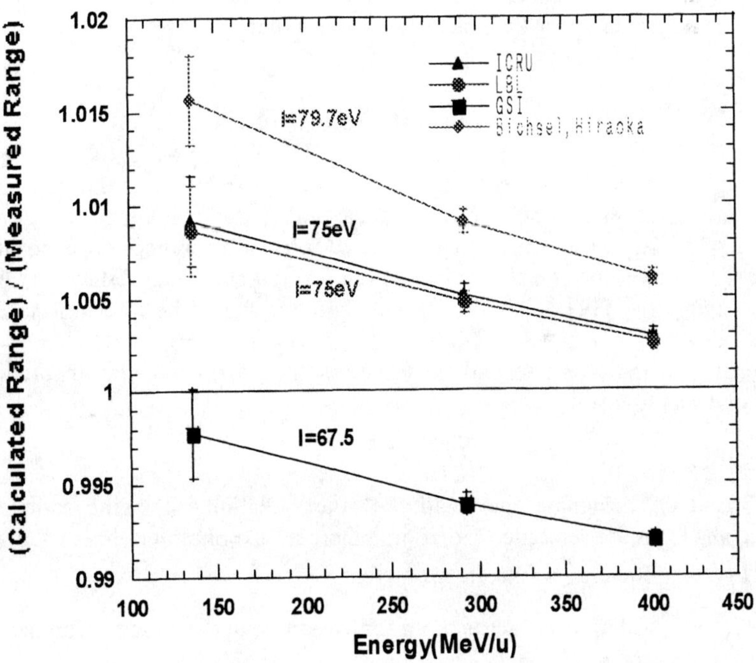

FIGURE 3. Range of high-energy carbon beams in water. The vertical axis is the ratio of the calculated range to the experimental range.

Ranges of Heavy Ions in Water

Usually, the stopping power of heavy ions in a material with water is fundamental quantity in the field of heavy-ion radiotherapy. An uncertainty of ±1 % of a range in water will be about ± 3 mm for heavy ions with 30 cm range. The error in the estimation of the water-equivalent length of a CT image and the uncertainty of the range in water becomes ± 6 mm for the worst cases. Then, it is very important to determine very accurately the range in water of heavy ion beams. Figure 4 shows tentative results of range measurements for high-energy carbon beams. The vertical axis of the figure is the ratio of the calculated range to the measured range. The triangle, circle, square and diamond marks show the results calculated by various authors. Triangles are the results being converted from the ICRU calculation for the proton and alpha particles using simple scaling.

The variations of the results of the calculations are almost dependent only on the averaged excitation values, <I>, adopted by those authors [7,8,9,10]. From this tentative experiments, the <I> value should be close to 70 eV for a water material. As shown in the figure, there is still a systematic contradiction between the experiments and the calculation. More detailed experiments or calculations are required in order to determine ranges in water to an accuracy of less than 1%.

Heavy-Ion Dosimetry

Although there is a very large uncertainty in the estimation of the constants in the formula, an ionization dosimetry is very stable and reproducible method for heavy-ion dosimetry [11, 12]. The dosimetry is based on an absorbed dose to the water calibration factor of an ionization chamber in a reference beam, which is taken to be the ^{60}Co gamma ray [12]. The ionization chamber should be traceable to a primary standard laboratory.

According to the IAEA formalism for heavy-ion dosimetry, the absorbed dose to water is given as follows:

$$D_{w,Q} = M_Q N_{D,w} k_{Q,Q_0} \tag{1}$$

where Q and Q_0 designate the quality of the radiation. Q is the radiation to be measured and Q_0 is the reference radiation, where a ^{60}Co photon field is used.

- M_Q Reading of the electrometer
- $N_{D,w}$ Calibration factor of an ionization chamber in the reference radiation field.
- k_{Q,Q_0} Beam Quality factor.

The beam quality factor $k_{Q,Q0}$ is given by,

$$k_{Q,Q_0} = \frac{(s_{w,air})_Q}{(s_{w,air})_{Q_0}} \frac{(W_{air})_Q}{(W_{air})_{Q_0}} \frac{p_Q}{p_{Q_0}} \tag{2}$$

Here, P_Q is a perturbation factor in the measurements of the dose in a Q radiation field. Usually, a heavy-ion radiation field contains various fragmented nuclei, and their energy spectra are not monochromatic. Thus, the stopping-power ratio and the w-value should be averaged over these whole spectra. The stopping-power ratio of water to air ($s_{w,air}$) is given by

$$s_{w,air} = \frac{\sum_i \int_0^\infty \Phi_i(E) \cdot (S_i(E)/\rho)_{water}\, dE}{\sum_i \int_0^\infty \Phi_i(E) \cdot (S_i(E)/\rho)_{air}\, dE} \tag{3}$$

where i stands for i-th particle in the radiation field; Φ_i shows energy spectrum of the i-th fragmented nuclei. The W value of air for heavy-ion beams is given by a dose-averaged value of the inverse of the W value, as follows:

$$\left(\frac{\overline{w}}{e}\right)_{HI} = \left(\sum_i \int_0^\infty \Phi_i(E) \cdot (S_i(E)/\rho)_{air}\, dE\right) \Big/ \sum_i \int_0^\infty \frac{\Phi_i(E) \cdot (S_i(E)/\rho)_{air}}{w_i(E)/e}\, dE$$
$$= \frac{D_{air}}{\sum_i \int \frac{e}{w_i(E)} \cdot d_{i,air}(E)\, dE} \tag{4}$$

In order to strictly calculate these stopping power ratios and W values, the complete constituents are requiredin the radiation fields of heavy-ion beam. Up to now, there is insufficient experimental and theoretical data concerning to these values. IAEA recommend 1.13 for the stopping power ratio and 34.5 eV for the W value, which is the same as the value for a proton beam.

The ionization dosimetry for heavy-ion radiation fields should be ascertained by other methods, such as the calorimetry or fluence method. Heavy-ion dosimetry using calorimeter has not been reported. The fluence of a heavy-ion radiation field can easily be measured compared to low-LET radiation fields. The energy loss of one heavy ion is very large compared to other radiation fields: photons, electrons, and proton beam radiation fields. Thus, the fluence of heavy-ion beam under therapeutic dose is much lower than those for other particles. It become possible to measure the fluence of mono-energetic heavy-ion beams using a CR 39 track detector or a plastic scintillator. The dose of a mono-energetic heavy-ion beam can be obtained by multiplying the fluence and stopping power of the heavy-ion in water.
Figure 5 shows an example of a comparison of two types of dosimetry: ionization

dosimetry and fluence dosimetry[13].

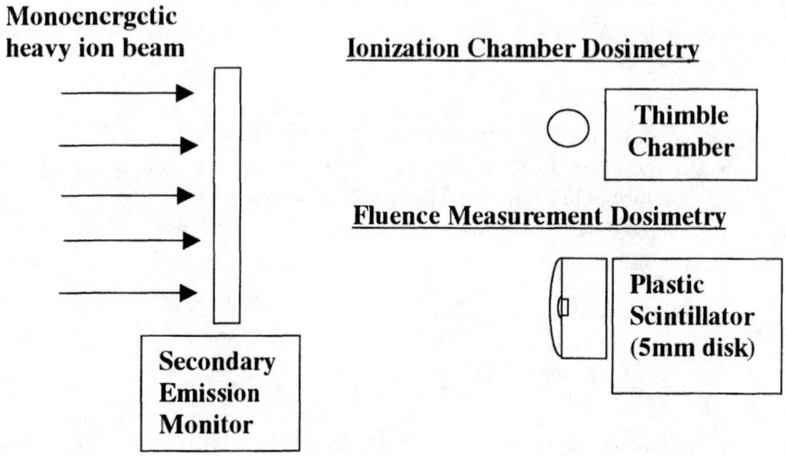

FIGURE 4. Comparison of fluence measurement dosimetry and ionization chamber dosimetry

TABLE 2. Experimental results of ionization chamber dosimetry and fluence dosimetry.

Ion	Stopping Power (KeV/μm)	Fluence (cm^2/SEM)	Dose (F) (Gy/SEM)	Dose (I.C.) (Gy/SEM)	Ratio D_F/D_{IC}
C135	23.60	3846.17 (±0.87%)	1.4540×10^{-4}	1.5056×10^{-4} ($\pm 1.07 \times 10^{-6}$)	0.966
C290	13.23	12738.5 (±2.6%)	2.700×10^{-4}	2.8570×10^{-4}	0.945
C350	11.76	9913.95 (±0.49%)	1.8679×10^{-4}	1.9415×10^{-4} ($\pm 0.93 \times 10^{-6}$)	0.962
C400	10.94	16736 (±0.67%)	2.9333×10^{-4}	3.0120×10^{-4} ($\pm 1.15 \times 10^{-6}$)	0.974
Ne230	44.74	2214.1 (±0.65%)	1.5870×10^{-4}	1.6203×10^{-4} ($\pm 1.07 \times 10^{-6}$)	0.979
Fe90	657.75	1860.5	1.9654×10^{-4}	1.9203×10^{-4}	1.023
Ar95	278.89	4055.4	1.8120×10^{-3}	1.8216×10^{-3}	0.995

A thimble ionization chamber, which is calibrated by a ^{60}Co radiation field, or a small plastic scintillator is placed at the center of the uniform irradiation fields of mono-energetic heavy ion beams. The irradiation dose is monitored by a secondary-emission monitor. A plastic scintillator, 5 mm in diameter and 3 mm in thickness, is embedded in Lucite light guide. From the reading of the electrometer and using equation (1), the heavy-ion dose per unit monitor can be obtained. On the other hand, the fluence per unit monitor can be directly obtained by the scintillator. Table 2 gives tentative results of a comparison of the two types of heavy-ion dosimetry. Figure 6 shows the ratio of the two dosimetry as a function of LET (Linear Energy Transfer). For high-energy carbon ions, the ratios of the fluence dosimetry to the ionization dosimetry were around 3% lower than one. Also with increasing the LET of the heavy ions, the ratio is increased. It may be possible to assume that the W value for high-energy heavy-ions is lower than the W value for a proton. The W value gradually increase as the LET of the heavy ions increase. Generally speaking, the W value for low-speed ions is larger than those for high-speed ions. In the heavy ion radiation field and proton radiation field, particles contributing to the radiation dose can be divided into two components: high-speed charged particles and low-speed charged particles. Primary particles and projectile fragments belong to the high-speed ions. Recoil particles and target fragments belong to the low-speed ions. When the radiation field is classified by these two components, equation (1) for the W value can be rewritten as follows:

$$\left(\frac{e}{w}\right)_{HI} = \sum_i \int \frac{e}{w_i(E)} \cdot \frac{d_i(E)}{D} dE = \left(\frac{e}{w}\right)_{high} f_{D,high} + \left(\frac{e}{w}\right)_{low} f_{D,low} \quad (5)$$

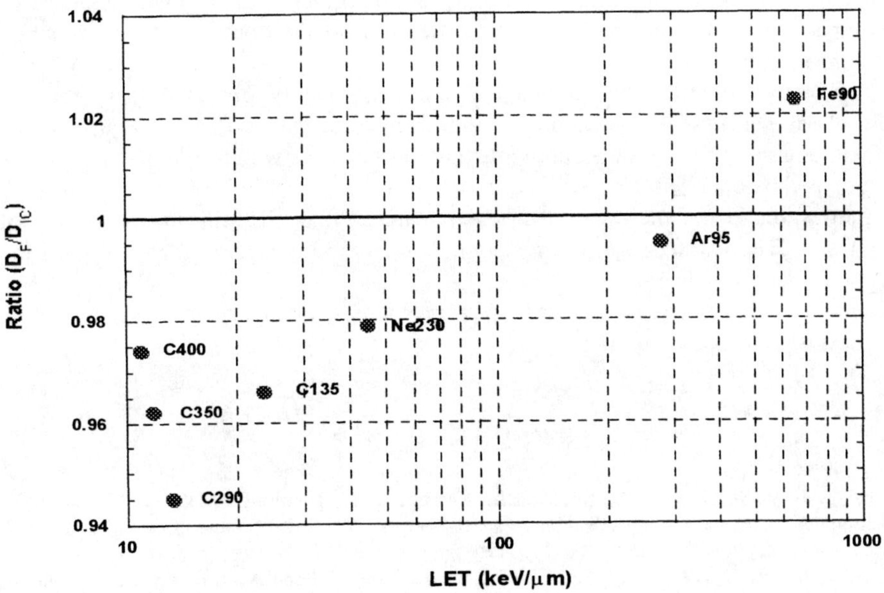

FIGURE 5. Comparison of the ionization chamber dosimetry and fluence dosimetry.

where $f_{D,low}$ and $f_{D,high}$ are the dose fractions of low-speed particles and high-speed particles, respectively. The inverse of the W value is calculated by averaging over the dose-weighted fraction of the spectrum.

The dose contribution by low-energy particles is very large for a proton beam radiation field. In the case of heavy-ion beams, the dose contribution by these low-energy particles may be low compared to the case of proton beams because the fluence of the heavy ions is much lower than that of the proton. Thus, W value for heavy-ion beam may be lower than the W value for the proton beam.

CONCLUSIONS

Since 1994, heavy-ion radiotherapy using high-energy carbon beams have been performed at the National Institute of Radiological Sciences, Japan. The physical frameworks for heavy ion radiotherapy were established using physical understanding of radiation physics. The obtained clinical result show that heavy-ion radiotherapy is very effective especially for advanced cancer.

In order to increase the accuracy of the heavy-ion radiotherapy, many physical problems should be solved.

The calculation result of depth dose distributions agrees with the experimental results for a 290 MeV/n carbon beam. For higher energies, calculation gradually over-estimate the dose contributed by fragmented particles. A complete analysis of the comparison should be made for these depth-dose distributions of high-energy heavy ions in water.

The electron density for patients can not be uniquely obtained from the CT numbers. For real tissues, the values of the electron density will scatter around the mean values up to ±1 %. An accurate method to obtain the water-equivalent length in patient should be developed.

In an estimation of the range in a water material, there is still an uncertainty of over ±1 %. The variations in the resultant range are almost dependent only on the averaged excitation values.

Heavy-ion dosimetry is still an open problem. Comparisons of completely different dosimetries are required.

REFERENCES

1. Kanai, T., Endo, M., Minohara S., Miyahara N., Koyama-Ito H., Tomura H., Matsufuji N., Futami Y., Fukumura A., Hiraoka T., Furusawa Y., Ando K., Suzuki M., Soga F., and Kawachi K., *Int. J. Radiation Oncology Biol. Phys.* 44, 201-210 (1999).
2. Hirao Y., Ogawa H., Yamada S., Sato Y., Yamada T., Sato K., Itano A., Kanazawa M., Noda K., Kawachi K., Endo M., Kanai T., Kohno T., Sudou T., Minohara M., Kitagawa A., Soga F., Takada E., Watanabe S., Endo K., Kumada M., Matsumoto S., Nucl. Phys. A538, 541c 0 550c (1992).
3. Tsujii H., Proc. Of the 6th International Meeting on Progress in Radio-Oncology ICRO/OGRO (edited by Kogelnik H. D., and Sedlmayer F.) Monduzzi Editore, Italy, '09 – 721 (1998)
4. Sihver L., Shadt D., and Kanai T., *Jpn. J. Med. Phys.* **18**, 1 – 21 (1998).

5. Chen G. T. Y., Singh R. P., Castro J. R., Lyman J. T., and Quivey J. M., *Int. J. Radiat. Oncol. Biol. Phys.* **5** 1809 – 1819 (1979).
6. Matsufuji N., Tomura H., Futami Y., Yamashita H., Higashi A., Minohara S., Endo M., and Kanai T., *Phys. Med. Biol.* **43** 3261 – 3275 (1998).
7. ICRU Report 49, 1993.
8. Salamon, M. H., *LBL Report* 10446, LBL, Berkeley (1980).
9. Heinrich W., Wiegel B., Kraft G., *GSI Report*, GSI – 91-30 (1991).
10. Hiraoka T., and Bichsel H., *Jpn. J. Med. Phys.* **15**, 91 – 100 (1995).
11. ICRU Report 59, 1998
12. To be published IAEA publication.
13. Kanai T., Kohno T., Minohara S., Sudou M., Takada E., Soga F., Kawachi K., and Fukumura A., *Radiat. Res.* **135**, 293 – 301 (1993).

Lamp Models Need Data

D. O. Wharmby

Technology Consultant, 55 Constable Road, Ilkley, West Yorkshire, LS29 8RW, UK
david@wharmby.demon.co.uk

Abstract. There are now very comprehensive computer models for the steady state operation of most lamp types of commercial importance. For historical reasons associated with computer power and availability of data most of these models arrived too late to assist the design of radically new lamp types, but they do provide excellent understanding of lamp operation. In the last decade models have often played a key role in lamp development. In the case of fluorescent lamps the collisional-radiative models are very complete, but accuracy still needs to be improved to turn them in to computer-aided design models for new lamp variants. The key to this is to improve the data, especially for ionization processes. In the case of high-pressure (HP) alkali arcs there has been remarkable success in calculating realistic spectra based on molecular energy levels. For HP metal-halide lamps energy balance models are very sophisticated but the data for calculating the radiation are far from adequate; semi-empirical treatment of radiation looks more hopeful. We are now entering an era in which computer power is adequate and *calculated* sets of self-consistent data for collision cross-sections will soon be available even for atoms with many electrons. Similarly good quality molecular energy levels and transition dipole moments for dimers containing heavy atoms look well within reach.

INTRODUCTION

The lamp industry has a venerable history of using theoretical models going back to at least to the time of Langmuir. In many cases the physics is highly non-linear and equations had to be evaluated by approximate methods using material functions that were often a rather poor representation of the data. Nevertheless the models provided important insights into the design of the earliest fluorescent lamps and high-pressure (HP) mercury discharges.[1-3]

As commercial computers became available, the equations could be solved numerically with realistic data insofar as it was available. To aid the development of fluorescent lamps a very complete numerical model of the low-pressure (LP) mercury rare-gas positive column was developed and extensively used at the AEI Research Laboratory in the UK in the 1950s. For commercial reasons the publications about this model were rather brief.[4] This work has formed the prototype for many of the current low-pressure discharge models; and as far as I know it was the very first computer model of a lamp.

Since that time computer models have been constructed for just about every type of lamp. However they have usually been completed well after products were first marketed. Despite this, the very attempt to construct computer models led to a remarkable increase in detailed understanding of lamps. There were two reasons for the delays in achieving useful models: lack of computer power and lack of data.

CP543, *Atomic and Molecular Data and Their Applications,* edited by K. A. Berrington and K. L. Bell

Nowadays there is no shortage of computer power and we seem to be entering an era in which large amounts of self-consistent data can be calculated quantum-mechanically. A good example of how the situation is changing is the work done by Sommerer[5] and colleagues on a possible alternative to the mercury in fluorescent lamps. Using a collisional-radiative (CR) model coupled with critical diagnostics they have been able to assess the commercial potential of LP Xe discharges without the need for the extensive Edisonian research that has been a feature of every other lamp development. As the commercial potential turns out to be rather unimpressive, this has probably saved many man-years of empirical research.

The rapidly improving state of calculated data was consistently underlined at the second ICAMDATA meeting as this volume shows. In particular calculations of collision cross-sections and molecular energy levels can now be done with impressive accuracy. There are still important areas for lamps that seem beyond reach; namely the properties of rare-earth elements and their halides.

In this paper I briefly describe three lamp types, outline the type of modelling that has been done and draw attention to some of the data issues. It is worth stating that the construction of physical and chemical models is often a helpful step in producing a new product. However, the *really difficult* technical barriers to making reliable and long-lived products are associated with the use of materials under extreme conditions; in this area modelling can offer little help and empiricism rules.

LOW-PRESSURE MERCURY DISCHARGES

There is no self-consistent model of the complete discharge but the mechanisms are well-enough understood.[6] Numerical models for the negative-glow have been published[7,8] but the bulk of the interest is in the positive column. This is where 185 and 254nm radiation from Hg atoms is produced with an efficiency of up to 80%. A phosphor is used to convert the UV radiation into visible light that has the colour properties that user needs. A review of this type of discharge is given by Maya and Lagushenko.[9]

Positive-Column Models

The operation of the positive column is clearly set out by Winkler *et al.*[10] Electrons gain energy in the electric field. They make elastic collisions with rare-gas atoms and Hg atoms. The rare-gas controls the electron mean free path so that the average electron energy is adjusted to give the optimum UV efficiency. The electrons form a near Maxwellian distribution with a deficiency of electrons at energies above 4.7 eV, the threshold for the first excited state of mercury. The electron temperature is about 12000K, whilst the gas temperature in the region of 40 celsius. The high-energy electrons excite a number of states of Hg. Transitions from Hg 6^3P_1 and 6^1P_1 levels give rise to the two UV lines. Radiation at these wavelengths is strongly trapped in the Doppler broadened lines. The UV efficiency depends on the density of Hg atoms. At low Hg number densities the radiation trapping is weak, but there are few atoms to radiate and the efficiency is low. At high mercury densities, transport of the photons out of the discharge involves so many emissions and re-absorptions that the excitation energy may be lost in a non-radiative events. The optimum conditions correspond to a mercury partial pressure of about 5-10 mtorr depending mainly on tube diameter. The pre-dominant ionization process is in two steps: excitation from ground to the 6^3P states followed by ionization from the 6^3P states. For this reason the metastable 6^3P_0 and 6^3P_2 states of Hg play a crucial part in achieving high efficiency. Direct ionization is relatively unimportant.

In their CR model, Winkler *et al.*[10] solve the Boltzmann equation for the electron energy distribution function (EEDF), but most other authors use a variety of approximations which have been compared by Yousfi *et al.*[11] Knowing the EEDF, the rates of excitation and de-excitation can be calculated from the electron collision cross-sections. The most complete procedure is to solve the continuity equations for the rates of gain and loss for each species at each radial node[12] and for the EEDF by full calculation or by approximation.[9,11,13] Of course, the radiative rates of gain and loss from excited states must also be included. Other versions of the model include time dependence.[13]

Data sources

Radiation data for the Hg lines has been very well documented.[14] The issues concerning radiation are related more to finding efficient algorithms for calculating the radiation transport that do not have the rather restrictive assumptions of the Bibermann-Holstein approach. In particular the assumptions concerning complete frequency redistribution and uniform lower state density are certainly not valid in important cases.[14,15] Recent approaches involve Monte Carlo simulations to give analytical equations that can be used in the models.[16]

As for collisional data, good summaries have been given by Waymouth[17] and Winkler *et al.*[10] Much of the data can be traced back to Kenty's 1948 paper[2] where many cross-sections were estimated. Elastic and ground state inelastic collision cross-sections for Hg have been deduced from electron transport data by Rockwood[18] and also Zissis *et al.*[19] Cross-sections for collisions to higher states are mostly estimated. At this meeting, however, I was made aware of important calculated cross-sections by Märk, Bartschat and Bray (this volume).

The major defects in the data used in the models seem to be related to the ion-ization processes. The very important cross-sections for ionization from the metastable ^3P levels of Hg are based on rather simplified quantum-mechanical calculations. There are many possible ionization processes that have now been incorporated in the models. These are various forms of chemi-ionization and associative ionization[20,21] that lead to Hg^+ or Hg_2^+ ions, ladder ionization from highly excited levels of Hg[22] and Penning ionization by metastable Ar atoms. Finally in some cases rare-gas ionization via metastable levels is significant. With several ion types present the ambipolar diffusion equations must be suitably modified.[24]

Although contemporary models include many of the same processes, individual authors make use of data adjusted in different ways. At least in some cases these data were adjusted in the early models that assumed Maxwellian EEDFs. For inelastic processes generally this procedure is likely to have led to significant errors in the adjusted data. For this reason individual models tend to have had their data tuned to give good agreement with reality over a limited range of input parameters.

Applications

The models fall short of a full computer-aided design tools. Nevertheless they have proved useful in engineering design. One example of this the link between lamp and circuit. Many new types of lamp are driven from electronic control gear. The lamp is a very complex load and electronic engineers have always found it difficult to understand its behaviour as a circuit element. One approach which could be very fruitful is to produce a time-dependent lamp model which can be incorporated as a circuit element into the electronic engineer's circuit simulator package.[25] Because of constraints on run time for a SPICE circuit simulator the lamp circuit element must be built up from a simplified time-dependent positive column model: great skill is required to judge what can be left out the lamp model.

In another development the 1D positive column model of a Hg rare-gas discharge has been applied to an inductively coupled discharge (ICD) lamp.[26] The discharge vessel contains a toroidal space.[27] The toroidal discharge is driven by the azimuthal electric field of a coil in a re-entrant in the centre of the torus; the plasma therefore forms the secondary of a transformer. The model[28] consists of an electromagnetic part which describes the RF fields produced by the coil as modified by the conductivity of the plasma; and a plasma part which is assumed only to depend on radius, and not on length. The 1D plasma model calculates properties per unit length and the effective length is then calculated by comparison with experiment on actual discharges. The output of this model is the complex impedance of the coil, UV output from the Hg lines, discharge current, voltage around the plasma loop, and the radial variation of electron temperature, excited state densities etc. For engineering development the information from the model of how the complex impedance of the driving coil depended on the plasma conditions was crucial. This model played an important part in the development of the GE Genura lamp®, and is one of few cases in the lamp industry that modelling arrived early enough to assist development of a product.

HP DISCHARGE OPERATION

High-pressure *sodium* discharge lamps provide one of the most efficient, long-lived sources for street lighting and for lighting warehouses, sports halls and other similar buildings.[29] The golden yellow light is a direct result of some intricate detail in the broadening of the sodium D lines as explained in the following section. Another important HP discharge is the *metal halide* type[30] that is used to produce high efficiency white light with excellent colour properties for commercial interiors. Again radiation data is the key issue as I discuss in the final section.

In contrast to the LP discharges described above, HP discharges are in Local Thermal Equilibrium (LTE)[31], or close enough to it for practical purposes. Collision rates are so high that the gas temperature is approximately equal to the electron temperature. The electric field accelerates electrons to a Maxwellian EEDF at 4000-

6000K depending on the ionization potential of the most easily ionised atoms. Heat is conducted to walls that are typically at 1000K, so a temperature gradient forms. In a sodium arc for example the electron and Na^+ densities in the centre regions which are at 4000K reaches $\sim10^{22}$ m^{-3}, whilst farther out where the temperature is 2000K, the densities are only $\sim10^{18}$ m^{-3}.[32] The current therefore flows mainly through the hot centre region of the plasma. The populations of excited states depend exponentially on local temperature through Boltzmann factors. This is why HP discharges are corded in appearance.

Once the spatial temperature profile is known from a solution of the energy balance equation including radiation, the properties of any LTE arc can be calculated, as the classic work by Lowke demonstrates[33] (see also reference 29 chapter 4). The calculation of the spectrum allowing for self-absorption is relatively straightforward as long as data for the absorption coefficient of the vapour is known; a matter I shall develop in detail in the following sections.

Densities in this type of lamp are so high that many spectral lines and molecular emission features suffer self-absorption. The intensity of the spectrum at each frequency ν (in W m^{-2} sr^{-1}) can be evaluated using the radiation transport equation[34] along a line of sight of length $2L$

$$I_\nu = \int_{-L}^{L} \kappa(\nu,T)B_\nu(T)\exp\left(\int_{x}^{L} \kappa(\nu,T)dx'\right)dx \qquad (1)$$

where $\kappa(\nu,T)$ is the absorption coefficient and $B_\nu(T)$ is the full radiator intensity. The expression under the first part of the integral is the local emission at point x (using Kirchhoff's law in LTE[31]), whilst the expression under the second integral shows the amount of attenuation of the ray as it travels out from the local position x to the wall. This absorption ensures that the spectral intensity never exceeds the full radiator intensity. A simple physical picture showing how resonance broadened sodium D lines become self-reversed is given in reference 32 p. 113.

Sodium has rather ideal properties for HP discharges, but the colour is not satisfactory for many indoor applications. Are there any other elements that give rise to excellent white colour spectra? The answer is yes, many. In particular scandium, titanium and many of the rare-earth elements have ideal spectra. Unfortunately all the elements of interest are so involatile that they cannot be used. However their halides and especially their iodides are much more volatile. For example, a lamp containing ScI_3 which is operated at a high enough wall temperature to evaporate the halides (~1000K) produces a high quality white light of great commercial importance.[6] The process is as follows. ScI_3 evaporates from the wall into the hot HP arc where it dissociates into radicals ScI_2 and ScI, and finally in the hottest parts of the arc into Sc and iodine. The hot parts radiate the Sc spectrum; there is little radiation from I. In the *metal halide* (MH) lamp the halide serves to introduce otherwise involatile metal atoms into the hot vapour near the centre of the arc where they can radiate their characteristic spectra.

Much effort has been expended in trying to find optimum combinations of metal halides.[6] MH halide lamps can be modelled as HP arcs, but the modelling has had limited success because of lack of radiation data as I describe in the final section. Another matter is of importance is the thermochemical calculation of the densities of the radiating species. This can readily be done for an LTE arc providing the Gibbs free energy of formation of the each chemical species is known or can be estimated. Much work has been done over the last 20 years to obtain this data. The situation is well reviewed by Hilpert in this volume. Equilibrium chemical composition programs linked to massive databases are available from a number of sources.

SPECTRA OF HP SODIUM ARCS

A standard HP sodium lamp contains sodium vapour at a pressure of about 0.1 atmosphere. The sodium D lines are so strongly broadened that the details of line broadening at wavelengths between 1 and 100nm from line centre have a dominating influence on emitted radiation. This is far beyond the range of most line broadening studies. The broadening of the far wings of spectral lines are usually treated by the quasi-static theory in which the radiator is supposed to be at rest relative to the perturber. The radiator and perturber interact and the emitted radiation will probably be at a different frequency from that at the line centre. For like atoms dipole-dipole interactions give rise to $1/r^3$ potentials where r is the atom separation; this is called resonance broadening. For the conditions in a HP sodium arc the atoms approach each other so closely that resonance broadening gives a very poor representation of broadening in the extreme wings of the line. The real molecular states must be used in the calculations.[34]

Fortunately an approximate calculation is relatively straightforward if the relevant molecular energy levels and dipole transition moments are known.[34] The basis is the classical Franck-Condon principle (CFCP) which assumes for electronic transitions of molecules that the vibrational and rotational levels can be ignored and that the electronic transitions are vertical; that is, they happen in a time so short that the nuclei do not move. The CFCP treats on an equal footing the cases of normal molecular spectra (bound-bound emission), the excimer case (bound-free emission), molecules with unstable upper states (free-bound emission) and colliding atoms (free-free emission). All these processes and their Kirchhoff related absorptions are important in the spectra of HP sodium arcs.

For a pair of atoms number densities [A] and [B] having a lower state potential $V_l(R_c)$at separation R_c and $V_l(\infty)$ at infinite separation, the absorption coefficient $\kappa(\nu,T)$ for the vertical transition to the upper state at separation R_c is proportional to (complete equation in reference 34)

$$\propto \frac{1}{\nu^2} R_c^2 (D_{ul}^2)_c \frac{1}{\left(\dfrac{d\nu}{dR}\right)_c} [\mathrm{A}][\mathrm{B}]\exp\{-(V_l(R_c) - V_l(\infty)/kT)\} \qquad (2)$$

where ν is the frequency, T the temperature, k Boltzmann's constant and D_{ul} the dipole transition moment at separation R_c. The derivative is the proportional to the slope of the potential difference between upper and lower states. It is obviously possible for the states to be parallel for a number of values of R_c. This corresponds to a classical satellite or 'head of heads'. The semi-classical theory of Szudy and Baylis[35] can be employed to remove the divergence and give an approximation to the satellite shape.

satellite shape. The total absorption coefficient is then the sum over all the Condon points c for which the difference potential is $h\nu$. If there are several pairs of states that can contribute to the absorption coefficient at ν, then the sum over these is also taken. It is of course possible to do the full quantum mechanical calculation and this reveals the rotational and vibrational structure and eliminates the need for the Szudy-Baylis approximation. Full calculations and measurements of absorption coefficients reported at this conference by Hyun-Kyung Chung, Shurgalin and colleagues are therefore of considerable interest.

By using the CFCP estimates of absorption coefficients in the radiation transport equation 1, Schlejen and colleagues have secured excellent agreement with calculated spectra over a wide range of conditions.[36,37] They have used as input data a wide range of quantum-mechanical calculations of energy levels, references to which can be found in their publications. Examples of spectra reproduced from Schlejen's thesis are shown in the following figures. The ordinates are not in the same units so the comparison is relative. However separate comparisons which I have done show very good agreement between the calculated and measured magnitudes of the D line self-reversal maxima.

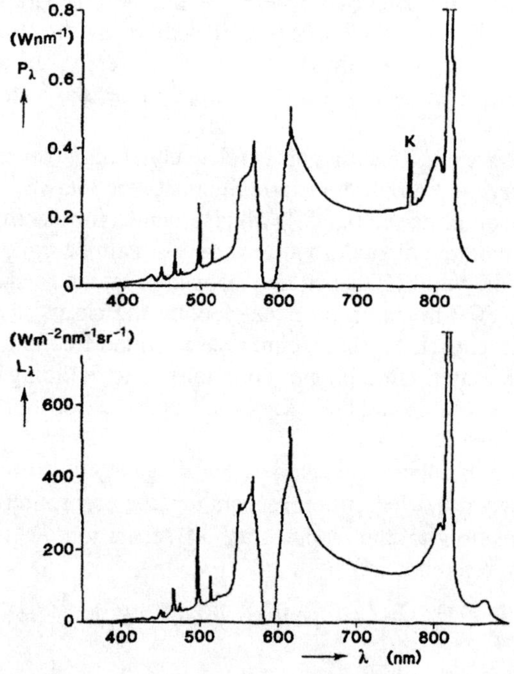

FIGURE 1. Measured (upper) and calculated (lower) spectra for a sodium vapour pressure 60 kPa (450 torr) and a tube radius of 3.8 mm. The line marked K is from a potassium impurity. Figure reproduced from Schlejen's thesis[36] with permission.

Figure 1 shows the extremely good agreement between measurement and calculation. Nevertheless there are differences. For example the magnitude of the infra-red continuum is too low in the calculations. The emission in this region is mostly from the transitions between the Na_2 $^1\Sigma_g^+$ and $^1\Sigma_u^+$ states. Although calculated dipole transition moments are available they were not included in the calculation of figure 1 and this may explain the small discrepancies.

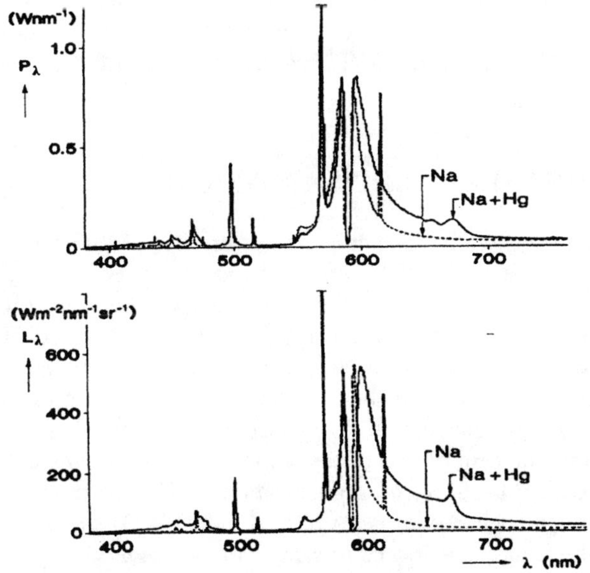

FIGURE 2. Showing the influence of Hg at a pressure of 225 kPa (1700 torr) on the spectrum of Na at a pressure of 15kPa (110 torr). Tube radius is 2.4 mm. Also shown (dotted line) is the data for the case with no Hg. For the calculation the temperature profile was parabolic with a centre temperature of 4000K and a wall temperature of 1500K. Reproduced from Schlejen's thesis[36] with permission.

Figure 2 shows similar calculations for a sodium discharge with added mercury, which has the benefit of lowering the thermal conductivity so that heat loss is minimised and radiated efficiency is increased. The ionization and excitation potentials of Hg are so high that very little Hg line radiation can be seen in the spectrum. Nevertheless the influence is seen in the extra red-wing broadening of the Na D line through van der Waals broadening and emission from NaHg excimer molecules. The agreement here is not so good, mainly because of the rather incomplete knowledge of the states of the NaHg dimer.

These results show that good quality molecular state and transition moment calculations can be used accurately to predict spectra under extreme conditions. It is worth emphasising that even for such a well-investigated molecule as Na_2 hardly any of the data needed has been *measured.* Most is not easily accessible to experiment, particularly for transitions between high-lying states and transitions that involve unstable states. Calculations of data are essential.

The corollary of this is that in certain cases high-pressure lamps offer an excellent way to test potential curves. Take for instance the case of the thallium Tl 535nm resonance line broadened by Hg. This is strongly broadened to the long wavelength side and in emission shows signs of the structure that is must be related to details of the Tl-Hg potentials. An HP discharge operated at 50 Hz has a maximum arc temperature that varies between about 5000K and 3500K. The time dependence of the emission line therefore involve emission from a wide range temperatures, or molecular separations through the Boltzmann factor in expression (2). Details of the equipment and methodology have been given by Wharmby *et al.*[38] where we show how to deduce high-temperature absorption coefficients from the time-varying emission spectra. This same approach can be used to validate molecular state calculations.

METAL-HALIDE LAMP MODELLING

Metal-halide lamps generally contain a high pressure of Hg to reduce the thermal conduction losses. Therefore the LTE energy balance can be solved for the spatial temperature profile, which in view of the other complications is fortunate indeed. Metal halide lamps are often short, squat cylinders or ellipsoids; so 2D or even 3D models are needed. Convection is usually a significant feature leading to special effects such as axial segregation of atomic species in vertical arcs and bowing in horizontal arcs. Of course convection also affects the energy balance by transferring energy from the radiating centre parts of the arc to other parts of the arc tube. A very good summary showing the level of sophistication in HP arc models, including an approximate treatment of electrodes is given by Fischer.[39] This paper deals with HP Hg arcs in which the radiation is treated in a very simple manner.

In metal halide arc models the major challenge is to calculate the spectrum. The problem is in the lack of the line strength and line broadening data for atoms and molecules. Dakin and colleagues [40-42] have taken this process about as far as it will go in a series of papers concerned with the modelling of the commercially important scandium-sodium iodide lamp. Even with many extra measured transition probabilities, a very complete model does not predict well how the critical lighting measures of colour and efficacy vary with the various input parameters. In other words the model helps in general understanding, but is little help in guiding the engineering design process.

My personal revelation concerning this issue came from a dissertation by Jun-Qin Lu[43] concerned with rare-earth halide arcs. He shows spectra of DyI_3-Ar and TmI_3-Ar arcs measured in the lines of sight through the centre of the arc and and halfway to the wall. In the former, the side-on intensity is dominated by the hottest regions, whilst in the latter it is dominated by the radiation from much cooler parts. From the hotter spectrum, there is thick forest of rare-earth atomic lines, but in the hotter spectrum there is also very strong continuum. This same continuum dominates the cooler spectrum, where atomic lines are much weaker. This means that the continuum must be molecular in origin. Since the plasmas have very simple compositions, the only possible conclusion is that this radiation is from DyI and TmI molecules, with perhaps some absorption from the di-iodides and tri-iodides. Most practical metal halide lamps have continua which, by analogy, are emitted by unidentified molecules; in some case it accounts (see below) for the major part of the visible spectra. What hope is there of calculating realistic realistic spectra? Another approach is needed.

Let us take as an example the ceramic metal halide lamps that have recently become so important commercially.[44,30] These lamps have a translucent alumina arc tube of diameter and length ~10mm and contain Na, Tl, Dy, Hg and I. The ceramic arc tube allows operation at higher power density than in fused silica with the result that the partial pressure of the constituents is higher and the spectrum more continuous. Moreover reaction rates with the arc tube are lower. For the user the benefits are better and more constant colour properties initially and through life.

Clearly the representation of the continuum in the model is a major challenge. Our approach is based[45] on an empirical representation of the rare-earth contribution obtained from measurement of the emission spectrum. We have used a measurement at 70W (figure 3) to explore the effect of changing power input on the colour properties.

We can use the data for Hg, Tl and Na radiation (including dimer effects described above) to calculate the properties of the arc using Dakin's 2D code.[44] This code includes thermochemistry, conduction, convection and 3D radiation transport. We then assume that most of the visible radiation from the rare-earth atoms is from upper states radiating to the ground state; the relatively low arc temperature that is a feature of arcs containing Na weights the emission to that from low lying energy levels. Let us assume therefore that the emission coefficient (measured in W m^{-3} sr^{-1}) at each wavelength λ can be expressed by

$$\varepsilon_\lambda = \frac{1}{4\pi}[\mathrm{R}]\exp(-\frac{hc}{\lambda kT})\frac{hc}{\lambda}A_\lambda^* \qquad (3)$$

where [R] is the concentration of rare-earth atoms, h and k are Planck and Boltzmann constants and c is the velocity of light and A^*_λ is the unknown spectral transition probability (measured in s^{-1} nm^{-1}) which is to be determined. For the purpose of calculation we divide the spectrum into 5nm bands over which A^*_λ is assumed constant. In LTE the absorption coefficient $\kappa(\lambda, T)$ for use in the radiation transport calculation (equation 3) in the model is calculated using Kirchhoff's law.

First we do a full model calculation of the arc properties assuming that A^*_λ is independent of wavelength. This yields a spectrum with the Na, Tl and Hg self-reversed lines on top of an arbitrary continuum. In each 5nm band we take the ratio of the measured spectrum at 70W to the calculated spectrum; scale A^*_λ according to this ratio; and finally recalculate the arc properties. After 3 iterations this process has converged to yield a set of values of A^*_λ in each 5nm band which represent the radiation not accounted for by the Na, Tl and Hg atoms.

When the power is changed in the calculation this affects the temperature of the molten mixed halide dose, thereby changing the partial pressure of each constituent of the dose. The full energy balance is therefore calculated for the new vapour composition using the A^*_λ values determined as above to give the temperature profiles, arc power and spectra. Figure 3 shows the measurements in 5nm bands in the range 60 to 90W, the 70W spectrum being used to establish the values of A^*_λ. Figure 3 also shows the calculated spectra of the arc for the same range of powers. For simplicity we did not convolve the calculated spectra with the instrument function of the measuring spectrometer, so calculated lines are appear narrower than the measured lines; the issue here was the calculation of the continuum where the agreement is rather impressive. Colour properties calculated were much more accurate than for other MH models, and with accuracy good enough to act as an aid to engineering design.

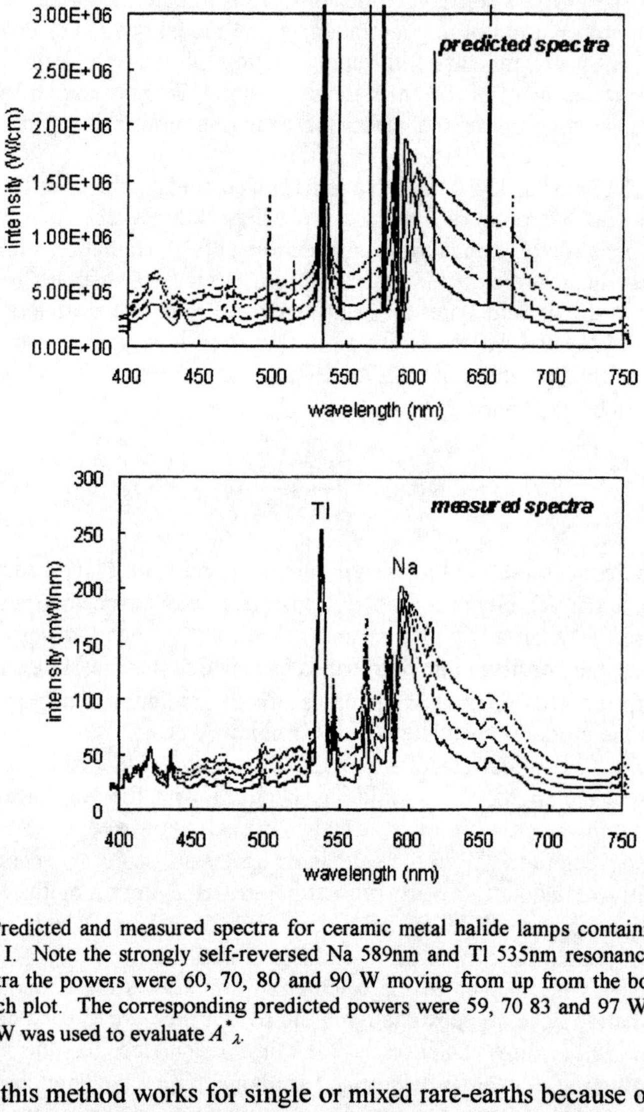

FIGURE 3. Predicted and measured spectra for ceramic metal halide lamps containing Na, Tl, rare-earths, Hg and I. Note the strongly self-reversed Na 589nm and Tl 535nm resonance lines. For the measured spectra the powers were 60, 70, 80 and 90 W moving from up from the bottom line to top spectrum in each plot. The corresponding predicted powers were 59, 70 83 and 97 W. The measured spectrum at 70W was used to evaluate A^*_λ.

Note that this method works for single or mixed rare-earths because of their similar thermochemistry. It also subsumes the radiation from other molecules that might be present such as rare-earth-iodide dimers, TlHg etc. There are clearly ways to improve this procedure; (i) by extending equation 3 to include more species; (ii) by including the possibility of transitions terminating in an effective energy level which is not the ground state (see reference 3 p. 53). By measuring time-dependent spectra from an arc operated at a low frequency, spectra representing a much wider range temperatures can be obtained; similar data can be obtained from side-on measurements but very high optical quality, not usually possible in lamps, is needed. The much wider range of temperatures allows the molecular and atomic effects can be discriminated[38] because molecular features dominate emission from the cooler regions.

CONCLUSIONS

This conference made it clear that very complete quantum-mechanical calculations of key data well tested by high quality experiments is the way forward. An important issue is that self-consistent sets of data are needed. This could open up a new phase of lamp models which can be used for computer aided design of new lamp types. Signs that this is happening are that in the USA, EPRI (the Electric Power Research Institute) and Osram Sylvania have co-funded a three year program of pre-competitive research, which has involved investigations on a number of lamp related topics. Whilst some of the information is regarded as company proprietary, a considerable amount of new data has been or will be published. Of particular interest in the context of this paper is work on electron impact cross-sections, chemi-ionization cross sections, branching ratios in rare-earth elements and improved methods of calculating radiation transport in LP discharges. It is to be hoped that the program will continue with increased input from lamp companies. The data needed for modelling the radiation from arcs containing rare-earths seems beyond reach at present.

ACKNOWLEDGMENTS

The mostly unpublished work on ICD lamps was done with former colleagues Graeme Lister, Steve Coe, Jim Dakin and Darryl Michael in a most enjoyable collaboration. I am very grateful for helpful discussions with Georges Zissis, Graeme Lister, and Tim Sommerer.

REFERENCES

1. Druyvesteyn, M.J. and Penning, F. M., *Reviews of Modern Physics* **12**, 87-174 (1940)
2. Kenty, C., *J. Appl. Phys.* **21**, 1309-1318, (1950)
3. Elenbaas, W., *The High Pressure Mercury Discharge*, Amsterdam: North-Holland, 1951
4. Cayless, M. A., *Brit. J. Appl. Phys.*, **14**, 863 (1963)
5. Sommerer T. J., "Atomic Data and Methods for Low-Pressure Lamps" in *Atomic and Molecular Data and Their Applications*, edited by P.J.Mohr and W.L.Wiese, AIP Conference Proceedings 434, American Institute of Physics, 1998, pp 295-313
6. Waymouth, J. F., *Electric Discharge Lamps*, Cambridge: MIT Press, 1971
7. Coe, S. E., Stocks, J. A. and Tambini, A. J., *J. Phys. D: Appl. Phys.*, 26, 1203-1210 (1993)
8. Ingold, J. H., *Phys. Rev.* A **43**, 3093, (1991)
9. Maya, J. and Lagushenko, R., "Progress in Low Pressure mercury-Rare Gas Discharge Research" in *Advances in Atomic Molecular and Optical Physics*, **26**, NewYork:Academic Press, 1990
10. Winkler, R. B., Wilhelm, J and Winkler, R, *Annalen der Physik* , **40**, 90-118 and 119-139 (1983)

11. Yousfi, M., Zissis, G., Alkaa, A. and Damelincourt J. J., *Phys. Rev.* **A 42**, 978-988 (1990)
12. Dakin, J. T., *J. Appl. Phys.*, **60**, 563-570 (1986)
13. Wani, K., *J. Appl. Phys.*, **68**, 5052 (1990)
14. van der Weijer, P. and Cremers, R. M. M., "Pulsed Optical Pumping in Low Pressure Discharges" in *Radiative Processes In Discharge Plasmas*, edited by J. M. Proud and L. H. Luessen, New York: Plenum Press, 1986
15. Molisch, A. F. and Oehry B. P., *Radiation Trapping in Atomic Vapours*, Oxford: Clarendon Press, 1998, chapter 18
16. Lawler J . E. and Curry J. J., *J. Phys. D: Appl. Phys.*, **31**, 3235-3242 (1998)
17. Waymouth, J. F., "Collision Phenomena in Electrical Discharge Lamps" in *Applied Atomic Collision Physics*, edited by H. S. W. Massey, E. W. Daniel and B. Bederson, New York: Academic Press, 1982, **5**, 331-347
18. Rockwood, S. D., *Phys. Rev.* A **8**, 2348-2358 (1973)
19. Zissis G., Benetruy, P. and Bernat I., *Phys. Rev.* A **45**, 1135-1148, (1992)
20. Vriens, L., Keijser, R. A. J. and Ligthart, F. A . S., *J. Appl. Phys.*, **49**, 3807-3813 (1978)
21. Lin, T., Goto, T., Arai, T. and Murayama, S., *J. Appl. Phys.*, **69**, 4201-4205 (1991)
22. Wani, K., *J. Appl. Phys.*, **75**, 4917-4926 (1994)
23. Karov, M. Rusinov, I. and Blagoev, A., *J. Phys. B: At. Mol. Opt.*, Phys., **30**,1361-68 (1997)
24. Lister, G. G. and Coe, S. E., *Computer Physics Comm.*, **75**, 160-184 (1993)
25. Vos, T., Ligthart, F., Hendrix, M., Stommen, T., Chittka, U., v.d. Mullen, J. and v.d. Bosch, W., "Modelling of Fluorescent Lamps for Power Supply Design" in *Proc. 8th Int. Symp. on Science and Technology of Light Sources*, edited by G. Babücke, Greifswald: Institute of Low Temperature Plasma Physics, 1998, pp. 288-9
26. Wharmby, D. O., Bull. Am. Phys. Soc., **39**, 1495 (1994)
27. Wharmby, D. O., "Electrodeless Lamps" in *Lamps and Lighting* 4th Edition, edited by J. R. Coaton and A. M. Marsden, London: Arnold, 1997, pp.216-225
28. Lister, G. G. and Cox, M., *Plasma Sources Sci. Technol.*, **1**, 67-73 (1992)
29. de Groot J. J. and van Vliet J. A. J. M., *The High –Pressure Sodium Lamp*, Deventer: Kluwer, 1986
30. Preston, B. and Odell, E. C., "Metal Halide Lamps" in *Lamps and Lighting* 4th Edition, edited by J. R. Coaton and A. M. Marsden, London: Arnold, 1997
31. Richter J. "Radiation from Hot Gases" in *Plasma Diagnostics* edited by W. Lochte-Holtgreven Amsterdam: North-Holland, 1968
32. Wharmby, D. O., "Radiation and Light Production" in *Lamps and Lighting* 4th Edition, edited by J. R. Coaton and A. M. Marsden, London: Arnold, 1997, pp.216-225
33. Lowke J. J., *J. Quant. Spectrosc. Radiat. Transfer*, **9**, 839 (1969)
34. Wharmby, D. O., "Molecular Spectral Intensities in Radiative Plasmas" in *Radiative Processes In Discharge Plasmas*, edited by J. M. Proud and L. H. Luessen, New York: Plenum Press, 1986, 147-170
35. Szudy, J. and Baylis, W., *J. Quant. Spectrosc. Radiat. Transfer*, **15**, 641 (1975)
36. Schlejen, J., *Spectroscopy of NaNa and NaHg diatomics in high-pressure vapours*, Thesis, University of Leiden, Holland, Amsterdam: Editions St.Jaques, 1987
37. de Groot, J. J., Schlejen, J., Woerdmann, J.P. and de Kievet, M .F. M., *Philips J. Res.*, **42**, 87-101 (1987)
38. Wharmby, D. O., Ede, N. R. and Meads J., J. Phys. D: Appl. Phys., **21**, 15-24 (1988)
39. Fischer, E., Philips J. Res., **42**, 59-85 (1987)
40. Dakin, J. T., Rautenberg, T. H. and Goldfield, E. M., J. Appl. Phys., **66**, 4074-4088 (1989)
41. Dakin, J. T., Gilliard R. P., J. Appl. Phys., **66**, 4074-4088 (1989)
42. Dakin, J. T., Gilliard R. P., J. Appl. Phys., **60**, 1281-1288 (1986)
43. Lu Jun-Qin, *Der Strahlungsmechanismus eines Seltenen-Erd-Halogenid Plasmas mit hohem Dampfdruck*, Thesis, Fakultät Universitat Elektrotechnik, Universität Fridericiana Karlsruhe, Germany, 1983, pp. 97-100
44. Seinen, P. A., "High Intensity Discharge Lamps with Ceramic Envelopes" in *Proc. 7th Int. Symp. on Science and Technology of Light Sources*, edited by R. Itatani and S. Kamiya, Tokyo: Illuminating Engineering Society of Japan, 1995, pp.101-109
45. Sommerer T. J. and Wharmby D. O., Bull. Am. Phys. Soc., **43**, 1453 (1998)

Modeling Network for Argon Glow Discharges: The Output Cannot Be Better Than The Input

Annemie Bogaerts and Renaat Gijbels

Department of Chemistry, University of Antwerp, Universiteitsplein 1, B-2610 Wilrijk-Antwerp, Belgium

Abstract. A modeling network has been developed for Ar glow discharges with Cu cathode, consisting of several submodels, i.e., Monte Carlo simulations, fluid approaches and collisional-radiative models. The plasma species described in the models, include the electrons, Ar atoms, Ar ions, fast Ar atoms, Ar atoms in various excited levels, sputtered Cu atoms and Cu ions, both in the ground state and in various excited levels. In this paper, the models are described, and the various input data needed for the calculations are discussed. Typical results of the models include the electrical characteristics, the potential and electric field distribution, the densities, fluxes and energies of the various plasma species, information about collision processes in the plasma and about sputtering at the cathode, optical emission intensities, etc. Some of these results will be presented here, to illustrate the possibilities of the model.

INTRODUCTION

Glow discharge plasmas are used in a large number of application fields, such as in the semiconductor industry (for the deposition of thin layers and the etching of surfaces), in materials technology (for the deposition of hard coatings), in the laser and light industry, in flat plasma display panels, and also for the spectrochemical analysis of (mainly) solid materials [1]. In the latter application, the material to be analyzed is used as the cathode of the glow discharge, which is sputter-bombarded by plasma species. The sputtered cathode atoms arrive in the plasma, where their concentration can be measured with atomic absorption or fluorescence spectrometry (glow discharge atomic absorption spectrometry or fluorescence spectrometry; GD-AAS or GD-AFS). Moreover, the atoms can become ionized or excited in the plasma. The corresponding ions can be detected in a mass spectrometer (glow discharge mass spectrometry; GDMS), whereas the excited atoms and ions emit characteristic photons which can be detected with optical emission spectrometry (GD-OES).

In order to improve the results of glow discharge analytical spectrometries, we have developed a number of models to describe the behavior of various plasma species. The models can, however, also be useful for other application fields.

Different kinds of models have been used in the literature to describe glow discharge plasmas, i.e., analytical approaches, fluid models, Boltzmann equations, Monte Carlo methods and particle-in-cell simulations. All these models have their specific advantages and disadvantages. Therefore, we use a hybrid model, which is a combination of (some of) the above models for the various plasma species, in order to

CP543, *Atomic and Molecular Data and Their Applications,* edited by K. A. Berrington and K. L. Bell

combine the advantages and avoid the disadvantages of the individual models. In the following, we will give a brief description of the models and show some typical results.

DESCRIPTION OF THE MODELS

The hybrid model network is developed for an Ar plasma with a Cu cathode. The species assumed to be present in the plasma are electrons, Ar gas atoms, Ar ions, fast Ar atoms, Ar atoms in various excited levels, sputtered Cu atoms and Cu ions, in the ground state and in excited levels. These species are described with a combination of Monte Carlo simulations, fluid approaches and collisional-radiative models. Briefly, the Monte Carlo simulations are used for the "fast" plasma species, which are not in equilibrium with the electric field (such as fast electrons); the fluid approach is applied for the "slow" plasma species, which can be considered in equilibrium with the electric field (like slow electrons or Ar ions); and the collisional-radiative models describe the behavior of the atoms and ions in excited levels. An overview of the various plasma species and the models used to describe them is given in Table 1.

The Monte Carlo models are all developed in three dimensions, whereas the fluid and collisional-radiative models are two-dimensional. Indeed, the model network is generally applied to cylindrically symmetrical glow discharge cells, so that the three dimensions can be reduced to two dimensions. The model network has mainly been applied to direct current (dc) glow discharges (see e.g., [2] and references therein), but capacitively coupled radio-frequency (rf) discharges have been investigated as well ([3] and references therein), and the modeling of microsecond pulsed discharges is in progress [4]. Below, each of the models will be described in some more detail.

TABLE 1. Overview of the Different Plasma Species and the Models Used to Describe Them.

Plasma Species	Model
Ar gas atoms	no model (uniformly distributed + at rest) or: gas heating model (dc case)
fast electrons	Monte Carlo model
slow electrons	fluid model
Ar^+, Ar^{2+}, Ar_2^+ ions	fluid model
Ar^+ ions in CDS	Monte Carlo model
fast Ar atoms in CDS	Monte Carlo model
Ar atoms in various excited levels	collisional-radiative model
sputtered Cu atoms: thermalization	Monte Carlo model
Cu atoms and ions in ground state + excited levels	collisional-radiative model
Cu^+ ions in CDS	Monte Carlo model

Monte Carlo Model for the Fast Electrons

It should be mentioned that the electrons are split up in two groups, i.e., the fast and slow electrons, depending on their energy. Indeed, the electrons are called "fast" when they have enough energy to produce inelastic collisions. Since excitation and ionization of Ar ground state levels are the most frequent collision types, the threshold is in practice mostly defined at 11.55 eV (which is the lowest excited Ar level).

The fast electrons are simulated with a Monte Carlo model. A large number of electrons are followed, one after the other, during successive time-steps. Their trajectory during one time-step (Δt) is calculated with Newton's laws:

$$
\begin{aligned}
z &= z_0 + v_{z_0}\Delta t + \frac{qE_{ax}(z,r,t)}{2m}(\Delta t)^2 \\
x &= x_0 + v_{x_0}\Delta t + \frac{qE_{rad}(z,r,t)\cos(\alpha)}{2m}(\Delta t)^2 \\
y &= y_0 + v_{y_0}\Delta t + \frac{qE_{rad}(z,r,t)\sin(\alpha)}{2m}(\Delta t)^2 \\
v_z &= v_{z_0} + \frac{qE_{ax}(z,r,t)}{m}\Delta t \\
v_x &= v_{x_0} + \frac{qE_{rad}(z,r,t)\cos(\alpha)}{m}\Delta t \\
v_y &= v_{y_0} + \frac{qE_{rad}(z,r,t)\sin(\alpha)}{m}\Delta t
\end{aligned}
\qquad (1)
$$

where z_0, x_0, y_0 and z, x, y are the position coordinates before and after Δt, $v_{z_0}, v_{x_0}, v_{y_0}$ and v_z, v_x, v_y are the velocities before and after Δt, E_{ax} and E_{rad} are the axial and radial electric field, as a function of axial and radial position and time (obtained from the Ar ion - slow electron fluid model, see below), α is the azimuthal angle of the radial position (i.e. the angle of the radial position coordinates with respect to the x-axis), and q and m are the electron charge and mass, respectively.

The probability of a collision during that time-step, $Prob_{coll}$, is calculated and compared with a random number between 0 and 1:

$$Prob_{coll} = 1 - \exp(-\Delta s \Sigma(n\sigma_{coll}(E))) \qquad (2)$$

where Δs is the distance traveled during Δt; n and $\sigma_{coll}(E)$ are the densities of the target particles and the cross sections of the different collision types of the electron with energy E. If the probability is lower than the random number, no collision occurs, and the Monte Carlo solver continues with the next electron during that time-step. If the probability is higher, a collision takes place. The collisions taken into account in the model, are elastic collisions with Ar ground state atoms, electron impact ionization, excitation and de-excitation for all Ar atom levels, Cu atom and Cu ion levels, and also electron-electron Coulomb scattering. Elastic collisions and electron impact excitation and ionization from the Ar ground state are the most frequent processes. Their cross sections are adopted from ref. [5]. The cross sections used for detailed electron impact ionization, excitation and de-excitation between various excited Ar atom and Cu atom and ion levels are explained in detail in refs. [6,7]. Some of these cross sections are based on experimental data, others on quantum-mechanical calculations, but most of them are obtained from empirical formulas (e.g., depending on the energy difference between the levels), and are subject to considerable uncertainties. Nevertheless, it appeared that they have only minor effect on the final calculation results, because they determine only the level populations of the excited levels (see below) and not the overall electron behavior. Finally, the cross section for electron-electron Coulomb scattering is taken from ref. [8]. The accuracy of this cross section is again not so important for the general electron behavior when only electrons with energy above 11.55 eV are simulated with the Monte Carlo model. Indeed, the cross section rises with decreasing energy (e.g., it is $3x10^{-17}$ cm^2 at 1000 eV and rises to 10^{-7} cm^2 at 0.01 eV), but since the density of the target particles (i.e., electrons) is 4-5 orders of magnitude lower than the Ar gas atom density, the process is, in practice, only important at electron energies of a few eV and lower [6].

To determine which collision takes place, the partial collision probabilities of the various collisions are calculated, and the total collision probability (which is equal to one, since it is the sum over all partial collision probabilities) is subdivided in intervals with lengths corresponding to these partial collision probabilities. A second random number between 0 and 1 is generated, and the interval in which the random number falls, determines the collision that takes place. Then, the new energy and direction after collision are also defined by random numbers, based on energy and angular differential cross sections [9-11] (see also refs [12-15] for more details).

This procedure is repeated for all electrons during that time-step. Then, the Monte Carlo procedure is repeated during the next time-step, again for all electrons, and so on, until (periodic) steady state is reached. However, the electrons can also be removed from the Monte Carlo model, when they undergo recombination at the cell walls, or when they are transferred to the slow electron group, which happens when their energy drops below 11.55 eV (see above). Indeed, these "slow" electrons cannot give rise to the most frequent inelastic collisions anymore, i.e., excitation and ionization from the Ar ground state; their most important role is carrying the electrical current and providing negative space charge, which can as well be described with a fluid model (see below), to save calculation time. However, when we want to calculate the detailed excitation and de-excitation between the various excited Ar and Cu levels (for the collisional-radiative models; see below), all electrons, also the slow ones, are simulated with the Monte Carlo model, because low energy electrons can cause de-excitation or excitation to nearby levels. More information about this model, developed for the dc, rf and pulsed mode, can be found in refs. [4,12-16].

Fluid Model for the Slow Electrons and Ar ions

The slow electrons are described, together with the Ar ions, in a fluid model. It is based on thermal equilibrium with the local electric field, which means that the energy gained from the electric field is locally balanced by the energy lost due to collisions. The fluid model consists of the continuity equations for ions and electrons, the transport equations for ions and electrons, an electron energy balance equation for the electrons (only used in the model when it is applied to the rf discharge, see [15,16]), and Poisson's equation to obtain the electric field distribution. Not only Ar^+ ions, but also Ar^{2+} and Ar_2^+ ions are considered in the model. This leads to 10 coupled differential equations:

$$\frac{\partial n_{Ar^+}(z,r,t)}{\partial t}+\overline{\nabla}\cdot\overline{j_{Ar^+}}(z,r,t)=R_{prod,Ar^+}(z,r,t)-R_{loss,Ar^+}(z,r,t)$$

$$\frac{\partial n_{Ar^{2+}}(z,r,t)}{\partial t}+\overline{\nabla}\cdot\overline{j_{Ar^{2+}}}(z,r,t)=R_{prod,Ar^{2+}}(z,r,t)-R_{loss,Ar^{2+}}(z,r,t)$$

$$\frac{\partial n_{Ar_2^+}(z,r,t)}{\partial t}+\overline{\nabla}\cdot\overline{j_{Ar_2^+}}(z,r,t)=R_{prod,Ar_2^+}(z,r,t)-R_{loss,Ar_2^+}(z,r,t)$$

$$\frac{\partial n_{e,slow}(z,r,t)}{\partial t}+\overline{\nabla}\cdot\overline{j_{e,slow}}(z,r,t)=R_{prod,e,slow}(z,r,t)-R_{loss,e,slow}(z,r,t)$$

$$\overline{j_{Ar^+}}(z,r,t)=\mu_{Ar^+}n_{Ar^+}(z,r,t)\overline{E}(z,r,t)-D_{Ar^+}\overline{\nabla}n_{Ar^+}(z,r,t)$$

$$\overline{j_{Ar^{2+}}}(z,r,t)=\mu_{Ar^{2+}}n_{Ar^{2+}}(z,r,t)\overline{E}(z,r,t)-D_{Ar^{2+}}\overline{\nabla}n_{Ar^{2+}}(z,r,t) \qquad (3)$$

$$\overline{j_{Ar_2^+}}(z,r,t)=\mu_{Ar_2^+}n_{Ar_2^+}(z,r,t)\overline{E}(z,r,t)-D_{Ar_2^+}\overline{\nabla}n_{Ar_2^+}(z,r,t)$$

$$\overline{j_{e,slow}}(z,r,t)=-\mu_{e,slow}n_{e,slow}(z,r,t)\overline{E}(z,r,t)-D_{e,slow}\overline{\nabla}n_{e,slow}(z,r,t)$$

$$\frac{\partial w_{e,slow}(z,r,t)}{\partial t}+\overline{\nabla}\cdot\left(\frac{5}{3}\varepsilon_{e,slow}(z,r,t)\overline{j_{e,slow}}(z,r,t)+\overline{q}(z,r,t)\right)=$$

$$-e\overline{j_{e,slow}}(z,r,t)\cdot\overline{E}(z,r,t)-R_{w,e,slow}(z,r,t)$$

$$\nabla^2V(z,r,t)+\frac{e}{\varepsilon_0}\left(n_{Ar^+}(z,r,t)+n_{Ar^{2+}}(z,r,t)+n_{Ar_2^+}(z,r,t)-n_{e,slow}(z,r,t)\right)=0$$

n and j are the densities and fluxes, respectively, of the Ar ionic species and electrons. R_{prod} and R_{loss} are the production and loss rates. Production of Ar^+ ions is due to electron impact ionization, which is calculated in the electron Monte Carlo model above, as well as Ar^{2+}-electron recombination. Loss of Ar^+ ions is due to Ar^+-electron recombination, atomic to molecular ion conversion from Ar^+ to Ar_2^+, and electron impact ionization from Ar^+ to Ar^{2+}. The production processes for the Ar^{2+} ions include electron impact ionization from Ar^0 and from Ar^+ (calculated in the Monte Carlo model above). The loss processes are Ar^{2+}-electron recombination and two-electron asymmetric charge transfer with Cu^0 (which is a resonant process [17]). Production of Ar_2^+ ions is caused by Hornbeck-Molnar associative ionization of Ar atoms, as well as by atomic ion to molecular ion conversion (see above). Loss of Ar_2^+ ions is assumed to occur entirely due to dissociative recombination. Finally, production of the slow electrons is due to electron transfer to the slow electron group (calculated in the above electron Monte Carlo model), whereas loss of these electrons is due to various electron-Ar ion recombination mechanisms. The data needed to calculate the above production and loss rates can be found in ref. [18]. Further, $\overline{E}$ is the electric field and V is the electric potential. D and μ are the diffusion coefficients and mobilities, respectively, of the Ar ionic species and electrons, for which the values are given in ref. [13]. Finally, in the electron energy balance equation: $w_{e,slow}$ is the slow electron energy density ($=n_{e,slow}\varepsilon_{e,slow}$, where $\varepsilon_{e,slow}$ is the mean electron energy), q is the heat flux [15], the first term in the right-hand side describes the energy gain by the electric field (i.e., Ohmic heating) and the second term, $R_{w,e,slow}$, represents the electron energy loss due to collisions. It should be mentioned that this electron energy balance equation is not included in our fluid model for the dc or pulsed discharge. Indeed, once the electrons are transferred to the slow electron group, which are described in the fluid model, their exact energy is not so important, because they cannot give rise to inelastic collisions (e.g., ionization) anymore. However, in the rf discharge, the slow electrons can become heated again by the oscillating rf electric field, and they can produce again ionization. Therefore, in the rf discharge, it is necessary to calculate their energy with this energy balance equation in the fluid model, to compute the ionization rate due to these electrons.

The four transport equations can be inserted into the four continuity equations, leading to a set of five (or six) coupled differential equations (i.e., including Poisson's equation, and, in the rf case, the electron energy balance equation). Due to the severe non-linearity and strong coupling of the equations, solving this model is a difficult numerical problem. The method we used was developed by Goedheer and coworkers [19], and is based on the Scharfetter-Gummel exponential scheme for the transport equations [20,21]. The basic idea is that the particle fluxes are assumed constant between mesh points, instead of the densities. The advantage of this scheme is its ability to switch between situations where either the migration component or the diffusion component of the particle flux is dominant (i.e. high and low electric field, cathode dark space (CDS) and negative glow (NG), respectively). More details about this model can be found in refs. [13-16,18].

Monte Carlo Model for the Fast Ar Ions and Atoms in the CDS

The Ar ions are not really in equilibrium with the strong electric field in the CDS (this is the region adjacent to the cathode, characterized by a strong electric field), and the fluid model is, therefore, only an approximation for the Ar ions in this region. Therefore, the Ar ions are also simulated with a Monte Carlo method in this region, which enables us to calculate the Ar ion energy distribution at the cathode, needed to calculate the amount of sputtering (see below). Only the Ar^+ ions are treated with this Monte Carlo model, because the Ar^{2+} and Ar_2^+ ions have a lower density and flux, and they play only a minor role in the sputtering process [18]. However, beside the Ar^+ ions, also the fast (i.e., non-thermal) Ar atoms (Ar^0_f), which are created from collisions of the Ar ions, are followed with this Monte Carlo model, since it was found that they play a dominant role in the sputtering process [12].

The Ar ion and fast Ar atom Monte Carlo model is similar to the electron Monte Carlo model. During successive time-steps, the trajectory of the ions and atoms is calculated by Newton's laws, and the occurrence of a collision, the nature of the collision and the new energy and direction after collision are determined by random numbers. The collision processes taken into account are elastic scattering collisions with Ar ground state atoms, for both ions and atoms, symmetric charge transfer for Ar ions (which is actually also a form of elastic collisions, because there is no change in kinetic energy [22]), and ion and atom impact ionization, excitation and de-excitation for all Ar atom levels. The cross sections for elastic collisions (with isotropic scattering and with backscattering, i.e., so-called charge transfer [22]) are adopted from [22]. The cross sections for ionization and excitation from the Ar ground state are taken from [23]. These processes are generally not taken into account in plasma models, but they were found to be important at the high discharge voltages of 1000 V encountered in analytical glow discharges, since the cross sections increase with rising energy, reaching a maximum around 1000-10000 eV. Indeed, we found that these processes had to be incorporated in our models, in order to predict the correct current-voltage-pressure characteristics in dc analytical glow discharges [24]. Finally, the cross sections for ionization, excitation and de-excitation from excited Ar levels are described in [6]. Due to lack of data, we had to use empirical formulas to calculate these cross sections, but again, this determined only the excited level populations (to a small extent) and had no effect on the overall fast Ar ion and atom behavior. More information about this Monte Carlo model can be found in refs. [12,24,25].

Heat Transfer Model for the Ar Gas Atoms

In most cases, we have assumed that the Ar gas atoms are at rest, uniformly distributed throughout the discharge, and no specific model is applied to describe their behavior. However, recently we have developed a model for the dc glow discharge to calculate gas heating, and consequently the gas temperature distribution, which yields, when the gas pressure is constant, a non-uniform gas density distribution. The gas temperature is calculated with the heat conduction equation:

$$\frac{\partial^2 T_g}{\partial z^2} + \frac{1}{r}\frac{\partial}{\partial r}\left(r \frac{\partial T_g}{\partial r} \right) = -\frac{P}{\kappa} \tag{4}$$

where T_g is the Ar gas temperature (function of z and r position), P is the power input and κ is the thermal conductivity (= 1.8×10^{-4} W cm^{-1} K^{-1} for Ar). The power input in the Ar gas is calculated in the ion and atom Monte Carlo models, based on collisions and subsequent energy transfer of the Ar ions, fast Ar atoms and Cu atoms (see below) to the Ar gas atoms. A detailed explanation of this model, as well as the required input data (such as the cathode temperature and thermal accommodation coefficients at the cell walls), can be found in ref. [26].

Collisional-radiative Model for the Ar Excited Atoms

A collisional-radiative model is actually a kind of fluid model, specifically applied to atoms and ions in excited levels. It consists of a set of coupled balance equations (one for each level) with various production and loss terms. Since the production and loss processes are either due to collisions or radiative decay (see below), this model is called “collisional-radiative model”.

Figure 1 shows a schematic energy diagram of the Ar atomic levels taken into account in this model. 64 Ar atomic excited levels are considered; most of them are effective levels, i.e., a group of individual levels with comparable excitation energy and quantum numbers. The four 4s levels (i.e., two metastable levels and two resonant levels) are, however, treated separately.

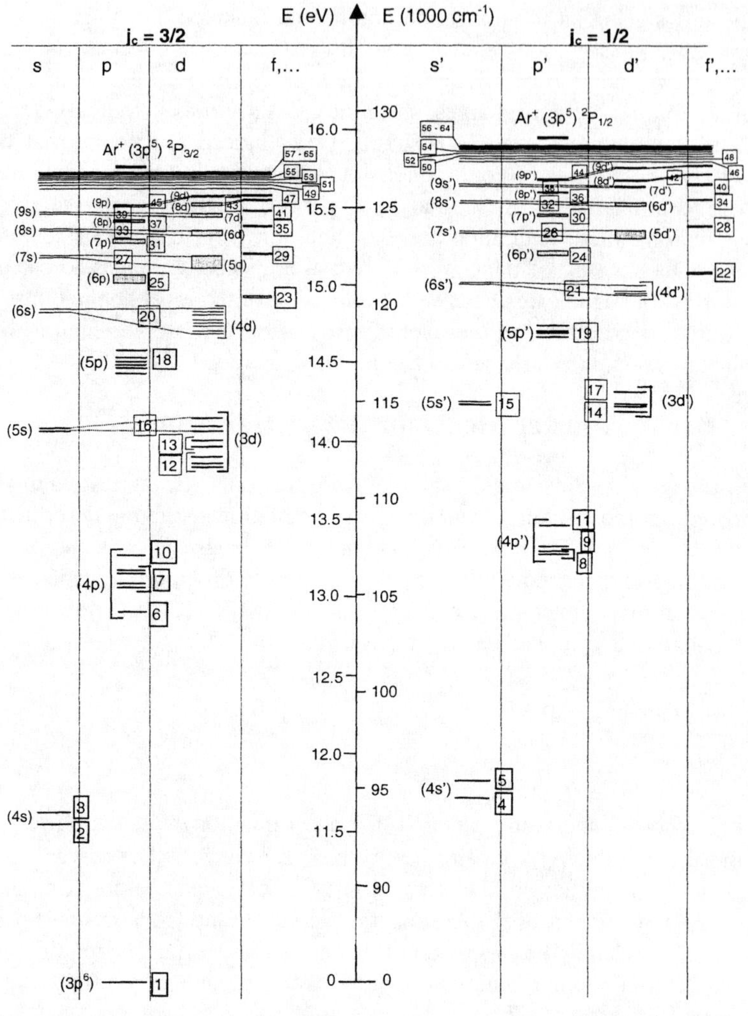

FIGURE 1. Energy level scheme of the Ar atoms, illustrating all the effective levels incorporated in the collisional-radiative model [6].

The behavior of the levels is described with 64 coupled balance equations, taking into account a large number of populating and depopulating processes:

$$\frac{\partial n_{Ar*}(z,r,t)}{\partial t} - D_{Ar*}\frac{1}{r}\frac{\partial}{\partial r}\left(r\frac{\partial n_{Ar*}(z,r,t)}{\partial r}\right) - D_{Ar*}\frac{\partial^2 n_{Ar*}(z,r,t)}{\partial z^2} = R_{prod}(z,r,t) - R_{loss}(z,r,t) \qquad (5)$$

The production and loss processes taken into account are: electron, Ar ion and atom impact ionization from all levels, excitation and de-excitation between all the levels, and electron-ion three-body and radiative recombination to all levels, as well as radiative decay between the levels and Hornbeck-Molnar associative ionization (for Ar* levels with excitation energy above 14.7 eV). Moreover, some additional processes are incorporated for the 4s metastable levels, i.e., metastable atom - metastable atom collisions, Penning ionization of the sputtered Cu atoms, and two-body and three-body collisions with Ar ground state atoms. The cross sections and rate coefficients of all these processes are discussed in detail in ref. [6]. As mentioned before, the cross sections for electron, ion and atom impact ionization and excitation are mainly determined from empirical formulas. The cross sections for electron impact de-excitation and for electron-ion recombination are obtained from the corresponding electron impact excitation and ionization cross sections, based on the principle of detailed balancing. The Einstein transition probabilities for the 4s-4p and 4s-5p transitions (which dominate the spectrum [27]) were obtained from [28], whereas the values for other (less important) transitions were calculated on the basis of intermediate and (j,K) coupling [29]. Hornbeck-Molnar associative ionization was initially not incorporated in the model [6], but was added later on (when the behavior of Ar_2^+ ions was described; see details in [18]), and it was found to be important for the levels above 14.71 eV [30]. Finally, the additional data for the processes concerning the 4s metastable levels, i.e., metastable atom – metastable atom collisions, two-body and three-body collisions with Ar atoms, and Penning ionization, are taken from the literature [31-34]. It should be mentioned that the rate coefficient for Penning ionization between Ar metastable atoms and sputtered Cu atoms was difficult to find in the literature. We used an empirical formula [33], which was fitted to some experimentally obtained cross sections for certain elements (e.g., Zn, Cd and Mg) [33,34] in order to arrive at approximate values for other elements, like Cu. It is clear that most of these input data in the model are subject to large uncertainties, which will be reflected in the excited level populations. However, the optical emission intensities which we calculated from the excited level populations appeared to be in reasonable agreement with experimental data (see [27]) which suggests that the above data are not so critical for a realistic description of the excited Ar atoms.

Transport of the atomic levels occurs, in principle, by diffusion. However, the latter plays only a role for the 4s levels, because the higher excited levels decay more rapidly to the ground state by emission of radiation, than they could move due to diffusion. Furthermore, when the two non-metastable 4s levels decay to the ground state, a large fraction of the emitted radiation is again re-absorbed by the ground level, leading again to formation of this 4s level. This phenomenon of "radiation trapping" is accounted for by means of "escape factors" which express the fraction of photons that can really escape without being re-absorbed [35,36]. The escape factors we used were calculated, based on combined Doppler and collisional line broadening [6].

The 64 balance equations are coupled to each other, because higher and lower levels affect each other due to radiative decay, excitation and de-excitation. The boundary conditions for these equations are $n_{Ar*}=0$ at all walls, because the excited levels will de-excite upon collision at the walls. More information about this model can be found in refs [6,30] for the dc and the rf case, respectively.

Cathode Sputtering and Thermalization of the Sputtered Atoms

The flux of sputtered Cu atoms, J_{sput}, is calculated from the energy distribution functions of the Ar ions, fast Ar atoms and Cu ions (see below) bombarding the cathode ($f_0(E)$; calculated in the Monte Carlo models), multiplied with an empirical formula for the sputtering yield as a function of the bombarding energy (Y(E) [37]):

$$J_{sput} = -\int_E \left\{ Y_{Ar-Cu}(E) \left[f_{0,Ar^+}(E) + f_{0,Ar_f^0}(E) \right] + Y_{Cu-Cu}(E) f_{0,Cu^+}(E) \right\} dE \qquad (6)$$

When the Cu atoms are sputtered from the cathode, they have energies in the order of 5-10 eV. However, they lose these energies almost immediately in the first mm's from the cathode, by collisions with Ar gas atoms, until they are thermalized. This thermalization process is described with a Monte Carlo model, similar to the electron Monte Carlo model (see above), except that the electric field does not come into play for the neutral atoms, and that only elastic collisions with Ar atoms are incorporated. Indeed, collisions with other plasma species could be neglected, due to the lower densities of these species. This Monte Carlo model is run until all atoms are thermalized, and it results in a so-called thermalization profile, F_T, i.e., the number of atoms thermalized as a function of position from the cathode. More details about the collision cross sections and the scattering formulas can be found in ref. [38].

The product of J_{sput} and F_T will be used as source term for the Cu atoms, described in the following model.

Collisional-radiative Model for the Cu Atoms and Ions

The further behavior of the thermalized Cu atoms (i.e., transport, ionization and excitation), and the behavior of the excited Cu atoms and of the Cu ions (in the ground state and in excited levels) is described with a collisional-radiative model. Eight Cu atom levels, seven Cu^+ ion levels and the Cu^{++} ions are considered (see the energy level scheme in figure 2). Some of the Cu atom and ion levels are grouped into effective levels. The behavior of all the levels is again described with a set of coupled balance equations with various production and loss terms, i.e., electron and atom impact ionization from all levels, excitation and deexcitation between all levels, radiative decay between all levels, electron-ion three-body recombination to the upper Cu atom and Cu ion levels, Penning ionization by Ar metastable atoms, and asymmetric charge transfer between Cu atoms and Ar ions. Moreover, an additional production term for the Cu ground state atoms is the product of J_{sput} and F_T, as is described above.

The transport occurs by diffusion for the atoms, and by diffusion and migration for the ions. The equations are also coupled due to the effect of higher and lower levels on each other. More information about this model can be found in refs. [7,39], as well as about the input data needed to calculate all the production and loss processes (which are mainly taken from the literature concerning Cu vapor lasers; see e.g., [40,41]).

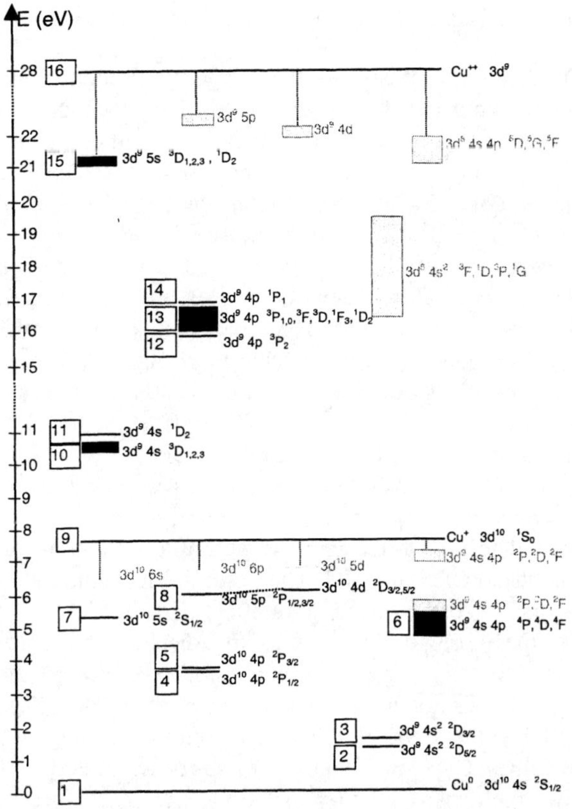

FIGURE 2. Energy level scheme of the Cu atoms and ions [7]. The levels considered in the model are presented in black. Reprinted from [7] with permission of Elsevier Science.

Monte Carlo Model for the Cu Ions in the CDS

Finally, the Cu ions are also treated with a Monte Carlo model in the CDS, because they are not in equilibrium with the strong electric field in this region. The procedure is again comparable to the electron Monte Carlo model (i.e., calculation of the trajectory by Newton's laws, and treatment of the collisions by random numbers); see refs. [42,43] for more information.

Coupling of the Models

All the models are coupled to each other due to the interactions between the various plasma species, i.e., the output of one model is used as input for the other models. The models are solved iteratively until final convergence is reached, to obtain an overall picture of the glow discharge. A schematic picture of the coupling procedure is given in ref. [44,45]. The whole calculation procedure takes several days on a unix professional workstation.

TYPICAL RESULTS OF THE MODELS

Table 2 presents an overview of typical results obtained with the models. All these plasma quantities can be calculated for different discharge conditions (voltage, current, power, pressure), operation modes (dc, rf or pulsed) and cell geometries (see e.g., [46-48]). To test the validity of the models, the calculated quantities must be compared with experimental data. Such a comparison has been carried out for some quantities (when experimental data were available; see Table 2), and in general, reasonable agreement between experimental and calculated results was reached.

Some of the characteristic calculated plasma quantities are illustrated in Figure 3, for a simplified analytical glow discharge cell geometry of 1 cm length, at typical discharge conditions used for GDMS (i.e., dc discharge at 1000 V, 3 mA, 0.5 Torr). The discharge consists of three regions (see part a): a cathode dark space (CDS, adjacent to the cathode), a negative glow (NG, which fills the main part of the discharge), and an anode dark space (ADC, adjacent to the anode).

The calculated potential and electric field distributions are shown in part (b). The potential (solid line, left axis) is -1000 V at the cathode and increases to zero at about 0.24 cm from the cathode. The position where the potential crosses the zero-line is defined as the interface between CDS and NG. The potential is slightly positive in the NG (i.e., a few V; which is called the plasma potential) and decreases to zero at the anode. Hence, the NG is the most positive part of the discharge. The electric field (dashed line, right axis) is extremely negative at the cathode (-8 kV/cm) and increases almost linearly in the CDS. It does not cross the zero-line at the CDS-NG interface but bends off to a small negative value in the NG. It goes through zero at about 0.7 cm and then takes small positive values. Close to the anode it rises to about 200 V/cm.

Parts (c) and (d) present the density profiles of the various plasma species. The Ar metastable atom density (part (c), dashed line, left axis) reaches a maximum of about 10^{12} cm^{-3} rather close to the cathode and decreases to low values in the rest of the discharge. The Ar ion density (part (c), solid line, left axis) is low and more or less constant in the CDS and reaches a broad maximum of about $5x10^{11}$ cm^{-3} halfway the discharge, whereafter it decreases to a low value at the anode. The slow electron density (part (c), solid line, left axis) is zero in the CDS and close to the anode and it is almost equal to the Ar ion density in the NG. This gives rise to a net positive space charge in the CDS and close to the anode and nearly charge neutrality in the NG. The latter defines the characteristic potential and electric field distributions presented in part (b). The fast electron density (part (c), solid line, right axis) reaches a maximum in the beginning of the NG; it is, however, four orders of magnitude lower than the Ar ion and slow electron densities, and hence does not contribute to the space charge. The sputtered Cu atom density (part (d), left axis) is at its maximum (ca. $6x10^{12}$ cm^{-3}) close to the cathode and decreases almost linearly towards the anode. The Cu ion density (part (d), right axis) shows the same profile shape as the Ar ion density but is about two orders of magnitude lower. From part (d) it can also be deduced that the ionization degree of sputtered Cu atoms (i.e., the ratio of Cu ion to Cu atom density) is of the order of 0.1 %. This calculated value should not be considered too strictly; it depends strongly on the discharge conditions, cell geometry and kind of cathode material; for other conditions, values in the order of a few % were calculated [46,55].

TABLE 2. Overview of the Typical Results Obtained with our Models, and Comparison with Experimental Data, if Available.

Calculated Quantities (+ references for more information)		**References for Comparison with Experiment**
Electrical characteristics:		
Current as function of voltage and pressure (dc)	24,44,46	24,44
Rf amplitude and dc bias voltage (rf)	16	49
Voltage, current, power as function of time (pulsed)	4	50
Potential, electric field distributions:		
3D potential distributions	4,13-16,44	-
3D axial and radial electric field distributions	4,13-16,44	-
Lengths of the different regions (CDS, NG)	4,13-16,44	51 (Aston formula)
3D density profiles of:		
* Ar atoms (gas heating)	26	-
* Ar^+, Ar^{2+} and Ar_2^+ ions	4,13-16,18,44	-
* fast Ar atoms	12,44	-
* Ar metastable atoms	6,30,43,44,52	53 (LIF)
* other Ar excited levels	6,30	-
* fast electrons	12-14,44	-
* thermal electrons	4,13-16,44	54 (Langmuir probe)
* atoms of the cathode material	7,39,42-44	55 (LIF)
* ions of the cathode material	7,39,42-44	55 (LIF)
* cathode atoms + ions in excited levels	7,39	
Ion fluxes of various Ar and cathode ions at the exit slit of the cell to the mass spectrometer	18,47,48	47,56 (ratios in glow discharge mass spectra)
Ionization degrees of Ar and cathode atoms	42-44	55 (based on LIF results)
3D energy distributions and mean energies of:		
* electrons	12,13,15,44	-
* Ar ions	12,25,44	57 (at cathode)
* fast Ar atoms	12,25,44	-
* cathode ions	42,44	57 (at cathode)
Information about collision processes:		
* collision rates of the various collision processes of electrons, Ar^+ ions and fast Ar atoms	4,12-16,24,44	-
* rates of Penning ionization, asymmetric charge transfer and electron impact ionization of sputtered atoms; relative contributions to the total ionization	7,39,42-44	-
* rates and relative contributions of the various populating and depopulating processes (see text) of the metastable and other excited Ar levels	6,30,43,44,52	-
* rates and relative contributions of the various populating and depopulating processes (see text) of the excited Cu atom + ion levels	7,39	-
Information about sputtering:		
* Sputtering (erosion) rates at the cathode	39,44,46,58	59,60
* Thermalization profiles of the sputtered atoms	38,44	-
* Amount of redeposition on the cathode by backscattering or backdiffusion	38,44,58	-
* Relative contributions of Ar ions, fast Ar atoms and cathode ions to the sputtering process	12,39,42-44	-
* 2D crater profiles due to sputtering at the cathode	58	59
Emission spectra and spatial distributions of emission intensities for Ar and Cu atoms + ions	27,61,62	27,62
Effect of cell geometry	47,48	-

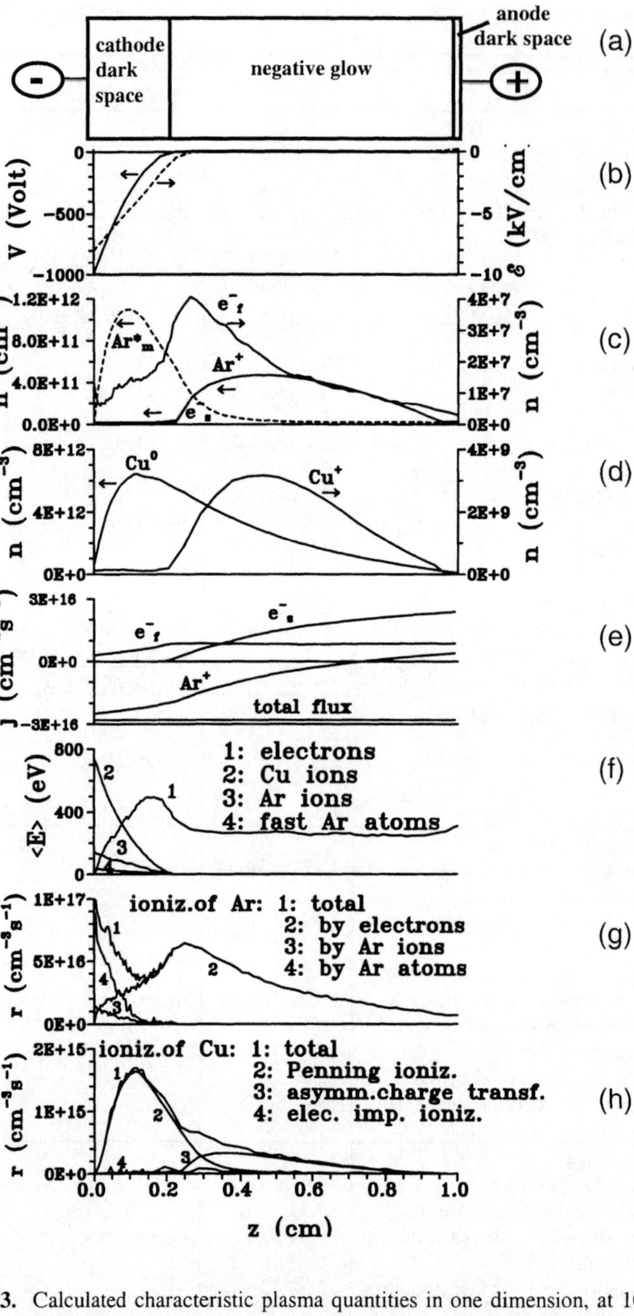

FIGURE 3. Calculated characteristic plasma quantities in one dimension, at 1000 V, 3 mA and 0.5 Torr (dc conditions). (a) discharge geometry; (b) potential and electric field distributions; (c) number density profiles of Ar metastable atoms (Ar_m*), Ar^+ ions and slow (e^-_s) and fast (e^-_f) electrons; (d) number density profiles of sputtered Cu^0 atoms and Cu^+ ions; (e) fluxes of Ar^+ ions, fast and slow electrons, and total flux flowing through the discharge; (f) mean energies of electrons, Cu ions, Ar ions and fast Ar atoms throughout the discharge; (g) ionization rates of Ar atoms (total + individual processes); (h) ionization rates of Cu atoms (total + individual processes).

The fluxes of the dominant current carriers are indicated in part (e). In the CDS most of the current is carried by the Ar ions, which are directed towards the cathode by the strong electric field in front of it. The fast electron flux at the cathode is about an order of magnitude lower than the Ar ion flux (determined by the ion induced secondary electron emission coefficient, which is in the order of 0.1) and it is in the opposite direction (i.e., away from the cathode). It increases in the CDS and remains nearly constant in the NG. The slow electron flux, on the other hand, is zero in the CDS. Indeed, slow electrons are assumed not to be present in this region since they would immediately be accelerated by the electric field and they would not remain in the slow group. Their flux increases, however, considerably in the NG, as more and more fast electrons are slowed down by collisions and are transferred to the slow electron group. In the NG, most of the current is carried by the (slow) electrons. The Ar ion flux changes sign in this region; hence Ar ions will bombard the anode as well.

Part (f) shows the mean energies of electrons, Cu ions, Ar ions and fast Ar atoms, as a function of distance from the cathode. The electrons (no. 1) start at the cathode with rather low energies, and they gain energy in the CDS from the electric field. They lose, however, also energy by collisions, so that their mean energy at the CDS-NG interface is not equal to the discharge voltage, but is about 50 % of this value. The electron energy decreases in the NG, since the electrons do not gain energy from the electric field anymore (which is very small, see part (b)) but they lose their energy more efficiently due to collisions. Further in the NG, the electron energy remains more or less constant since the electrons travel back and forth in this region. The Cu ions, Ar ions and fast Ar atoms are more or less thermalized in the NG, but their energy increases as they move towards the cathode. The mean Cu ion energy (no. 2) at the cathode is about 700 eV (hence about 70 % of the discharge voltage), since the Cu ions do not lose their energy very efficiently by collisions. Indeed, they are only subject to elastic collisions with Ar atoms; asymmetric charge transfer collisions with Ar atoms have a lower cross section, and symmetric charge transfer with Cu atoms is of low probability due to the much lower Cu atom density compared to Ar atom density (the latter is ca. 10^{16} cm^{-3}). The Ar ions (no. 3), on the other hand, reach a maximum mean energy of only about 150 eV (i.e., 15 % of the discharge voltage) at the cathode, since these species lose the energy they gained from the electric field very efficiently due to symmetric charge transfer collisions with Ar atoms. The mean energy of the fast Ar atoms (no. 4) at the cathode is only in the order of 30 eV (ca. 3 % of the discharge voltage). Indeed, these species are created from the Ar ions with energies corresponding to the Ar ion energies, but they cannot gain more energy from the electric field. It should be mentioned that the term "fast" Ar atoms is used for those atoms which are not thermalized, and have energies higher than ca. 0.05 eV.

From the energies and the fluxes of the species bombarding the cathode (i.e., Ar ions, fast Ar atoms and Cu ions), the amount of sputtering can be calculated. It is generally found that the flux of fast Ar atoms bombarding the cathode is definitely higher than the fluxes of Ar ions and Cu ions. Therefore, the fast Ar atoms play a dominant role in sputtering [12]. However, the efficiency of sputtering increases with the energy of the bombarding particles and with their mass [37]; therefore, it is expected that the contribution of Cu ions to sputtering (so-called self-sputtering) is non-negligible, in spite of their lower total flux, and this contribution rises significantly with pressure and voltage. It was calculated that the fast Ar atoms, Ar ions and Cu ions contribute for about 30-70 %, 20-30 % and 0.1-50 % to the sputtering, respectively, at the typical discharge conditions used in analytical glow discharges (i.e., 600-1200 V, 1-50 mA, 1-5 Torr) [44,46].

Finally, parts (g) and (h) of Figure 3 present the ionization rates of Ar atoms and Cu atoms, respectively. The Ar atoms are mainly ionized by electron impact ionization, but fast Ar ion and atom impact ionization play also a role in the CDS, especially close to the cathode where the Ar ions and atoms reach their maximum energy (see part (f)). Integrated over the entire discharge region, electron, fast Ar ion and atom impact ionization contribute for about 90 %, 2 % and 8 %, respectively, to the ionization of Ar. A more or less similar figure can be made for the excitation rate of Ar atoms, with a maximum adjacent to the cathode due to fast Ar ion and atom impact excitation and a second peak in the beginning of the NG caused by electron impact excitation. This corresponds well with the luminous intensity throughout the glow discharge: the NG is the most luminous part of the discharge, but close to the cathode, a bright layer is often observed, called the cathode glow. For the ionization of sputtered Cu atoms, Ar ion and atom impact ionization are not included in the model, basically because no cross sections are available in the literature and also because these processes are probably not important for the general ionization of Cu atoms. Indeed, besides electron impact ionization, two other processes come into play for the sputtered atoms, i.e., Penning ionization by Ar metastable atoms and asymmetric charge transfer with Ar ions. Both these processes seem to be more important than electron impact ionization. The relative contributions of Penning ionization, asymmetric charge transfer and electron impact ionization, integrated over the entire discharge region, were calculated to be about 60 %, 36 % and 4 %, respectively, at the discharge conditions of 1000 V and 3 mA. These values are only approximate, because the rate coefficients of Penning ionization and especially of asymmetric charge transfer are subject to uncertainties. Moreover, these calculated contributions depend strongly on the discharge conditions, cell dimensions and especially kind of sputtered material. Nevertheless, we found for all conditions investigated that Penning ionization and asymmetric charge transfer were more important than electron impact ionization. Since the latter process is the dominant one for Ar atoms (see above), it follows that the sputtered atoms are more efficiently ionized than the Ar atoms. Indeed, we calculated typical ionization degrees of 10^{-5}-10^{-3} for Ar, and in the order of 10^{-4}-$5\text{x}10^{-2}$ for the sputtered atoms (Cu, Ta, …).

CONCLUSION

A hybrid modeling network has been developed, consisting of Monte Carlo, fluid and collisional-radiative models to describe the various species in glow discharge plasmas. The different submodels are explained, and the data needed to solve the models are discussed. It is clear that the modeling network requires a large number of input data. Some of them are rather well-known (e.g., electron impact ionization and excitation cross sections from the Ar ground state), but most of them are subject to large uncertainties (e.g., the cross sections for excitation and ionization from various

Ar and Cu excited levels). Fortunately, the latter data do not affect the overall model calculations to a large extent; they determine only the excited level populations. The data which have the largest influence on the overall model calculations are the electron, Ar ion and atom impact ionization cross sections from the Ar ground state (of which the first is well-known, but the two others are subject to larger uncertainties), as well as the gas pressure and temperature (of which the latter is also generally unknown, and can be significantly higher than room temperature [26]). Indeed, these data determine the number of ionization collisions, and hence the creation of electrons and Ar ions, and therefore also most other collision rates, and the densities, fluxes and energies of the various plasma species, as well as the total electrical current. Nevertheless, it has been shown that the calculated current is in reasonable agreement with experimental values, which suggests that the input data used in the model are sufficiently reliable.

In order to show the possibilities of the model network, an overview is given of the typical results that have been calculated, and some of them are illustrated. Reasonable agreement has been reached between the model results and experimental data, which suggests that our models present a realistic picture of the analytical glow discharge.

ACKNOWLEDGMENTS

A. Bogaerts is indebted to the Fund for Scientific Research (FWO) Flanders for financial support. The authors also acknowledge financial support from the Federal Services for Scientific, Technical and Cultural Affairs (DWTC/SSTC) of the Prime Minister's Office through IUAP-IV (Conv. P4/10). Finally, they would also like to thank W. Goedheer, J. Vlcek and R. Carman for their cooperation in developing the models.

REFERENCES

1. Marcus, R.K. (ed.), *Glow Discharge Spectroscopies*, New York: Plenum Press, 1993.
2. Bogaerts, A., and Gijbels, R., *J. Anal. At. Spectrom.* **13,** 945-953 (1998).
3. Bogaerts, A., and Gijbels, R., *J. Anal. At. Spectrom.* **15**, (in press: 2000).
4. Bogaerts, A., and Gijbels, R., *J. Anal. At. Spectrom.* (submitted).
5. Phelps, A.V., and Petrovic, Z. Lj., *Plasma Sources Sci. Technol.* **8**, R21-R44 (1999).
6. Bogaerts, A., Gijbels, R., and Vlcek, J., *J. Appl. Phys.* **84**, 121-136 (1998).
7. Bogaerts, A., Gijbels, R., and Carman, R. J., *Spectrochim. Acta B* **53,** 1679-1703 (1998).
8. Hashiguchi, S., *IEEE Trans. Plasma Sci.* **19**, 297-300 (1991).
9. Boeuf, J.P., and Marode, E., *J. Phys. D* **15**, 2169-2187 (1982)
10. Surendra, M., Graves, D.B., and Jellum, G.M., *Phys. Rev. A* **41**, 1112-1125 (1990).
11. Hasted, J.B., *Physics of Atomic Collisions*, London: Butterworth, 1972.
12. Bogaerts, A., van Straaten, M., and Gijbels, R., *Spectrochim. Acta B* **50**, 179-196 (1995).
13. Bogaerts, A., Gijbels, R., and Goedheer, W. J., *J. Appl. Phys.* **78,** 2233-2241 (1995).
14. Bogaerts, A., Gijbels, R., and Goedheer, W. J., *Anal. Chem.* **68,** 2296-2303 (1996).
15. Bogaerts, A., Gijbels, R., and Goedheer, W. J., *Jpn. J. Apl. Phys.,* **38,** 4404-4415 (1999).
16. Bogaerts, A., Yan, M., Gijbels, R., and Goedheer, W. *J., Appl. Phys.,* **86**, 2990-3001 (1999).
17. Jogwich, M., Huber, B.A., and Wiesemann, K., *Z. Phys.D*, **17,** 171-179 (1990).
18. Bogaerts, A., and Gijbels, R., *J. Appl. Phys.,* **86**, 4124-4133 (1999).
19. Passchier, J.D.P., and Goedheer, W.J., *J. Appl. Phys.* **73**, 1073-1079 (1993).

20. Gummel, H. K., *IEEE Trans. Plasma Sci,* **11**, 455-465 (1964).
21. Scharfetter, D. L., and Gummel, H. K. *IEEE Trans. Electron. Devices* **16,** 64-77 (1969).
22. Phelps, A.V., *J. Appl. Phys.* **76**, 747-753 (1994).
23. Phelps, A.V., *J. Phys. Chem. Ref. Data* **20**, 557-573 (1991).
24. Bogaerts, A., and Gijbels, R., *J. Appl. Phys.* **78,** 6427-6431 (1995).
25. Bogaerts, A., and Gijbels, R., *IEEE Trans. Plasma Sci.* **27**, 1406-1415 (1999).
26. Bogaerts, A., Gijbels, R., and Serikov, V.V., *J. Appl. Phys.* **87**, (in press: 15 June 2000).
27. Bogaerts, A., Gijbels, R., and Vlcek, J., *Spectrochim. Acta B* **53,** 1517-1526 (1998).
28. Wiese, W.L., Brault, J.W., Danzmann, K., Helbig, V., and Kock, M., *Phys. Rev. A* **39**, 2461-2471 (1989).
29. Vlcek, J., *J. Phys. D* **22**, 623-631 (1989).
30. Bogaerts, A., and Gijbels, R., *Spectrochim. Acta B* **55**, 263-278 (2000).
31. Ferreira, C.M., Loureiro, J., and Ricard, A., *J. Appl. Phys.* **57**, 82-90 (1984).
32. Tachibana, K., *Phys. Rev. A* **34**, 1007-1015 (1986).
33. Riseberg, L.A., Parks, W.F., and Schearer, L.D., *Phys. Rev. A* **8**, 1962-1968 (1973).
34. Inaba, S., Goto, T., and Hattori, S., *J. Phys. Soc. Jpn.* **52**, 1164-1167 (1983).
35. Holstein, T., *Phys. Rev.* **83,** 1159-1168 (1951).
36. Walsh, P. J., *Phys. Rev.* **116,** 511-515 (1959).
37. Matsunami, N., Yamamura, Y., Itikawa, Y., Itoh, N., Kazumata, Y., Miyagawa, S., Morita, K., Shimizu, R., and Tawara, H., *Atom. Data Nucl. Data Tables* **31**, 1-80 (1984).
38. Bogaerts, A., van Straaten, M., and Gijbels, R., *J. Appl. Phys.* **77,** 1868-1874 (1995).
39. Bogaerts, A., and Gijbels, R., *Spectrochim. Acta B* **55**, 279-297 (2000).
40. Carman, R.J., Brown, D.J.W., and Piper, J.A., *IEEE J. Quant. Electron.* **30**, 1876-1895 (1994).
41. Carman, R.J., *Opt. Lett.* **21**, 872-874 (1996).
42. Bogaerts, A., and Gijbels, R., *J. Appl. Phys.* **79,** 1279-1286 (1996).
43. Bogaerts, A., and Gijbels, R., *Anal. Chem.* **68,** 2676-2685 (1996).
44. Bogaerts, A., *Ph.D. Dissertation*, University of Antwerp (1996).
45. Bogaerts, A., and Gijbels, R., *Plasma Phys. Reports* **24**, 573-583 (1998).
46. Bogaerts, A., and Gijbels, R., *Spectrochim. Acta B* **53,** 437-462 (1998).
47. Bogaerts, A., and Gijbels, R., *J. Anal. At. Spectrom.* **12,** 751-759 (1997).
48. Bogaerts, A., and Gijbels, R., *J Am. Soc. Mass Spectrom.* **8,** 1021-1029 (1997).
49. Hoffmann, V., private communication.
50. Yang, C., and Harrison, W.W., private communication.
51. Aston, F. W., *Proc. Roy. Soc. London, Ser. A* **79,** 80 (1907).
52. Bogaerts, A., and Gijbels, R., *Phys. Rev. A* **52**, 3743-3751 (1995).
53. Bogaerts, A., Guenard, R.D., Smith, B.W., Winefordner, J.D., Harrison, W.W., and Gijbels, R., *Spectrochim. Acta B* **52,** 219-229 (1997).
54. Bogaerts, A., Quentmeier, A., Jakubowski, N., and Gijbels, R., *Spectrochim. Acta B* **50,** 1337-1349 (1995).
55. Bogaerts, A., Wagner, E., Smith, B.W., Winefordner, J.D., Pollmann, D., Harrison, W.W., and Gijbels, R., *Spectrochim. Acta B* **52,** 205-218 (1997).
56. Kennis, P., *Ph.D. Dissertation*, University of Antwerp (1996).
57. van Straaten, M., Bogaerts, A., and Gijbels, R., *Spectrochim. Acta B* **50,** 583-605 (1995).
58. Bogaerts, A., and Gijbels, R., *Spectrochim. Acta B* **52,** 765-778 (1997).
59. Jonkers, C., *Ph.D. Dissertation*, University of Antwerp (1995).
60. Bengtson, A., and Lundholm, M., *J. Anal. At. Spectrom.* **3,** 879-882 (1988).
61. Bogaerts, A., and Gijbels, R., *J. Anal. At. Spectrom.* **13,** 721-726 (1998).
62. Bogaerts, A., Donko, Z., Kutasi, K., Bano, G., Pinhao, N., and Pinheiro, M., *Spectrochim. Acta B*, (submitted).

Charge Exchange at Solid Surfaces for Medium Energy He Ions

Yoshiaki Kido

Department of Physics, Ritsumeikan University, Kusatsu
Shiga-ken 525-8577, Japan

Abstract. Charge exchange process for medium energy He ions backscattered from solid surfaces was studied using a high-resolution toroidal electrostatic analyzer ($\Delta E / E = 9 \times 10^{-4}$). He ions backscattered from topmost atoms have a high fraction of nonqeuilibrium singly-ionized charge state and thus it generates a surface peak in the backscattering spectrum from an amorphous solid. The height and width of the surface peak provide the information about the electron capture and loss cross sections together with the ion-surface interaction region expanded from the top atomic plane toward the vacuum side. Dependence of the surface peak profile upon energy, exit-angle and materials of the surface is discussed in detail in the context of the ion-surface interactions.

1. INTRODUCTION

Medium energy ion scattering (MEIS) provides a powerful tool to determine elemental compositions and atomic configurations at surfaces and interfaces[1]. A high-resolution electrostatic or magnetic spectrometer allows a layer-by-layer analysis of near surface regions[2-4]. Using the ion shadowing/blocking effect makes it possible to determine the displacements of atoms near a surface with an accuracy of better than 0.1 Å. In the case of the elemental analysis, we need the knowledge of the fraction of ions emerging from a surface, which depends upon the energy, exit angle, and material of the surface, because the above spectrometers detect only ions. Unfortunately, there are few reports on the energy, exit-angle, and materials dependence of the fractions of ions in the low and medium energy regimes.

In the previous paper[5], we reported the anomalous surface peaks observed with a toroidal electrostatic analyzer (ESA) for medium energy He ions backscattered from amorphous Si (a-Si) and SiO_2 (a-SiO_2) films. The spectrum height is proportional to the fraction of the singly ionized He^+ ions emerging from the solid surface. The surface peak was explained quantitatively by considering nonequilibrium charge exchange processes near the surfaces. In order to reproduce the surface peak, it was needed to extend to some extent the ion-electron interaction region from the top atomic plane toward the vacuum side.

In this study, we measured the backscattering spectra from graphite, a-SiO_2, a-Si, a-N and a-NiSi thin films for scattered He energies from 25 up to 85 keV. Th amorphous and clean surfaces were checked by reflective high energy electro

Tel: +81-77-561-2710; Fax: +81-77-561-2657; E-mail: ykido@se.ritsumei.ac.jp

CP543, *Atomic and Molecular Data and Their Applications,* edited by K. A. Berrington and K. L. Bell

diffraction (RHEED) and Auger electron spectroscopy (AES), respectively. The surface peaks were observed in the energy spectra for He ions backscattered from C, O, and Si atoms but not from Ni atoms. Dependence of the surface peak profiles upon the energy, exit-angle, and materials of the surface is studied systematically in the context of the ion-surface interactions.

2. EXPERIMENT

The clean surface of pyrolytic graphite films was prepared by stripping its flake with an adhesive tape in the air and by subsequent annealing at 300℃ for 10 min in an ultrahigh vacuum (UHV: 2×10^{-10} Torr). The graphite is c-axis oriented with a domain size of about 1000 Å. Thin SiO_2 layers were formed by thermal oxidation of Si(001) substrates at 800℃ in a wet atmosphere and then annealed at 800℃ for *1 h* in a dry N_2 ambiance. Prior to the MEIS measurement, the samples were annealed at 200℃ for 10 min in UHV. The AES observation confirmed the clean surfaces without oxygen and carbon for the graphite and the SiO_2/Si(001), respectively.

Thin a-Si, a-Ni, and a-NiSi layers were formed on SiO_2(70 Å)/Si(001) substrates at

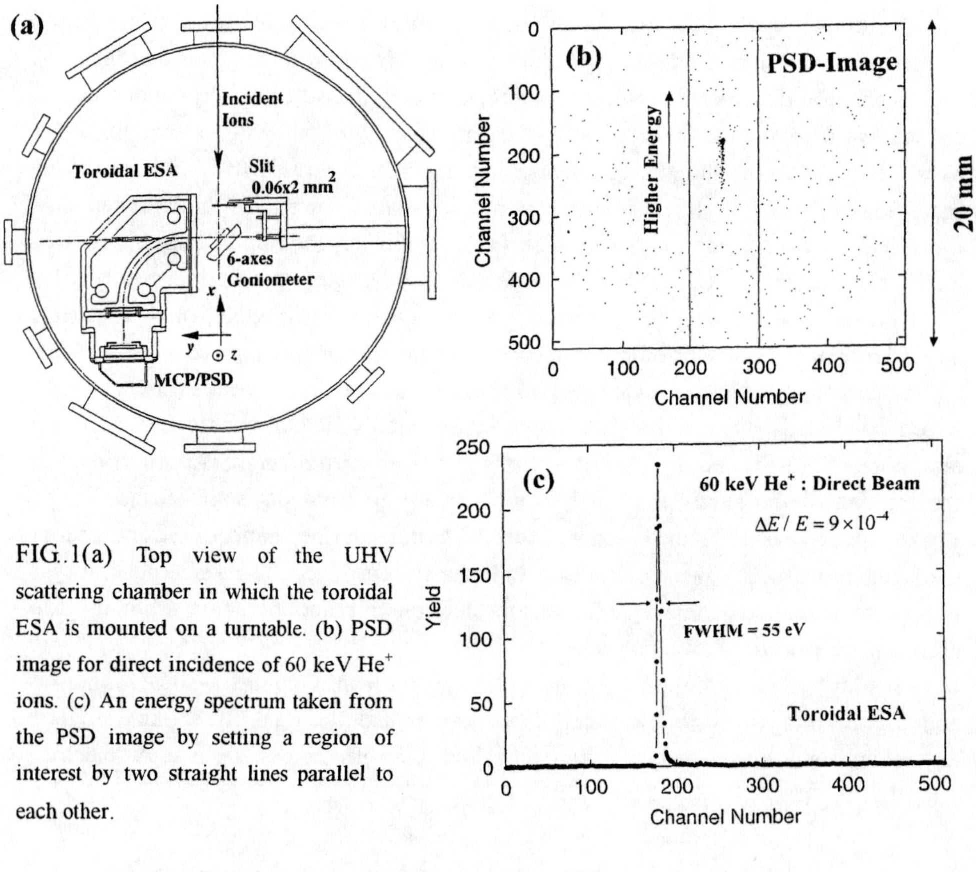

FIG.1(a) Top view of the UHV scattering chamber in which the toroidal ESA is mounted on a turntable. (b) PSD image for direct incidence of 60 keV He^+ ions. (c) An energy spectrum taken from the PSD image by setting a region of interest by two straight lines parallel to each other.

room temperature (RT) with Knudsen cells under a UHV condition (base pressure: 7 $\times 10^{-11}$ Torr) In the case of a-NiSi films (30-50 Å), Ni and Si were deposited simultaneously with two Knudsen cells at a same deposition rate of 1.5 MLs/min(ML: monolayer). The RHEED and AES observation confirmed the amorphous and clean surfaces without contamination such as C and O. It must be noted that the word 'amorphous ' used here means no diffraction spots in the RHEED images.

The MEIS experiments were performed at 'SORIS' beam line[6] in Ritsumeikan SR Center. The new toroidal ESA[4] was mounted on a turntable in a UHV scattering chamber (2×10^{-10} Torr)(see Fig. 1(a)). Our toroidal ESA covers a broad energy range of 10 % of the pass energy at a fixed applied voltage. The energy resolution ($\Delta E/E$) was estimated to be 9×10^{-4} by direct incidence of 60 and 80 keV He^+ ions (Figs. 1(b) and (c)). Such an excellent energy resolution allows a layer-by-layer analysis and makes it possible to observe the surface peaks clearly. In the present MEIS measurement, in order to avoid the deterioration of the surfaces such as radiation damage, we shifted the beam spot position after the integrated beam current of 1 μC.

3. ANOMALOUS SURFACE PEAKS

Anomalous surface peaks were observed in the backscattering spectra from graphite, a-SiO_2, and a-Si thin films but not from a-Ni films for medium energy He ions in the energy range from 25 to 85 keV. On the other hand, we found no surface peaks for 60 keV H^+ incidence on SiO_2(46Å)/Si(001). In the case of a-NiSi films, the surface peaks appeared in the backscattering spectra from Si atoms but not from Ni atoms.

Figs. 2(a), (b), and (c) show the MEIS spectra from graphite, a-Si(17Å)/Si(111), and a-Ni(67Å)/SiO_2(70Å)/Si(001), respectively for 80 keV He^+ incidence at a same scattering angle of 90° under similar scattering geometry. In the case of (a), the He^+ ions were randomly incident on the c-axis oriented graphite. On the other hand, the He^+ ions were incident along the major crystal axis of the Si substrates ((b): [100], (c): [110]). It is clearly seen that the surface peaks appeared for graphite and a-Si but not for a-Ni films. The relative surface peak intensity tends to increase with decrease in Z_2 (atomic) number of the target.

Energy dependence of the surface peak intensity was measured for graphite, a-SiO_2, and a-Si films. Figure 3 shows the MEIS spectra from Si of a-SiO_2 films for 60, 90, and 110 keV He^+ incidence under a same scattering condition. Increase in He energy reduces the surface peak intensity and the surface peak is not significant for

4. ORIGIN OF SURFACE PEAKS

It is clear that the surface peak is ascribed to a high fraction of singly-ionized charge state for He ions backscattered from the topmost atoms compared with that for those from deeper layers. Here, it must be noted that the toroidal ESA detected only the IIe$^+$ ions. Such a high Hc$^+$ fraction for the scattering component from the topmost layer reflects a nonequilibrium charge exchange process near the surface. We make the following speculation to explain the origin of the surface peak quantitatively. In the previous papers [7,8], we proposed a simple formula for surface stopping power, which was assumed to be equal to the bulk stopping power in a surface region up to λ above the surface plane. Here, a bulklike cutoff λ is defined by $\sigma_S / 2\rho_B$, where σ_S and ρ_B are the areal density of the surface plane and the atomic density of the bulk, respectively. The real surface stopping power would be position-dependent ($dE/dx(z)$) and not constant [9](see Fig. 4). The constant surface stopping power multiplied by the cutoff length λ should be equal to the total stopping power integrated along the surface normal from the top atomic plane toward the vacuum side. Thus, it is reasonable to assume that the charge exchange region is also extended from the top atomic plane toward the vacuum side up to λ . Under this assumption, we introduce the electron capture(σ_{10} : $He^+ + e^- \rightarrow He^0$) and loss(σ_{01} : $He^0 \rightarrow He^+ + e^-$) cross sections available also in the extended surface region. In the present medium energy regime (25 – 85 keV), the fraction of He^{2+} is neglected [14,15].

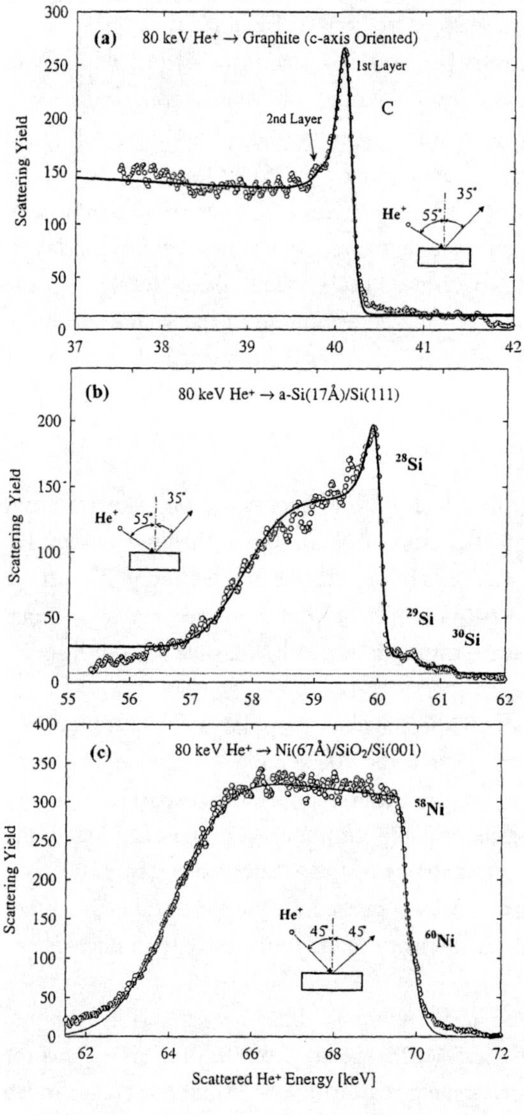

FIGs. 2. MEIS spectra for 80 keV He$^+$ ions incident on graphite (a), a-Si (b), and a-Ni (c). The insets indicate the scattering geometry.

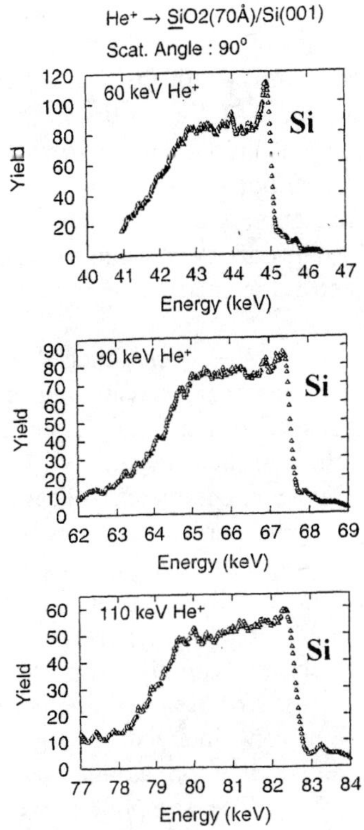

FIG. 3. Surface peak profiles for 60, 90, and 110 keV He^+ incident on a-SiO_2(70 Å)/Si(001) and backscattered to 90° from Si atoms of SiO_2

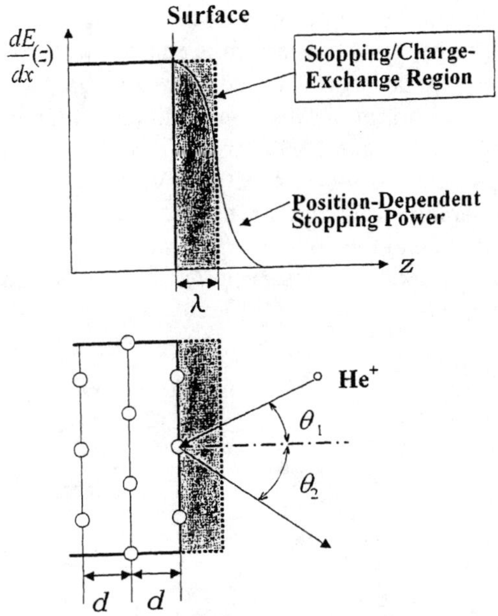

FIG.4. Schematics of the position -dependent stopping power (*dE/dx(z)*) and theion-electron interaction (surface-stopping/charge-exchange) region which is extended by λ from the top atomic plane toward the vacuum side.

Now, we have the following rate equations for the charge fractions ($Y_0(x)$, $Y_+(x)$) of He^0 and He^+:

$$\Delta Y_0 = Y_+(x)\sigma_{10}\Delta x - Y_0(x)\sigma_{01}\Delta x, \quad (1)$$

$$\Delta Y_+ = Y_0(x)\sigma_{01}\Delta x - Y_+(x)\sigma_{10}\Delta x, \quad (2)$$

$$Y_0(x) + Y_+(x) = 1, \quad (3)$$

where x denotes the outgoing path length in the ion-electron interaction region. This rate equation can be easily solved, and we obtain

$$Y_0(x) = \sigma_{10}/(\sigma_{10}+\sigma_{01}) + [Y_0(0) - \sigma_{10}/(\sigma_{10}+\sigma_{01})]\exp[-(\sigma_{10}+\sigma_{01})x], \quad (4)$$

$$Y_+(x) = \sigma_{01}/(\sigma_{10}+\sigma_{01}) + [Y_+(0) - \sigma_{01}/(\sigma_{10}+\sigma_{01})]\exp[-(\sigma_{10}+\sigma_{01})x]. \quad (5)$$

Figure 5 shows the charge fractions ($Y_0(x)$ and $Y_+(x)$) for 80 keV He^+ ions incident on graphite and backscattered to 90° as a function of outgoing path length x. Here, we assumed the best-fitting condition that $Y_+(0) = 1.0$, $\sigma_{10} = 2.1\times10^{-16}$ cm^2/atom, and the equilibrium charge fraction of He^+ is 0.2 for the MEIS spectrum (see Fig. 2 (a)). The charge state equilibrium is attained within the path length of 1.5×10^{16} atoms/cm^2. Apparently, a larger capture cross section gives a shorter path length for the charge state equilibrium. This nonequilibrated high charge fraction of He^+ gives the surface peak. In the case of H ions, the equilibrium average charge is more than $+0.5\,e$ in the energy range from 30 to 100 keV[10]. In addition, the stopping powers for H are considerably smaller than those for He. Small stopping powers degrade the depth resolution and thus broaden and lower the surface peak. They are the reasons why no surface peaks are observed in the backscattering spectra for medium energy H^+ incidence.

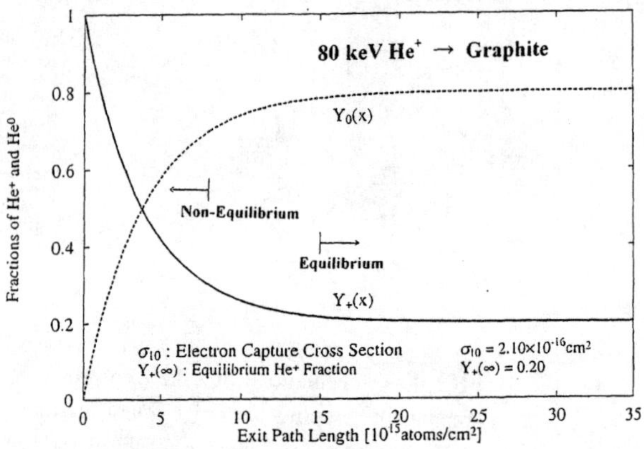

FIG. 5. Charge fractions of He^0(neutral) and He^+ for 80 keV He^+ incident randomly on graphite and backscattered to 90°. It was assumed that the equilibrium charge fraction of He^+ ($Y_+(\infty)$) is 0.20, the initial charge state (Y_+0)) is $+e$, and $\sigma_{10} = 2.0\times10^{-16}$ cm^2/atom. This assumption reproduced well the observed surface peak (see Fig. 2(a)).

We reproduce the surface peaks considering the charge exchange process near the surface region. The relevant parameters generating the surface peak are the average charge state immediately after the large-angle back- scattering, the electron capture and loss cross sections, and the ion-electron interaction region above the top atomic plane (λ). The stopping powers were measured in advance using the thin films deposited on C(100 Å)/Si substrates whose thickness was determined by Rutherford back-scattering (RBS) with 1.5 MeV He^+ beams. The energy straggling was estimated from the slopes of the front and rear edges in the MEIS spectra. The ratio of the yield of He^+ to the total yield of He^+ to the total yield (He^0(neutral) + He^+) coincides with $\sigma_{01}/(\sigma_{01}+\sigma_{10})$. The total yield was calculated from the integrated beam current of the incident He^+ ions and from the solid angle subtended by the toroidal ESA. The detection efficiency of the toroidal ESA was estimated to be 0.50 ± 0.03 for He with energies more than 1 keV[11]. Thus, the equilibrium charge fractions of He^0 and He^+ together with the ratio of σ_{10} to σ_{01} are determined experimentally.

5. CHARGE EXCHANGE AT SURFACES

As mentioned in the previous paper[5], in order to reproduce consistently all the surface peaks, it is needed to assume that the charge state of He ions immediately after the large-angle backscattering was almost $+e$ (singly ionized, e: electronic charge). Figure 6 shows the schematic of a large-angle scattering at a top surface. For a medium energy He ion, its trajectory is deflected in a large-angle collision within the spherical region whose radius is about 0.2Å. In this region, the He ion is first decelerated and then accelerated by the Coulomb repulsion force of the target nucleus. The acceleration is completed within the distance about 0.2Å from the target nucleus. The collision time is roughly $2 - 4\times10^{-17}$ s and thus the energy uncertainty is estimated to be 150 – 200 eV. The binding energy of the 1s electron of He^0 and He^+ is 23 and 54 eV, respectively. Therefore, the state of the bound electrons of the He after the large-angle scattering are independent of those of the incident He. In order to reproduce the surface peak, we must assume that the average charge of He immediately after the large-angle scattering is high enough compared with the equilibrium one. For simple and quantitative analysis, the charge state of He ions immediately after the large-angle scattering is assumed to be $+e$. Thus, the fitting parameters to reproduce the surface peak are reduced to only (1) electron capture (or loss) cross section and (2) the outgoing path length from the position immediately after the complete acceleration in the large-angle scattering. The increase in capture cross section decreases both the peak height and width, while the increase in outgoing path length decreases the peak height and lowers slightly the peak energy. It indicates that the electron capture cross section and the outgoing path length are the independent fitting parameters[5].

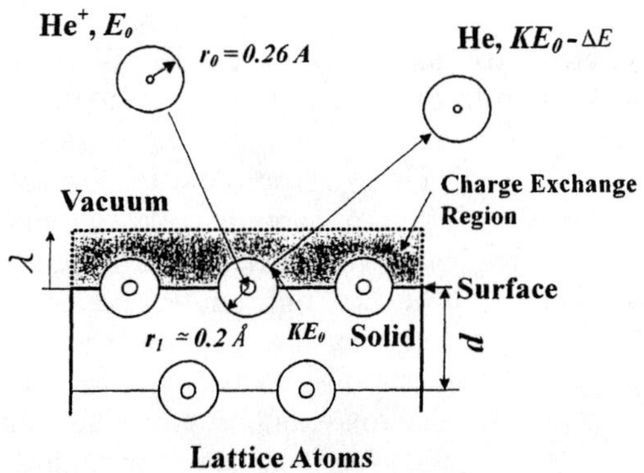

FIG. 6. Schematic of a large-angle collision between He^+ and topmost atom. Hatched region denotes the charge exchange region from the top atomic plane toward the vacuum side.

The electron capture and loss cross sections derived from the surface peak analysis for graphite are plotted in Fig. 7 and compared with the data obtained for gas targets(CH_4, H_2, CO_2, O_2)[12,13]. Here, we assumed the additive rule to deduce the cross sections for He in C. The electron capture cross sections derived here are comparable with the data obtained for the gas targets. In addition, the energy dependence of the electron capture and loss cross sections is also consistent with the data for gas targets. Such consistency supports our simple model. It must be noted that the charge exchange considered here is limited to the region far from the target nucleus typically more than 0.2 Å. For impact parameters more than 0.2Å, medium energy He ions undergo only small angle scattering.

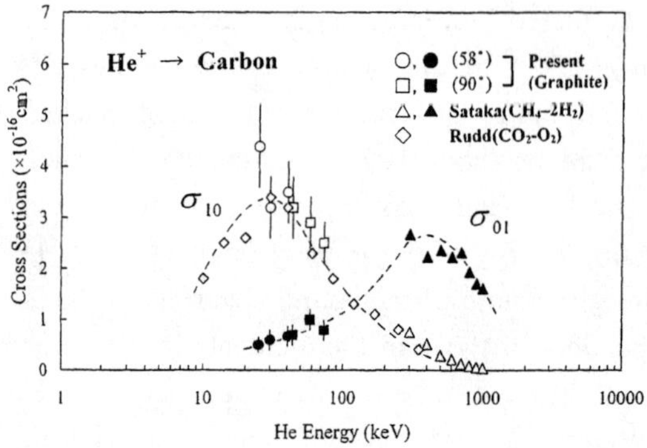

FIG. 7. Electron capture (open) and loss (full) cross sections (10^{-16} cm^2/atom) for He ions passing through carbon medium. Squares and circles denote the present data for scattering angle of 90° and 58°, respectively. Triangles and rhombi denote the data reported by Sataka et al. (CH_4 and H_2 gases)[12] by Rudd et al.(CO_2 and O_2 gases)[13], respectively. For gas targets, the additive rule was used.

Figure 8 shows the path length (equilibrium thickness: T_{eq}) to attain a charge state equilibrium for He ions. In the medium energy region (≲ 0.02 MeV/amu, amu: atomic mass unit), the major charge states are $+e$ and neutral and thus we defined the T_{eq} value as the inverse of $(\sigma_{10}+\sigma_{01})$. On the other hand, in the high energy region (≳ 0.2 MeV/amu), the dominant charge states are He^+ and He^{2+}. Therefore, the T_{eq} value was defined by the inverse of $(\sigma_{21}+\sigma_{12})$ for He energies larger than 0.2 MeV/amu. The T_{eq} values in the high energy region were calculated from the cross section data measured for gas targets using the additive rule[14-16]. Apparently, the T_{eq} value increases pronouncedly with increase in He energy.

Here, we define the relative surface peak intensity (I_S) as the ratio of the height of the surface peak to that of the scattering yield for He ions backscattered from a deeper region whose charge state is equilibrated. In the case of the a-SiO_2 films, the I_S value for the He ions backscattered from O atoms is considerably larger than that from Si atoms. The large I_S value is responsible for a small electron capture cross section. Such a Z_2-dependence suggests that not only outer (valence) electrons but also inner (core) electrons contribute to the electron capture cross section and thus it is large for high-Z_2 atoms because of larger number of the bound electrons. According to our model, decrease in surface peak intensity with increase in He energy is ascribed to the fact that the equilibrium average charge increases and the electron capture cross section decreases with increase in He

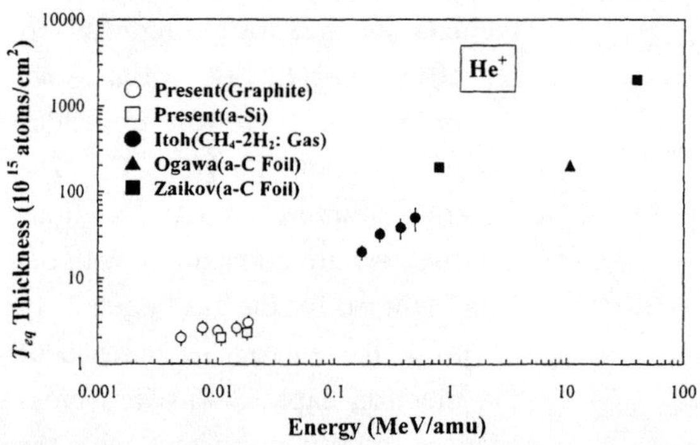

FIG. 8. Thickness to attain a charge state equilibrium for He ions. Open circles and squares denote the values derived from the present data for graphite and a-Si, respectively. Full circles and triangle are the T_{eq} values estimated from the data for gas targets(CH_4, H_2)[14] and for C-foil[15], respectively. Full square denote the estimate by Zaikov et al.[16].

energy (see Fig. 7).

From our model, it is expected that the surface peak intensity decreases with increase in exit angle measured from the surface normal. Exit-angle dependence of the surface peak profile was measured for graphite and a-Si. Figures 9 (a) and (b) show the exit-angle dependent outgoing path length from the top atomic plane derived by best-fitting the simulated surface peaks to the observed ones for graphite and a-Si, and a-Si, respectively. The present results support the existence of the extended charge exchange region from the top atomic plane toward the vacuum side by a half to one of the interplanar distance of the major crystal plane. However, they do not indicate clearly the sharp cutoff of the charge exchange region above the surface. The long outgoing path lengths in the interaction region for graphite are probably due to the fact that the π bond electrons are largely expanded toward the vacuum side (interlayer distance: 3.4Å), in spite of low concentrations. In fact, the electron capture and loss cross sections derived for graphite are considerably smaller than those for a-Si and a-SiO_2.

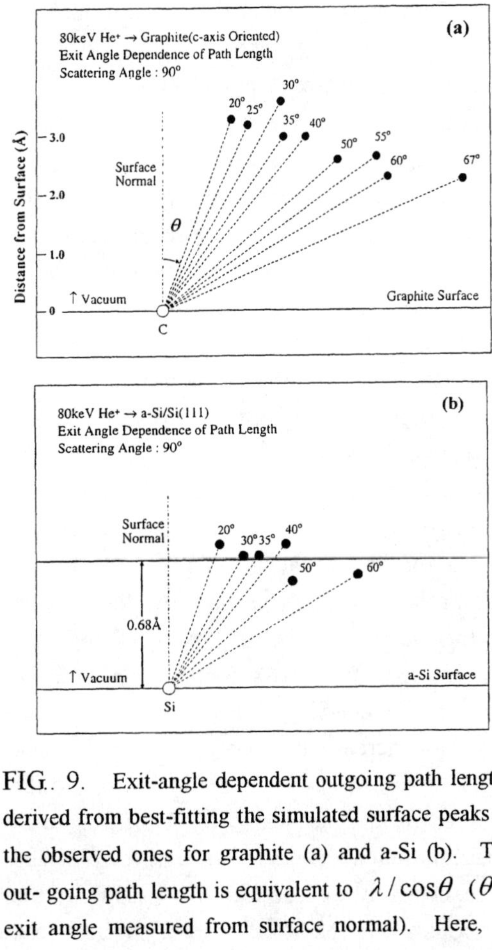

FIG. 9. Exit-angle dependent outgoing path lengths derived from best-fitting the simulated surface peaks to the observed ones for graphite (a) and a-Si (b). The out- going path length is equivalent to $\lambda / \cos\theta$ (θ : exit angle measured from surface normal). Here, 80 keV He^+ ions were incident and the scattering angle was fixed to 90°.

The present analysis does not support the sharp cutoff of charge exchange region above the top atomic plane, which was successfully introduced to explain quantitatively the exit-angle dependence of the energy loss of medium energy H ions traversing the Si(001)2 × 1-Sb surface[7]. The weak dependence of exit-angle upon the surface peak profile is possibly due to the fact that the amorphous surfaces are not flat and locally rumpled in an atomic scale. The exit-angle dependence of the energy loss of H ions is pronounced for glancing emergence condition (>85°) [7]. Under this condition, the H ions travel a long path almost parallel and close to the flat surface of the Si(001)2 × 1-Sb.

6. SUMMARY

Anomalous surface peaks were observed in the backscattering spectra measured with the toroidal ESA for medium energy He^+ incidence on low-Z_2 materials such as graphite, a-SiO_2, and a-Si films. The surface peak is explained quantitatively by considering the charge exchange near the surface regions. The surface peak is originated from the high fraction of nonequilibrium singly-ionized charge state of He ions backscattered from the topmost atoms. The physical quantities to reproduce the surface peak are reduced to two independent parameters, (1) electron capture (or loss) cross section and (2) outgoing path length in the ion-electron interaction region. In order to reproduce the surface peak systematically, it is essential to assume the charge state of the He ions immediately after the large-angle scattering is almost $+e$.

The electron capture and loss cross sections determined by best-fitting the simulated surface peak to the observed one are comparable with those measured for gas targets. The energy dependence of the capture and loss cross sections derived here is also consistent with that for gas targets. In order to reproduce the surface peak, the ion-electron interaction region must be extended toward the vacuum side from the top atomic plane by a half to one of the interplanar distance of the major crystal plane.

With increase in exit-angle, the relative peak intensity is significantly reduced and the peak width is broadened, although the change is not so pronounced. The present results support the existence of the extended charge exchange region from the top atomic plane toward the vacuum side. However, they do not indicate clearly the sharp cutoff of the charge exchange region above the surface. The long outgoing path lengths above the graphite surface are probably due to the fact that the π bond electrons are largely expanded toward the vacuum side, in spite of the low concentrations. In fact, the electron capture and loss cross sections derived for graphite are considerably smaller than those for a-Si and a-SiO_2

The surface peak intensity is reduced with increase in Z_2-number. In fact, no surface peak appeared in the backscattering spectra from a-Ni ($Z_2 = 28$) films. In the case of a-NiSi, strong surface peaks were seen in the backscattering spectra from Si but not from Ni. This surface peak intensity is much stronger than that for a-Si and a-SiO_2. The strong surface peak was successfully reproduced by assuming the surface segregation of Si with thickness about 2 MLs. Thus, such a surface peak provides the quantitative information about the elemental composition near the top surface region of amorphous compounds.

ACKNOWLEDGMENTS

The author would like to thank Dr. T. Nishimura and Mr. F. Fukumra for their assistance in the MEIS measurement and also for valuable discussion on how to analyze the surface peaks. Special thanks are also due to Mr. Y. Hoshino, Mr. Y. Taki, Mr. H. Toi and Mr. E. Toyota for their assistance in the sample preparation and the MEIS measurement.

REFERENCES

1. J.F. van der Veen, Surf. Sci. Rep. **5** (1985) 199.
2. J. Vrijmoeth, P.M. Zagwijn, J.W.M. Frenken and J.F. van der Veen, Phys. Rev. Lett. **67** (1991) 1134.
3. K. Kimura, K. Ohshima and M. Mannami, Appl. Phys. Lett. **64** (1994) 2232.
4. T. Nishimura, A. Ikeda and Y. Kido, Rev. Sci. Instrum. **69** (1998) 1671.
5. Y. Kido, T. Nishimura and F. Fukumura, Phys. Rev. Lett. **82** (1999) 3352.
6. Y. Kido, H. Namba, T. Nishimura, A. Ikeda, Y. Yan and A. Yagishita, Nucl. Instrum. Methods **B136-138** (1998) 798.
7. K. Sumitomo, T. Nishioka, A. Ikeda and Y. Kido, Phys. Rev. **B56** (1997) 7011.
8. A. Ikead, K. Sumitomo, T. Nishioka, T. Yasue, T. Koshikawa and Y. Kido, Surf. Sci. **385** (1997) 200.
9. Y. Fujii, S. fujiwara, K. Narumi, K. Kimura and M. Mannami, Surf. Sci. **277** (1992) 164.
10. J.B. Marion and F.C. Young, *Nuclear Reaction Analysis-Graphs and Tables* (North-Holland, Amsterdam, 1968).
11. K. Tobita, H. Takeuchi, H. Kimura, Y. Kusama and M. Nemoto, Jpn. J. Appl Phys. **26** (1987) 509.
12. M. Sataka, A. Yagishita and Y. Nakai, J. Phys. **B23** (1990) 1225.
13. M.E. Rudd, T.V. Goffe, A. Itoh and R.D. DuBois, Phys. Rev. **A32** (1985) 829.
14. A. Itoh, Doctor Thesis (Kyoto University, 1979).
15. H. Ogawa, I. Katayama, H. Ikegami, Y. Haruyama, A. Aoki, M. Tosaki, F. Fukuzawa, K. Yoshida, I. Sugai and T. Kaneko, Phys. Rev. **B43** (1991) 11370.
16. V.P. Zaikov, E.A. Kralkina, N.F. Vorobjev, I.S. Dmitriev, V.S. Nikolaev and Ya. A. Teplova, Nucl. Instrum. Methods **B5** (1984) 10.

Atomic and Molecular Data Activities and Services

Interstellar and Circumstellar Chemistry: The UMIST Database

T J Millar

Department of Physics, UMIST, PO Box 88, Manchester M60 1QD, UK

Abstract. Molecules have been detected in interstellar and circumstellar clouds for over 60 years, but it has been only in the past 20 years or so that a significant effort has gone into chemical kinetic modelling of the processes by which they are formed and destroyed. In this paper, I review the physical conditions in the regions in which molecules are detected, and indicate the nature of the data required for a full interpretation of the observations. In particular, I describe the database for astrochemistry we have developed at UMIST, indicating areas for future development.

INTRODUCTION

To date, over 100 molecules have been detected in interstellar and circumstellar regions. These regions include objects as diverse as diffuse clouds, dust clouds, the envelopes around cool, late-type stars, proto-planetary nebulae, planetary nebulae, and even in the extreme environments around novae, supernovae and Wolf-Rayet stars. In the following Section, I shall discuss the physical properties of a number of these regions. In many respects it is these physical properties which determine the important classes of chemical reaction.

The first detections of interstellar molecules were made around 1937-1941 through optical absorption spectroscopy of simple diatomic species CH, CN and CH^+. For many years, it was thought that the low densities and high flux of ultraviolet photons in interstellar clouds would prevent large abundances of molecules being synthesised. In particular, gas-phase routes to form H_2 were known to be very inefficient and it was thought that the photodissociation rate of H_2 was very large, indicating that the simplest neutral molecule, containing the most abundant element in the Universe, would have a very low abundance.

This view began to change in the late-1960s when millimeter-wave telescopes detected polyatomic molecules such as NH_3 and H_2CO. Observations of several rotational transitions showed that the gas temperature was generally very low, $\sim$ 10-50 K, but that the molecules were found in clouds with number densities $\sim 10^4$ - 10^6 cm^{-3}, 2-4 orders of magnitude larger than those detected through optical spectroscopy. At the same time, it was recognised that the larger densities evident

CP543, *Atomic and Molecular Data and Their Applications,* edited by K. A. Berrington and K. L. Bell

in these clouds would help prevent rapid photodissociation through absorption of ultraviolet photons by dust grains, that these dust particles would provide an efficient catalytic surface for the recombination of H atoms to form H_2, and that the H_2 molecule was photodissociated via line absorption rather than through continuum absorption.

In the last 20 years, the development of sensitive millimeter and submillimeter detectors, the construction of large aperture telescopes at high mountain sites, and the development of infrared space telescopes have revolutionised our view of the molecular universe [1].

Table 1 lists the 118 molecules detected in interstellar and circumstellar regions. One should note that around 30 of the molecules listed as interstellar are also found in circumstellar regions. The molecules listed as circumstellar are those which are found only in this environment. It is likely that some of these, such as C_3 and C_5, are present but currently undetected in interstellar clouds. I consider the detection of all these species as secure; some other species have been identified tentatively, while there is clear evidence from infrared emission bands for a population of large polycyclic aromatic hydrocarbons, although individual species have not yet been identified.

It should be recognised that selection effects operate in the detection of interstellar molecules. For example, molecules without a permanent electric dipole moment, such as H_2, O_2, N_2, C_2H_2, and C_4, have no allowed rotational transitions and thus cannot be observed by the most sensitive technique of radio astronomy, which has the ability to detect species with fractional abundances relative to H_2 as small as 10^{-12}. These non-polar molecules can, in some cases be observed through ro-vibrational absorption in the infrared (C_2H_2 for example) or through infrared emission from hot gas (H_2), but such detections require regions with special properties and are therefore not necessarily useful probes of molecular abundances throughout the interstellar medium. Additionally, large non-linear species, such as branched chains and ring molecules, are difficult to detect because of partition function effects, that is, the energy of the molecule is spread out over very many more accessible transitions than occurs for a linear species of comparable complexity. This is one reason why so many linear carbon chain molecules are detected.

In addition to these species detected in the gas phase, there have been a number of molecules detected in solid-state ices through infrared absorption spectroscopy. The identifications have been aided greatly by use of the *Infrared Space Observatory (ISO)* and species such as H_2O, CO, CO_2, CH_3OH and CH_4 have been detected. Detailed comparison of the observed profiles with ice mixtures created in the laboratory enables information on the ice matrix and its thermal history to be determined. Note that infrared observations are incapable of identifying species whose fractional abundance relative to H_2 is less than about 10^{-6}-10^{-7} [2].

TABLE 1. Observed interstellar and circumstellar molecules: May 2000.

Number of atoms							
2	3	4	5	6	7	8	≥ 9
			Interstellar inorganic molecules (22)				
H_2	H_2O	NH_3					
OH	H_2S	H_3O^+					
SO	SO_2						
SiO	HNO						
SiS	NH_2						
NH	N_2H^+						
NO	N_2O						
NS	H_3^+						
PN							
HCl							
SO^+							
HF							
			Interstellar organic molecules (75)				
CH^+	HCN	H_2CO	HC_3N	CH_3OH	HC_5N	$HCOOCH_3$	HC_7N
CH	HNC	H_2CS	C_4H	CH_3CN	CH_3CCH	CH_3C_3N	CH_3OCH_3
CN	HCO	HNCO	CH_2NH	CH_3NC	CH_3NH_2	CH_3COOH	CH_3CH_2OH
CO	OCS	HNCS	CH_2CO	CH_3SH	CH_3CHO	H_2C_6	CH_3CH_2CN
CS	HCO^+	c-C_3H	NH_2CN	NH_2CHO	CH_2CHCN		CH_3C_4H
C_2	HOC^+	l-C_3H	HOCHO	HC_2CHO	C_6H		CH_3C_5N
CO^+	HCS^+	C_3N	c-C_3H_2	C_5H	c-C_2H_4O		CH_3COCH_3
	C_2H	C_3O	CH_2CN	H_2CCCC			HC_9N
	C_2O	C_3S	H_2CCC	HC_3NH^+			$HC_{11}N$
	C_2S	H_2CN	HCCNC				
	CH_2	CH_3	HNCCC				
	CO_2	C_2H_2	CH_4				
		$HOCO^+$	H_2COH^+				
		$HCNH^+$					
			Circumstellar molecules (21)				
CP	C_3	HCCN	C_5	C_2H_4		C_7H	C_8H
SiC	c-SiC_2	c-SiC_3	SiH_4	C_5N			
SiN	NaCN		SiC_4				
NaCl	MgCN						
AlCl	MgNC						
KCl							
AlF							

MOLECULAR ENVIRONMENTS

In this Section, I will briefly describe the physical and chemical characteristics of regions in which the bulk of molecules are found.

Diffuse Clouds

Diffuse clouds are so-called because they are penetrated by optical (and ultraviolet) radiation. The ability of clouds to exclude visible radiation can be parameterised by a quantity, denoted A_V, called *extinction* such that an extinction of 1 magnitude in the violet corresponds to a reduction in intensity of about a factor of e^{-1}. The magnitude of extinction at ultraviolet wavelengths, A_{UV}, which determines the importance of photoprocesses, is typically 2-3 times larger than A_V. Diffuse clouds typically have A_V less than 1. The general ultraviolet radiation field can be characterised by a blackbody function with an intensity which goes to zero below the hydrogen ionisation threshold wavelength at 91.2 nm; photons with energies greater than 13.6 eV are all absorbed close to the hot stars which produce them and do not penetrate the general interstellar medium. Typical time-scales for photodissociation of molecules in diffuse clouds are around 1000 years, compared with cloud lifetimes of 10^6 years.

Diffuse clouds have very low densities by interstellar standards, with the number of hydrogen nuclei cm^{-3} being less than 1000, and more typically 200-500. The material is mostly atomic, although some objects contain a significant fraction of H_2, perhaps 30 %, the number of H_2 molecules being determined by the nature of its photodissociation, in particular the ability of H_2 to self-shield.

The relatively large radiation flux and the low densities means that the clouds are relatively warm, around 100 K - heated by photoelectrons from dust grains - but chemically simple, mostly containing rather low fractional abundances of diatomic molecules, typically around 10^{-7}-10^{-9}. Some polyatomics are also detected, notably HCN, HNC, HCO^+, C_2H, H_2CO and C_3H_2.

Chemical models of these clouds give a very good agreement with observation for most species although the more complex molecules are more difficult to understand. Essentially, photoionisation sets the fractional ionisation at a high level, around 10^{-4}, which ensures that dissociative recombination of molecular ions with electrons is fast, and photodissociation controls the loss of molecules so that it is difficult to build the more complex species efficiently. Most models overcome these problems by requiring the diffuse clouds to contain a higher density core which both provides shielding against the photon flux and ensures larger collisional formation rates. For these clouds, detailed models require accurate dissociation and ionisation cross-sections for a range of simple atoms, radicals and molecules [3].

Dust Clouds

Dust, or dark, clouds have higher densities, typically 10^4-10^5 cm^{-3}, which ensure that their extinction is large, more than 5 magnitudes, and thus that interstellar photons do not influence the chemistry. The lack of photons also implies that the kinetic temperature is low, around 10 K. In these objects all hydrogen is molecular - in fact H_2 is undetectable, but measurements of the 21 cm line of atomic hydrogen show that this is indeed the case - and ionisation is provided by high energy cosmic ray particles which stream through the cloud. The cosmic ray ionisation rate is low and gives a fractional ionisation of around 10^{-7}-10^{-8}. Although this is small, it is very important in providing the basis for a rich chemistry driven in the main by ion-neutral reactions. The dust clouds are small, with sizes on the order of 1-5 pc and masses of 10-500 $M_{\odot}$. Observations of nearby dust clouds have detected over 50 of the 96 interstellar molecules, in particular all of the carbon chain molecules detected in interstellar clouds are observed in these sources. Although quite complex molecules exist, up to $HC_{11}N$, there is a distinct lack of saturated, large organic species, with low upper limits derived for the fractional abundances of species such as ethanol, dimethyl ether, and ethyl cyanide, this despite the overwhelming abundance of hydrogen. Very large enhancements in the abundances of deuterated molecules are observed, with abundance ratios of these species relative to their hydrogen-bearing analogues around 10^3-10^4 larger than the cosmic D/H ratio of 1-2 10^{-5}. These objects are of interest to astronomers because they are sites of low mass star formation, low mass here indicating stars similar in mass to the Sun. An overview of abundances in the TMC-1 dust cloud is given by Ohishi & Kaifu [4].

The chemistry of dust clouds is dominated by ion-neutral, neutral-neutral and dissociative recombination reactions at 10 K. Current models typically involve between 2000 and 6000 reactions. For the majority of ion-neutral reactions, theory is often sufficient to predict a rate coefficient at low temperature; for neutral-neutral reactions, theory can be used but there have been enough 'surprises' in measurements of rate coefficients at low temperature that care must be taken in adopting particular values. Note that while several thousand ion-neutral reactions have been measured at room temperature, with results that can often be extrapolated to low temperature, and confirmed by measurements at 80 K, only a few tens of neutral-neutral reactions have been measured below 200 K. For dissociative recombination reactions, branching ratios as well as total rate coefficients are needed.

Giant Molecular Clouds

These are the most massive objects in the Galaxy, with masses around 10^5-10^6 $M_{\odot}$ and sizes of 100-500 pc. They are very heterogeneous containing gas with densities ranging from 100 to 10^6 cm^{-3} with temperatures typically in the range 10-50 K. The gas is very clumpy and the clouds often contain many young, high mass stars. Because of their size, these clouds are detectable in many molecules. Furthermore, the process of star formation often involves powerful stellar winds and jets which drive shock waves through the medium and can heat the gas to high temperatures, typically 1000-3000 K. As a result, infrared emission from ro-vibrational transitions of H_2 can be detected and high temperature chemistry can drive reactions which do not occur at low temperatures.

Chemistry in these regions is primarily dominated by gas-phase reactions.

Hot Molecular Cores

Regions of active high-mass star formation often contain very small, $\sim$ 0.01-0.1 pc, high-density, 10^6-10^9 cm^{-3}, hot, 100-300 K, clumps of gas, which are heated either by a nearby massive star, or by an internal source. These objects are very opaque to ultraviolet radiation, with typical extinctions of 1000 magnitudes or so and contain very few ions. Observationally they contain very large fractional abundances of both small and large saturated molecules, such as H_2O, NH_3, CH_4, H_2S, CH_3OH, C_2H_5OH and $(CH_3)_2O$ with enhancements of more than a factor of 1000 over those in dust clouds, and contain only the smallest carbon-chain molecules. Despite their high temperatures, enhanced fractionation in deuterium is observed, indeed some doubly deuterated molecules have been detected. Since deuterium fractionation is known to depend on small zero-point energy differences between reactants and products, the degree of fractionation must be set at low temperature rather than at the high temperatures observed. This, together with the degree of saturation, is taken to imply that the molecular composition of hot molecular cores is set in a previous cold phase, and set most likely by chemistry in an ice mantle, which is then evaporated to the gas following heating, internally or externally, by a newly formed massive star.

For these regions, solid-state chemical reactions, about which relatively little is known, determine the initial chemical composition of the hot gas. Once evaporated, further evolution is determined by a neutral, radical chemistry [5].

Proto-Planetary Disks

Recently, the use of interferometers has allowed the study of a small number of simple molecules in the disks which accompany the formation of low-mass stars and which may provide the material for planet formation. Disks form during the collapse process and material falls slowly into the central object at a rate of about 10^{-7} $M_\odot$per year, taking about 10^6 years to fall in from a distance of 1000 AU. The density and temperature in the disk varies radially, ranging from 10^{11} cm^{-3}and 10K at 100 AU to 10^{16} cm^{-3} and 1500 K at 0.1 AU. At these densities, three-body collisions become important, and at the highest temperatures grain destruction can return refractory elements such as Si, Fe and Mg to the gas. Over much of the disk the density is large enough and the temperature low enough to ensure that the gas accretes on to the grains, so that surface chemistry is again expected to play a role in determining the chemical composition. Finally, the central object can be a source of both ultraviolet and X-ray photons so that the effects of these must be considered, especially on the chemistry at the disk surface.

The range of physical parameters in the disk means that a wide range of processes contributes to the chemistry [6–8].

Circumstellar Envelopes

The objects discussed so far are all associated with star formation, but stars approaching the end of their lives also exhibit a rich chemistry. The so-called *Asymptotic Giant Branch (AGB)* stars are relatively cool objects which loss mass at a rate of 10^{-7}-10^{-4} $M_\odot$per year. This mass loss is in the form of both gas and dust, indeed such stars are the sites of most of the solid particles found in the Galaxy. One well-studied object is CW Leo (IRC+10216), which is carbon-rich, i.e. the abundance of carbon is greater than that of oxygen, and is the closest AGB star to Earth at a distance of less than 200 pc. Although it is optically invisible, at near infrared wavelengths it is the brightest object in the sky after the Sun. It is losing mass at a velocity of 14 km s^{-1} and at a rate of 3 10^{-5} $M_\odot$ per year. Over 50 molecules have been detected, mostly at millimeter wavelengths, and mostly the same carbon chain species found in dust clouds [9]. Most of these molecules are found in the circumstellar envelope at a distance of 10^{16}-10^{17} cm from the stellar surface. Infrared spectroscopy probes the warmer material closer to the star and shows that in this region simple molecules such as CO, HCN, and C_2H_2 are present with abundances similar to those expected in local thermodynamic equilibrium. As the gas flows away from the star, it experiences a geometric dilution to give a density proportional to r^{-2}, where r is the radial distance from the star. This dilution in density allows photons from the interstellar radiation field to penetrate the outer envelope, photodissociating the parent molecules and driving a rich chemistry.

Oxygen-rich stars, having O/C greater than 1, also contain molecules but show less chemical complexity, mostly due to the fact that oxygen does not form chemical bonds as readily as does carbon.

Despite the relatively simple geometry, chemical models are difficult to construct since they need to include LTE chemistry with dust formation close to the stellar surface, where the density can reach 10^{14} cm^{-3} at a temperature of 1000-1500 K, shock chemistry since the stars pulsate with periods of around one year, the gas-dust interaction as material flows away form the stellar surface, photochemistry and gas-phase reactions. All this occurs in a time of 20,000 years which is the time it takes for material leaving the stellar surface to reach the interstellar medium [10].

Planetary Nebulae and Supernovae

The end-point of a star depends on its initial mass. If the mass is not too large, the AGB phase is followed by the collapse of the central core of the star to form a white dwarf while the envelope drifts out to form a thin shell of material. The white dwarf is hot and provides a source of highly energetic photons, and even X-rays, which ionise the gas and alters the chemistry in the shell. In this planetary nebula, the shell material becomes very clumpy and contains relatively simple molecules [11].

If the mass of the original star is large enough the mass loss process in the AGB phase is not large enough to allow the star evolve to a white dwarf. Instead, the star destroys itself in a supernova explosion which ejects several solar masses of material at a velocity of up to 10^4 km s^{-1} . Ionisation of the ejected material is provided by γ-rays released in the decay of ^{56}Co. Compton scattering degrades these to X-rays which ionise the atomic gas. The resulting chemistry is thus driven primarily by charge-transfer processes. The degree of ionisation differs from that in interstellar clouds, being around 10^{-2} in Type II supernovae and close to 1 in Type Ia supernovae. The ejecta range in density from 10^{11} cm^{-3} (Type II) to 10^6 cm^{-3} (Type Ia) with temperatures of a few thousand K. In addition, the original 'onion-like' structure of the star is reflected in the ejecta which is stratified into a layer containing H and He, and another containing C and O. Observations of supernovae, particularly SN 1987A, show the presence of CO, SiO and H_3^+ before dust condensation occurs, so that the initial chemistry must involve radiative association processes between atoms and atomic ions [12].

THE UMIST DATABASE FOR ASTROCHEMISTRY

The first detailed steady-state chemical model for interstellar chemistry was developed by Herbst & Klemperer [13] and contained 100 reactions among 37 species to describe the synthesis of 5 observed species in dust clouds. The largest model published to date is a time-dependent calculation by Bettens & Herbst [14] which contains about 10,000 reactions and 1000 species, including molecules with up to 64 carbon atoms. Chemical models for circumstellar matter are more difficult to calculate because of the dispersive flow which makes the ODEs very stiff. The early model of CW Leo by Nejad & Millar [15] included 270 reactions among 85 species. More recently, Millar, Herbst & Bettens [16] have studied the chemistry of this object with 4000 reactions and 400 species.

The ability of modern computers and codes to deal with solving large systems of coupled, non-linear ODEs means that one needs a huge range of reaction rate data. As is clear from the previous section, this data includes rate coefficients, branching ratios, and photo-cross-sections over a wide range of temperature, and, in certain situations, data on three-body interactions, and solid-state chemistry. Around 1990, the Astrophysics Group at UMIST decided to build a database for interstellar chemistry, partly driven by the need to ensure that all internal research projects were using consistent data and partly to make available data to the outside community. At this time it was also decided to make available additional software to write the chemical database as a FORTRAN subroutine as well as a complete suite of programmes to allow time-dependent kinetic models to be studied [17]. The release also contained electric dipole moments for neutral species, to enable the low-temperature effects of ion-dipolar collisions to be incorporated, as well as heats of formation. The data was, in some sense, critically evaluated in that each member of the team had to suggest a preferred rate coefficient and assess some degree of accuracy to it. Rate coefficients were given for 2880 reactions among 313 species.

Since 1992, PPARC have supported the gathering and evaluation of data through the provision of a PhD studentship. In 1997, a second release was made public [18], with data complete to the end of 1995, and contained 3864 gas-phase reactions and 395 species, including neutral molecules with up to 12 atoms. At this point, Bettens & Herbst [14] were already studying systems with 10,000 reactions, much larger than our database. We decided not to increase our database accordingly, since we wished to keep to a basic set of reactions which would reproduce well the chemistry in a variety of interstellar clouds. At the same time the data and codes were provided in such a way that researchers would be able to add particular reactions or classes of reaction to investigate new possible chemistries. We still prefer this approach today.

The latest version of the database, RATE99, has just been released (Le Teuff, Millar & Markwick [19]). For the first time, it contains information on three-body reactions and photo-cross-sections to reflect the growing interest in the chemical evolution of more extreme environments such as proto-planetary disks and supernovae, and a set of reactions which drive the fractionation in deuterium. The current release contains over 4000 reactions and 400 species.

With this release, we have created a Java applet with which the user can interrogate the database to search, for example, for all reactions involving a particular species. Highlighting a particular reaction on the database gives further information on the reaction type, rate coefficient, the temperature range over which it is applicable, the accuracy of the data and the reference from which it was taken, The rate coefficients of individual reactions can be plotted. In addition, there is software available which will enable the user to build his or her own dataset. These customised datasets can be saved to a local computer, together with associated software, to enable detailed models to be made. Since this step involves the applet writing to the user's local hard disk, the site incorporates a security certificate which needs to be accepted before any data can be written. Web access can be made via the home page of the UMIST Astrophysics Group at URL: http://saturn.phy.umist.ac.uk:8000/

The database and associated software has proven very useful to the community with between 10 and 20 groups using it to support modelling, including several who might not have got involved in this area otherwise. Our belief is that the new

WWW facilities will make the database even more useful to the community.

The future development of the database depends on continued support from PPARC. A major item for updating for the next release is to incorporate much more information on photo-cross-sections so that rate coefficients can be calculated directly for a variety of radiation fields. Another useful addition would be to compile a list of reactions likely to be important on grain surfaces. To do this, however, will require a much better level of agreement in the community on the important parameters, such as binding energies, many of which enter the rate equations through an exponential so that small differences between the values adopted can lead to large differences in abundances calculated.

ACKNOWLEDGMENTS

Astrophysics at UMIST is funded by PPARC. I would like to thank the organisers for support of my attendance at this conference.

REFERENCES

1. Hartquist, T. W., and Williams, D. A., eds., *The Molecular Astrophysics of Stars and Galaxies*, The Clarendon Press, Oxford, (1998).
2. Ehrenfreund, P., *Sp. Sci. Rev.* **90**, 233-238 (1999).
3. van Dishoeck, E. F., and Black, J. H., *Astrophys. J. Suppl.* **62**, 109-145 (1986).
4. Ohishi, M., and Kaifu, N., *Faraday Disc.* **109** , 205-216 (1998).
5. Millar, T. J., 'Models of Hot Molecular Cores' in *Molecules in Astrophysics: Probes and Processes,* edited by E. F. van Dishoeck, Kluwer, Dordrecht, 1997, pp. 75-88.
6. Bauer, I., Finocchi, F., Duschl, W. J., Gail, H.–P., and Schlöder, J. P., *Astron. Astrophys.* **317**, 273-289 (1997).
7. Willacy, K, Klahr, H H, Millar, T J, and Henning, Th, *Astron. Astrophys.* **338**, 995-1005 (1998).
8. Aikawa, Y., and Herbst, E., *Astron. Astrophys.* **351**, 233-246 (1999).
9. Olofsson, H., 'The Neutral Envelopes around AGB and Post-AGB Objects' in *Molecules in Astrophysics: Probes and Processes,* edited by E. F. van Dishoeck, Kluwer, Dordrecht, 1997, pp. 457-468.
10. Millar, T. J., 'Molecular Synthesis in the External Envelopes of AGB Stars' in *The Molecular Astrophysics of Stars and Galaxies*, edited by T. W. Hartquist and D. A. Williams, The Clarendon Press, Oxford, 1998, pp. 331-346.
11. Howe, D. A., and Williams, D. A., 'The Chemistry of Planetary Nebula Formation' in *The Molecular Astrophysics of Stars and Galaxies*, edited by T. W. Hartquist and D. A. Williams, The Clarendon Press, Oxford, 1998, pp. 347-370.
12. Clayton, D. D., Liu, W., and Dalgarno, A., *Science* **283**, 1290-1292 (1999).
13. Herbst, E., and Klemperer, W., *Astrophys. J.* **185**, 505-534 (1973).
14. Bettens, R. P. A., and Herbst E., *Int. J. Mass Spectrom. Ion Proc.* **149/150**, 321-343 (1995).

15. Nejad, L. A. M., and Millar, T. J., *Astron. Astrophys.* **183**, 279-286 (1987).
16. Millar, T. J., Herbst, E., and Bettens, R. P. A., *Mon. Not. R. Astr. Soc.* in press.
17. Millar, T. J., Rawlings, J. M. C., Bennett, A., Brown, P. D., and Charnley, S. B., *Astron. Astrophys. Suppl. Ser.*, **87**, 585-619 (1991).
18. Millar, T. J., Farquhar, P. R. A., and Willacy, K., *Astron. Astrophys. Suppl. Ser.*, **121**, 139-185 (1997).
19. Le Teuff, Y. H., Millar, T. J., and Markwick, A. J., *Astron. Astrophys. Suppl. Ser.* in press.

Atmospheric Molecules

Laurence S. Rothman* and Marina Zivkovic-Rothman†

*Harvard-Smithsonian Center for Astrophysics
Atomic and Molecular Physics Division, MS50
60 Garden Street, Cambridge MA 02138, USA

†188 Summit Ave, Brookline MA 02446, USA

Abstract. The history and development of the HITRAN molecular spectroscopic database is described. HITRAN was created as a compilation of spectroscopic parameters that could serve to provide the basic information to computer programs modeling atmospheric attenuation due to molecular gaseous absorption in the infrared. The first edition of the HITRAN database evolved from monographs on prominent bands of water vapor and carbon dioxide published by the National Bureau of Standards published in the 1960s. This first edition contained a limited set of parameters for bands of the seven most active infrared absorbers in the terrestrial atmosphere. HITRAN has since expanded to cover a broad spectral range from the millimeter through ultraviolet regions, to include absorption cross sections in addition to the discrete line transitions, to include information on about 60 different molecular species, to include more parameters for sophisticated remote-sensing retrieval programs, and has employed newer distribution methods.

EVOLUTION OF THE HITRAN DATABASE

The HITRAN spectroscopic database was created as a compilation of spectroscopic parameters that could serve to provide the basic information to computer programs modeling atmospheric attenuation due to molecular gaseous absorption in the infrared. The first edition [1] was made public in 1973. The work had evolved from several monographs produced in the 1960s at the National Bureau of Standards [2].

The first edition of HITRAN [1] was intended to include fundamental parameters for the strong infrared transitions of the principal atmospheric absorbers in the infrared. It covered the spectral region from about 1000 to 10,000 cm^{-1} (1 to 10 μm); it was comprised of the seven gases, H_2O, CO_2, O_3, N_2O, CO, CH_4, and O_2 including their most significant isotopomers; and it consisted of line parameters for the transitions of these gases that would satisfy the Lambert-Beers law of attenuation, namely the frequency of the discrete transition (the ro-vibration transition in vacuum wavenumbers, cm^{-1}), the intensity of the transition (in $cm^{-1}/(\text{molecule}\times cm^{-2})$), the air-broadened halfwidth (in cm^{-1}/atm), and the lower state energy of the transition (in cm^{-1}). Besides these four parameters from which one could calculate a simple absorption through an atmospheric path whose pressure and temperature were given, the HITRAN database gave the unique quantum identifications of each transition.

HITRAN is the international standard molecular spectroscopic database and has developed over time with respect to several key aspects: (1) The spectral range now spans millimeter through ultraviolet transitions (the database employs the units of

CP543, *Atomic and Molecular Data and Their Applications,* edited by K. A. Berrington and K. L. Bell

wavenumber in cm^{-1} throughout, which is proportional to energy and very convenient for most modeling codes). Most of the transitions included in HITRAN are due to the electric dipole moment of the molecules, although there are also magnetic dipole and electric quadrupole transitions (which, although weak, are of interest in the case of very abundant species such as O_2 and N_2). (2) The HITRAN compilation now includes absorption cross-sections in addition to the traditional line-by-line spectroscopic parameters. The cross-sections are used where the density of spectral features is very crowded (as is the case for "heavy" molecular species in the infrared such as the chlorofluorocarbons and especially for most bands in the UV). The cross-sections are given at different pressure-temperature pairs, which facilitates modeling of transmission or radiance in non-isothermal paths. (3) The number of gases covered has expanded from the original seven gases that were the principal atmospheric absorbers in the infrared to about 60 different molecules along with their isotopomers when significant to terrestrial atmospheric remote-sensing problems. From about 1978 until 1986, the additional gases were published separately from the HITRAN database, and was known as the Trace Gas Compilation [3]. The format of this compilation was identical to HITRAN (although the definition of the intensity cutoff was different). In 1986 they were merged together into the HITRAN database. (4) The quantity of parameters for individual transitions in the line-by-line portion of the HITRAN compilation has substantially increased. The original database contained information for the line position (transition frequency in vacuum wavenumbers), the intensity of the transition, the Lorentzian halfwidth (broadening of the spectral line due to collisions of the molecule with air), the lower-state energy of the transition, and the unique quantum number identification of each transition. These parameters were considered minimal for the calculation of molecular attenuation via the Lambert-Beers law. With the increased sophistication of high-resolution instrumentation, and the advances in retrieval algorithms, more fundamental parameters have been added, including transition probabilities, self-broadened halfwidths, temperature dependence coefficients for the halfwidth, and pressure-shift coefficients. (5) The media and software for HITRAN has also evolved. Originally released on magnetic tape, it is currently available on CD-ROM and more and more data is becoming accessible on the internet. The CD-ROM has certainly emphasized the archival nature of the database. HITRAN is not a program (although the MODTRAN band-model code has been supplied with the 1996 edition of HITRAN). However, software has been included that allows various database functions, such as filtering, sorting, plotting the data, and access to relevant references.

Figure 1 illustrates the growth in the number of transitions over time with each edition of the database. What this figure does not reveal is the increased information per transition nor the vastly improved accuracy of the data. A typical example is the case of the air-broadened halfwidths. In the first editions of HITRAN, the information was lacking for this parameter for many gases, and as a result a rough default value was often adopted. The situation has rapidly changed as laboratories throughout the world have applied high-resolution technology and techniques to measure this parameter over a wide range of energy differences and temperatures.

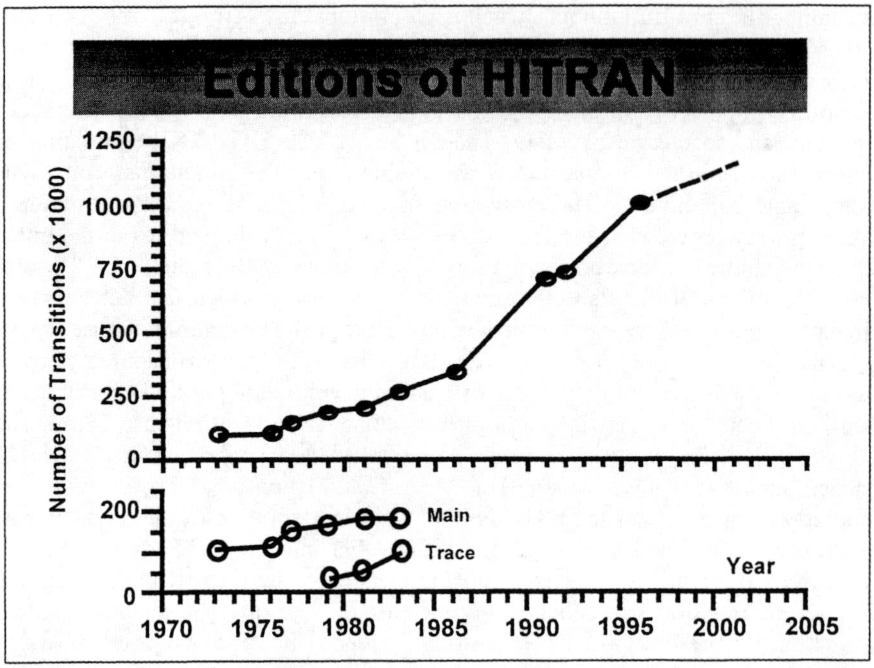

FIGURE 1. Increase in the number of line transitions with successive publications of HITRAN.

STATUS OF THE HITRAN DATABASE

The current edition of HITRAN is the edition of 1996 which is available on CD-ROM [4]. The total compilation is termed HAWKS (HITRAN Atmospheric Workstation) with the file structure shown in Fig. 2.

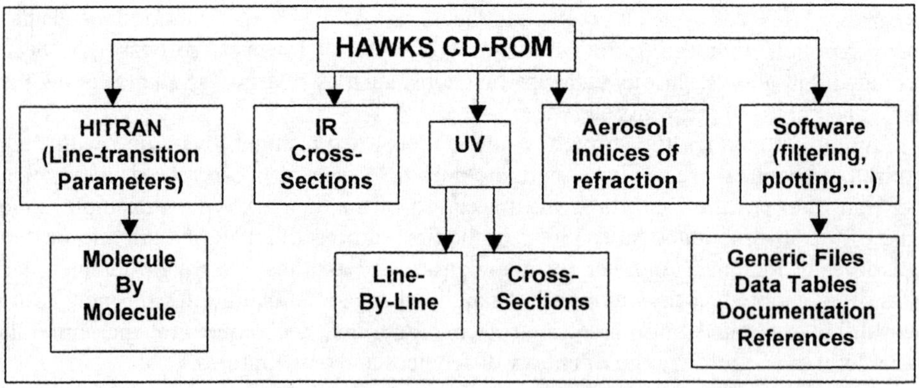

FIGURE 2. HAWKS file structure.

The molecules covered in the traditional line-by-line portion of the compilation are shown in Table 1. Column one gives the arbitrary number based on the chronological entry of the molecule into HITRAN. An intensity cutoff has been employed to limit the quantity of lines. This cutoff has been based on atmospheric transmission considerations; in practice the criteria has only been rigorously defined for uniformly mixed gases in the terrestrial atmosphere. Present difficulties with quantum mechanical calculations have also made the cutoff criteria difficult to reach for a few species.

TABLE 1. Summary of Molecules Represented with Line Transition in HITRAN96.

Mol. No.	Molecule	Number of Isotopomers	Number of Bands	Number of Lines	Spectral Coverage (cm^{-1})
1	H_2O	4	137	49 444	0 - 22657
2	CO_2	8	589	60 802	442 - 9649
3	O_3	5	106	275 133	0 - 4033
4	N_2O	5	164	26 174	0 - 5132
5	CO	6	47	4 477	3 - 6418
6	CH_4	3	51	48 032	0 - 6185
7	O_2	3	19	6 292	0 - 15928
8	NO	3	50	15 331	0 - 3967
9	SO_2	2	9	38 853	0 - 4093
10	NO_2	1	12	100 680	0 - 2939
11	NH_3	2	40	11 152	0 - 5295
12	HNO_3	1	13	165 426	0 - 1770
13	OH	3	103	8 676	0 - 9997
14	HF	1	6	107	41 - 11536
15	HCl	2	17	533	20 - 13458
16	HBr	2	16	576	16 - 9759
17	HI	1	9	237	12 - 8488
18	ClO	2	12	7 230	0 - 1208
19	OCS	4	7	858	0 - 2089
20	H_2CO	3	10	2 702	0 - 2999
21	HOCl	2	6	15 565	0 - 3800
22	N_2	1	1	120	1922 - 2626
23	HCN	3	8	772	2 - 3422
24	CH_3Cl	2	8	9 355	679 - 3173
25	H_2O_2	1	2	5 444	0 - 1500
26	C_2H_2	2	11	1 668	604 - 3375
27	C_2H_6	1	2	4 749	720 - 3001
28	PH_3	1	2	2 886	708 - 1411
29	COF_2	1	7	54 866	725 - 1982
30	SF_6	1	1	11 520	940 - 953
31	H_2S	3	15	7 151	2 - 2892
32	HCOOH	1	1	3 388	1060 - 1162
33	HO_2	1	4	26 963	0 - 3676
34	O	1	1	2	68 - 159
35	$ClONO_2$	2	3	32 199	763 - 798
36	NO^+	1	6	1 206	1634 - 2531
37	HOBr	2	2	4 358	0 - 316

Table 2 illustrates the parameters given for each transition in the line-by-line portion of the compilation. HITRAN originally was based on a FORTRAN card-image format and contained 80 characters per record (card). It was expanded in the 1986 edition to a 100 character per transition record, and it is now proposed to expand this format to a 150-character record. This increase is to accommodate new parameters, such as the upper- and lower-level statistical weights of the transitions, to include a flag for cases where line coupling is significant for the transition, and to allow enough space for representation of quanta.

TABLE 2. Proposed Format for the Next Edition of HITRAN.

Parameter	Field Length	Comment
Molecule number	2	Chronological assignment
Isotope number	1	Ordered by abundance
Frequency (cm^{-1})	12	Vacuum wavenumber
Intensity	10	At standard 296K
Einstein-A coefficient	10	Revised parameter
γ-air (Lorentzian halfwidth)	5	HWHM at standard 296K
γ-self	5	HWHM at standard 296K
E" (cm^{-1})	10	Lower-state energy
η	4	Temperature-dependence coefficient of γ-air
δ (cm^{-1})	8	Pressure shift in air
v′	15	Upper-state "global" quanta
v″	15	Lower-state "global" quanta
Q′	15	Upper-state "local" quanta
Q′	15	Lower-state "local" quanta
Error indices	6	Accuracy of critical parameters
Reference indices	12	References for critical parameters
Flag for line coupling	1	Pointer to algorithm
g′	2	Upper-level statistical weight
g″	2	Lower-level statistical weight

The infrared cross-sections cover additional molecular species to those given in Table 1. Table 3 presents a summary of the molecules that have been included since the 1992 edition of HITRAN as well as some that are planned for the next edition. The cross-sections are taken directly from high-resolution Fourier-transform spectrometer observations. Instrumental resolutions have ranged between 0.01 to 0.03 cm^{-1}. Some of the data have been extrapolated to zero pressure, whereas many of the more recent data have been taken at a variety of pressures. The experiments have generally been made at conditions representing the terrestrial atmosphere.

There are two species, sulfur hexafluoride (SF_6) and chlorine nitrate ($ClONO_2$), that have been represented with cross-sections and some line-by-line parameters. It has been recognized that for most modeling situations, the cross-sections actually are superior. There are several reasons for this situation. These molecules possess several very low frequency fundamental modes of vibration, thereby making the sequence of hot bands significant for most applications. The crowded spectra of these "heavy" molecules will often have effects such as line coupling. Cross-section data, being derived from observations, include these effects, whereas they are lacking from the line-by-line portion of HITRAN at this time.

TABLE 3. Summary of Molecules Represented by IR Cross-section Data in HITRAN.

Molecule	Common Name	Temperature Range (K)	Pressure Range (Torr)	Number of P,T sets	Spectral Coverage (cm^{-1})
SF_6	Sulfur hexafluoride	180-295	20-760	29	925-955
$ClONO_2$	Chlorine nitrate	213-296	0	2	740–840
		213-296	0	5	1240-1340
		213-296	0	2	1680-1790
CCl_4	Carbon tetrachloride	208-297	8-760	32	750-812
N_2O_5	Dinitrogen pentoxide	233-293	0	4	555-600
		233-293	0	4	720-765
		233-293	0	4	1210-1275
		233-293	0	4	1680-1765
HNO_4		268	0.7	1	770-830
CCl_3F	CFC-11	201-296	40-760	39	810-880
		201-296	40-760	39	1050-1120
CCl_2F_2	CFC-12	190-296	7.5-760	52	850-950
		190-296	7.5-760	52	1050-1200
$CClF_3$	CFC-13	203-293	0	6	765-805
		203-293	0	6	1065-1139
		203-293	0	6	1170-1235
CF_4	CFC-14	180-296	7.5-761	55	1250-1290
$C_2Cl_2F_3$	CFC-113	203-293	0	6	780-995
		203-293	0	6	1005-1232
C2C12F4	CFC-114	203-293	0	6	815-860
		203-293	0	6	870-960
		203-293	0	6	1030-1067
C_2ClF_5	CFC-115	203-293	0	6	955-1015
		203-293	0	6	1110-1145
		203-293	0	6	1167-1260
$CHCl_2F$	HCFC-21	296	1	1	785-840
$CHClF_2$	HCFC-22	181-297	0-764	29	760-860
		181-296	0-761	33	1070-1195
		253-287	0	3	1274-1380
$CHCl_2CF_3$	HCFC-123	253-287	0	3	740-900
		253-287	0	3	1079-1450
$CHClFCF_3$	HCFC-124	287	0	1	675-1430
CH_3CCl_2F	HCFC-141b	253-287	0	3	700-1500
CH_3CClF_2	HCFC-142b	253-287	0	3	650-1475
$CHCl_2CF_2CF_3$	HCFC-225ca	253-287	0	3	695-1420
$CClF_2CF_2CHClF$	HCFC-225cb	253-287	0	3	600-1500
CH_2F_2	HFC-32	203-297	0-750	17	995-1236
		203-297	0-750	17	1385-1475
CHF_2CF_3	HFC-125	287	0	1	700-1500
CHF_2CHF_2	HFC-134	203-297	0-750	9	600-1700
CFH_2CF_3	HFC-134a	253-287	0	3	815-1485
CF_3CH_3	HFC-143a	203-297	0-750	9	580-630
		203-297	0-750	9	750-1050
		203-297	0-750	9	1100-1500
CH_3CHF_2	HFC-152a	253-287	0	3	840-1490

The format of the cross-sections has been chosen in a rather compact manner. The absorption cross-sections are presented in a series of equally-spaced wavenumbers for each temperature-pressure pair of observations. A header at the beginning of every set provides information on the molecule, the wavenumber interval, the number of cross-sections, the pressure, the temperature, reference, and other pertinent qualifiers.

The software that is included in the 1996 edition runs in MS Windows and most UNIX systems. Generic files are also supplied in the c language that a user can compile and use on other systems, such as the Macintosh. The next edition of HAWKS is being written in the Java language and will run with the same full functionality on a Mac. An example of the plotting capabilities is given in Fig. 3, where the pull-down menu system is shown. This example shows both the capability of plotting the line-by-line portion of the HITRAN compilation and the cross-section portion (a feature to be introduced with the new Java version of the software).

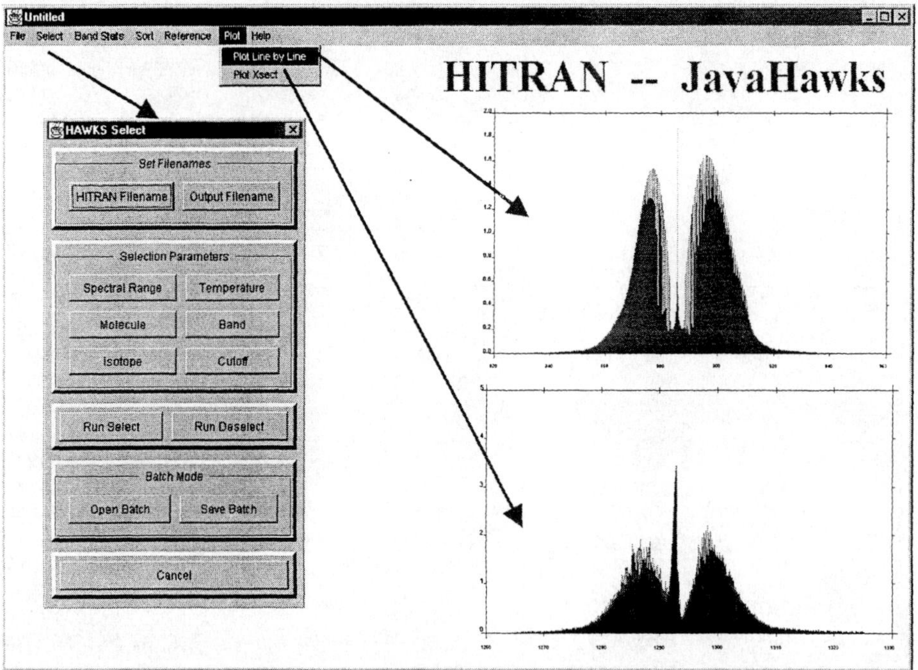

FIGURE 3. HAWKS software capabilities.

The principal function of the software is to provide the user with the ability to extract a smaller set of the line-by-line data in HITRAN for the particular problem at hand. The pull-down menu of SELECT allows one to specify the spectral range required and to select out molecules out of the total of 37 on the 1996 edition. The other four criteria of selection are less frequently employed. One can specify the set of isotopes (if more than one is available for a molecule). The temperature of the final filtered file can be scaled between 70 and 3000K (temperature dependent parameters in the HITRAN database, such as intensity and halfwidth, are given at a standard temperature of 296K). Selected vibration-rotation bands can be filtered out of the database. This function is mainly for theoretical applications, although it may have some use in non-local thermodynamic equilibrium (non-LTE) problems. Finally, a cutoff in the intensity is provided as a criterion for filtering; this choice is risky and presumably would only be used in cases where the transmission-model scenario is repetitive and well understood.

The HITRAN database was first distributed on 1600 bit per inch magnetic tape, and later on higher density 6200 BPI magnetic tape. The compilation is sent free of charge to qualified users in government agencies, industry, and academia. With the 1992 edition [5], the distributed media became CD-ROM. The CD-ROM has not only allowed storage of more information, but its archival nature has been a blessing for the managers of the database; they are no longer plagued by constant questions due to the passing of the database at facilities from one experienced user to a new operator. More recently, the internet has taken on the role of getting updates, corrections, and new data to the users. The database is still in ASCII form, so that at present very long data files, such as for molecules with dense spectra (ozone and nitric acid, for example) are not very practical for distribution via the web. However, we anticipate that having at least one dedicated server with the compilation and putting up the HAWKS software on the web will be the future method for accessing the compilation.

A web-site is maintained for the HITRAN database. Its address is: CfA-www.Harvard.edu/HITRAN. In addition to updates to parameters, this web-site contains relevant information and news about HITRAN, links to relevant spectroscopic meetings, documentation for HAWKS, links to related databases, and the questionnaire and request form for the CD-ROM.

Out of the surveys obtained from the web-site, we have been able to assess the distribution of the user community and their applications. Figure 4 shows the distribution of over 800 users who filled out the CD-ROM request form during a recent 12-month period with respect to their affiliation. This distribution is fairly representative of the approximately 3300 users on the HITRAN mailing list. A very high percentage of the users are from outside the US (about 33%).

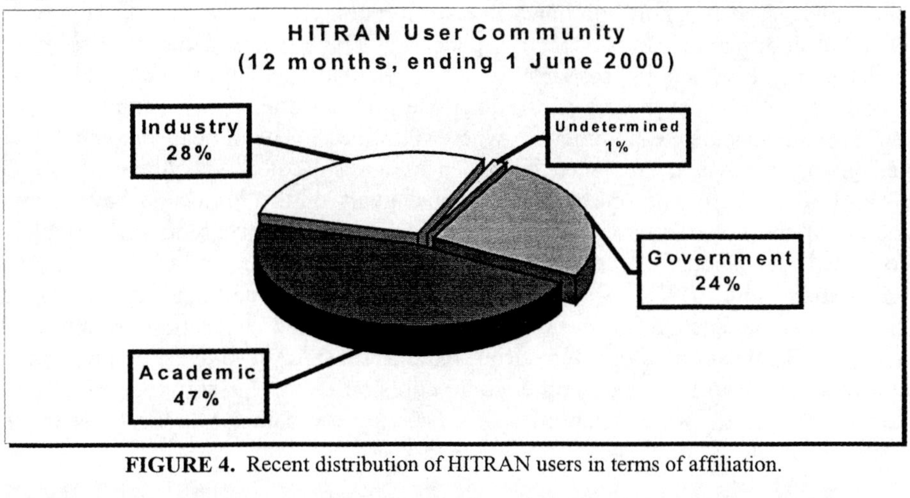

FIGURE 4. Recent distribution of HITRAN users in terms of affiliation.

The surveys also give a breakout of user applications. Table 4 is a representative sample of applications given by recent respondants. We allowed users to check off more than one application. The table illustrates the difficulty in creating these questionnaires: what had been assumed by us to be somewhat independent applications are not. We therefore have also examined the correlation among the various categories. This matrix is shown in Table 5.

As might be expected, the highest correlations occurred in three areas. Those users employing HITRAN for transmission or radiance simulations were highly likely to also be doing remote-sensing retrievals. Those people using HITRAN for lidar or laser propagation often considered their work to be simultaneously involved in remote sensing. And it should come as no surprise that investigators using HITRAN for plume, smoke stack emission, or combustion analysis would also check off band-model development. We had intended "band-model development" to mean employing the HITRAN database in the creation of band-model codes, not the separate use of these codes.

Clearly the surveys show that some refinement is necessary. These categories, which seemed almost orthogonal to the creators, certainly were not. Also included in the questionnaire were questions about the primary and secondary computers and operating systems being used with HITRAN. These have shown the expected trends: the rise in the use of UNIX systems, and the decline in the use of the VAX operating system among users.

Many interesting questions and comments were also obtained. This feedback is extremely useful in the development of the compilation to better serve the scientific community. The most common requests were for: high-temperature parameters (this difficult problem is being addressed by the HITEMP database [6]), higher precision with the parameters, and for additional molecular species. Many users have commented on the great service HITRAN has been for the scientific community, and also many have wished that the database could be made available via ftp.

The HITRAN molecular spectroscopic database serves a very broad user community. The program continues in its effort to provide the most accurate and complete set of parameters for spectral modeling. There is an international team that validates the data, and is responsible for particular specialties contained in the compilation. The contributors to the database provide both experimental data, and also provide calculated parameters where feasible. Theoretical treatments and methodologies have in fact been advanced as a result of the requirements of the HITRAN database. It is one of the principal goals of the project to have a great amount of the compilation result from self-consistent, convergent, and predictive theoretical calculations.

For the future, HITRAN will improve on the accuracy and extent of the spectroscopic parameters for absorption bands throughout the spectrum as applicable to diverse applications from terrestrial remote sensing to planetary atmospheric retrievals. There will also be parameters for collision induced absorption. Finally, we expect enhanced software capabilities of processing the data for applications via the internet.

TABLE 4. Representative Responses Giving Applications of HITRAN Users.

	1	2	3	4	5	6	7	8	9	10	11	12	13	14
1. Transmission/RadSimulation	190													
2. Climate Modeling	45	140												
3. Band-model Development	63	56	210											
4. Remote Sensing	80	33	56	156										
5. Surveillance	44	39	57	44	115									
6. Open-Path Monitoring	43	54	78	50	49	174								
7. Plumes/Combust Stack Gas	62	36	89	65	48	62	190							
8. Lidar/Laser	63	45	50	68	44	64	64	162						
9. Industrial Development	37	22	41	27	21	31	38	34	100					
10. Theory Development	37	30	58	33	35	44	49	41	42	149				
11. Laboratory Spectroscopy	44	21	32	26	14	25	33	31	31	34	119			
12. Fundamental Spectroscopy	11	3	5	9	5	3	9	8	6	7	0	24		
13. Planetary Atmospheres	20	3	7	12	0	3	7	10	1	4	12	1	32	
14. Other	6	14	21	17	19	11	26	22	11	26	15	5	2	92
Total	190	140	210	156	115	174	190	162	100	149	119	24	32	92

TABLE 5. Correlation Between Applications in Representative Responses of HITRAN Users.

	1	2	3	4	5	6	7	8	9	10	11	12	13	14
1. Transmission/RadSimulation	1.00													
2. Climate Modeling	0.22	1.00												
3. Band-model Development	0.29	0.37	1.00											
4. Remote Sensing	**0.59**	0.16	0.31	1.00										
5. Surveillance	0.28	0.35	0.47	0.37	1.00									
6. Open-Path Monitoring	0.17	0.43	0.52	0.31	0.45	1.00								
7. Plumes/Combust Stack Gas	0.33	0.17	**0.56**	0.44	0.39	0.40	1.00							
8. Lidar/Laser	0.39	0.28	0.24	**0.53**	0.35	0.44	0.40	1.00						
9. Industrial Development	0.20	0.05	0.28	0.11	0.08	0.18	0.25	0.19	1.00					
10. Theory Development	0.06	0.09	0.33	0.07	0.20	0.23	0.26	0.15	0.38	1.00				
11. Laboratory Spectroscopy	0.24	-0.03	0.06	0.04	-0.13	0.00	0.09	0.07	0.25	0.16	1.00			
12. Fundamental Spectroscopy	0.11	-0.27	-0.16	0.08	-0.13	-0.25	0.02	-0.02	-0.11	-0.11	-0.41	1.00		
13. Planetary Atmospheres	0.36	-0.24	-0.15	0.15	-0.35	-0.25	-0.10	0.00	-0.29	-0.27	0.17	-0.23	1.00	
14. Other	-0.30	-0.13	-0.04	-0.12	-0.02	-0.16	0.04	-0.05	-0.15	0.10	-0.10	-0.14	-0.33	1.0

ACKNOWLEDGMENTS

The HITRAN molecular database has been a project with strong international cooperation during its development. Laboratories throughout the world have contributed both experimental data and theoretical calculations. We in particular want to thank the major recent contributions of A. Barbe, D.Chris Benner, L.R. Brown, C. Camy-Peyret, C. Chackerian, K.V. Chance, C. Clerbaux, E.A. Cohen, V. Dana, V.Malathy Devi, H. Dothe, D.P. Edwards, J. Esmond, M.P. Esplin, J.-M. Flaud, R.R. Gamache, L.P. Giver, A. Goldman, D. Goorvitch, K.W. Jucks, R.L. Hawkins, M.L. Hoke, W.J. Lafferty, J.-Y. Mandin, S.T. Massie, K. Minschwaner, D. Nelson, V. Nemtchinov, D. Newnham, J. Orphal, A. Perrin, P. Pracna, C.P. Rinsland, J. Schroeder, M.A.H. Smith, K. Smith, R.A. Toth, Š. Urban, P. Varanasi, R.B. Wattson, M. Weber, K. Yoshino, and M. Zahniser. We wish to especially thank M.D. King, D.O'C. Starr, G.M. Stokes, and T.S. Cress for their support for this project.

The current effort has been funded by the NASA Earth Observing System (EOS), grant NAG5-8420; the Atmospheric Radiation Measurement (ARM) program of the Environmental Sciences Division, Office of Biological and Environmental Research, US Department of Energy; and the US Air Force Office of Scientific Research.

REFERENCES

1. McClatchey, R.A., Benedict, W.S., Clough, S.A., Burch, D.E., Calfee, R.F., Fox, K., Rothman, L.S., and Garing, J.S. "AFCRL Atmospheric Absorption Line Parameters Compilation," AFCRL-TR-0096 (1973).
2. Gates, D.M., Calfee, R.F., Hansen, D.W., and Benedict, W.S., "Line Parameters and Computed Spectra for Water Vapor Bands at 2.7μ," NBS Monograph 71 (1964); Calfee, R.F., and Benedict, W.S., "Carbon Dioxide Spectral Line Positions and Intensities Calculated for the 2.05 and 2.7 Micron Regions," NBS Technical Note 332 (1966); Benedict, W.S., and Calfee, R.F., "Line Parameters for the 1.9 and 6.3 Micron Water Vapor Bands," U.S. Dept. of Commerce, ESSA paper 2 (1967).
3. Rothman, L.S., Clough, S.A., McClatchey, R.A., Young, L.G., Snider, D.E., and Goldman, A., *Appl. Opt.* **17,** 507 (1978).
4. Rothman, L.S., Rinsland, C.P., Goldman, A., Massie, S.T., Edwards, D.P., Flaud, J.-M., Perrin, A., Camy-Peyret, C., Dana, V., Mandin, J.-Y., Schroeder, J., McCann, A., Gamache, R.R., Wattson, R.B., Yoshino, K., Chance, K.V., Jucks, K.W., Brown, L.R., Nemtchinov, V., and Varanasi, P., *J. Quant. Spectrosc. Rad. Transfer* **60,** 665-710 (1998).
5. Rothman, L.S., Gamache, R.R., Tipping, R.H., Rinsland, C.P., Smith, M.A.H., Benner, D.C., Devi, V.M., Flaud, J.-M., Camy-Peyret, C., Perrin, A., Goldman, A., Massie, S.T., Brown, L.R., and Toth, R.A., *J. Quant. Spectrosc. Rad. Transfer* **48,** 469-507 (1992).
6. Rothman, L.S., Wattson, R.B., Gamache, R.R., Schroeder, J., and McCann, A., *Proc. Soc. Photo-Optical Instrumentation Engineers* 2471 (1995).

Spectral Data for Atoms and Ions

Joseph Reader

National Institute of Standards and Technology
Gaithersburg, MD 20899, U.S.A.

Abstract. Progress continues on the development of databases for spectral properties of atoms and ions. Version 2.0 of the NIST Atomic Spectra Database (ASD) became available on the World Wide Web in March 1999. This version contains data for many more spectra than version 1.2, which it replaced, and the web interface is much more flexible in terms of presentation options for the user. We are now working to further improve the user interface, especially concerning ease of access to the references. The next version will provide hyperlinks from the spectrum page to references for the wavelengths and transition probabilities of that spectrum. As a continuing effort we are working to expand the coverage of spectra. The NIST Bibliographic Database on Atomic Transition Probabilities has been updated recently, and the NIST Bibliographic Database on Atomic Spectral Line Broadening will be updated sometime during 2000. Some uses of the databases for wavelength calibration of VUV lasers, online modeling of plasmas, and searches for the transition probabilities for all downward branches to specific energy levels of a particular atom for plasma diagnostics will be described.

INTRODUCTION

When first invited to present this paper I was asked to discuss "spectral line data." However, not having much experience with molecules I decided that a suitable title should be something like "spectral line data *for atoms*." Since then there have been some interesting applications of our databases to molecules, and I will in fact mention some of them.

Perhaps I should start with a question: Is there a need for spectral line data for atoms? You might just reflexively answer *yes* to this; everyone knows that atomic spectra are used in numerous ways. So it is certainly clear that it is useful to organize what is known in ways that make the data available to those who need them. Hence our atomic spectra databases, especially those that can be reached via the World Wide Web. The need is clear: since version 2.0 of the NIST Atomic Spectra Database came online one year ago, we have been receiving about 45000 requests per month. The

CP543, *Atomic and Molecular Data and Their Applications,* edited by K. A. Berrington and K. L. Bell
2000 American Institute of Physics 1-56396-971-8

requests originate from many sources, as I will show later. We also get a lot of mail - I will review some of the e-mail messages below. So this is not a problem to answer.

The harder question is: Do we need *more* atomic line data? The answer to this is not as clear. Some years ago, before we started supplying data for tokamak research to the fusion community, we had a seminar at NIST by someone from the U.S. Department of Energy. He told about materials that would be used in tokamaks - Mo, W, Ni, etc., and how their spectra would be used to diagnose the plasmas in these machines. At some point I asked (what I thought was rhetorically) as to whether there was a need for spectroscopic research on these elements to get the needed data. Whereupon he replied - Oh, no; we already have everything that we need - they are in tables that the Bureau of Standards (now NIST) gives us. Well, of course, that was *not true*. And we went on to carry out a very successful program to supply the data that really were in fact needed. This program still exists today, and there still is a need for various data in this area. Nevertheless, it is hard to overcome the impression that because there is so much data now, there is not a need for more. Those of us in the field know how much more needs to be done.

ATOMIC SPECTRA DATABASES

In what follows, I will describe some of the atomic spectra databases that are available, including our NIST Atomic Spectra Database, and look at some of the mail that we get. First I will summarize the principal atomic spectra databases:

1. Atomic Spectral Line Database. P. L. Smith, C. Heise, J. R. Esmond, and R. L. Kurucz, Harvard Smithsonian Center for Astrophysics and Deutsche Telekom AG; http://cfa-www.harvard.edu/amdata/ampdata/kurucz23/sekur.html
(good flexibility; no observed intensities; no separate energy level tables)

2. Atomic Spectra Database. D. E. Kelleher, W. L. Wiese, W. C. Martin, K. Olsen, A. Musgrove, J. R. Fuhr, L. Podobedova, E. B. Saloman, P. J. Mohr, J. Sugar, J. Reader, C. J. Sansonetti, and G. R. Dalton, National Institute of Standards and Technology; http://physics.nist.gov/asd
(good flexibility; observed intensities; separate energy level tables)

3. Vienna Atomic Line Data-Base. VALD-EMS: Vienna Atomic Line Data-Base - Electronic Mail Service, Vienna, Austria; e-mail to vald@galileo.astro.univie.ac.at
(*e-mail inquiries/replies only*)

4. Atomic Spectral Line List. R. Hirata and T. Horaguchi, Department of Astronomy, Faculty of Science, Kyoto University and National Science Museum, Hyakunin-cho, Shinjuku-ku, Tokyo; http://cdsweb.u-strasbg.fr/htbin/myqcat3?VI/69/
(gf-values and wavelengths for comparatively low-ionized species)

5. Databases for Atomic and Plasma Physics. http://plasma-gate.weizmann.ac.il/DBfAPP.html; *(listing of databases by country)*

6. Atomic Data for Astrophysics. D. Verner, U. of Kentucky
http://www.pa.uky.edu/~verner/atom.html; *(listing of databases)*

The database that is most similar to the NIST Atomic Spectra Database is the Atomic Spectral Line Database maintained at the Harvard-Smithsonian Center for Astrophysics, No. 1 above. Figs. 1 & 2 show a request form and retrieval for the case of Na I. Extensive references are also available.

The NIST Atomic Spectra Database (ASD) can be accessed through the URL given in No. 2 above or through the NIST Physics Laboratory home page (http://physics.nist.gov). By linking to Physical Reference Data, a page displaying the several databases on atomic spectroscopy that NIST maintains is obtained (Fig. 3). I will mention something about each one of these below. Ver. 2.0 of ASD came online in March 1999, replacing Ver. 1.2, and has been very popular. As mentioned, it is receiving about 45000 inquiries per month. Breakdown of requests by domain name is roughly 15% commercial(.com), 14% academic(.edu), 11% government or military(.gov or .mil), 39% non U.S. countries(.fr, .de, .ca, .jp, .uk, etc.), 5% network(.net), 1% organization(.org), and 15% unknown origin. The currently active contributors to ASD are J. R. Fuhr, D. E. Kelleher, W. C. Martin, P. J. Mohr, A. Musgrove, K. Olsen, L. Podobedova, J. Reader, E. B. Saloman, C. J. Sansonetti, and W. L. Wiese.

The "Spectrum of Platinum Lamp for Ultraviolet Spectrograph Calibration," has proved to be surprisingly popular. Originally developed for wavelength calibration on the Hubble Space Telescope and used on our website as a test case for display of graphic information, this atlas is receiving about 7000 requests per month. The applications to spectra of molecules that I will describe below are related to the availability of this atlas on the web.

The general contents of ASD are shown in Fig. 3. Some diagrams showing breakdown of the contents by element and stage of ionization for the transition and energy level data have been given by Kelleher *et al.* [1]. Of prime importance is the fact that all of these data have been critically evaluated by NIST scientists. For example, if several values for a particular wavelength or transition probability have appeared in the literature, only the value judged to be the most reliable by virtue of having met certain evaluation criteria is presented. An example of a retrieval of multiplet-ordered lines for Na I is given in Figs. 4 & 5. The first value for each multiplet is an average for the multiplet. Further details about the contents of ASD are given in an accompanying paper [2]. Some examples of e-mail messages that we have received follow.

"Hi. I am a student in Australia and I'm trying to find the ionization energy for the 1st and 2nd ionization of the elements of the second and third periods. Please send all the information that you have to me. Thank you.-AUSTRALIA" *(This is easily responded to by referring to our database on ionization energies.)*

"On examining the Cu I data listed by NIST (comparing them with the MIT Wavelength Tables) one finds that MIT has many more lines. Is the reason that you have fewer lines because the missing lines are less accurately known or have the other lines since been found to be due to impurities and therefore not belong to copper? Thank you.-FLORIDA, U.S.A." *(The problem here is that for many elements the list*

Na I Search Results

Wl_vac / nm	Wl_air / nm	log_gf	Element (Name)	E_lower_lev. / cm^(-1)	J lower	E_upper_lev. / cm^(-1)	J upper	Ref.
568.4210	568.2633	-0.700	Na I	16956.172	0.5	34548.766	1.5	KP
568.9772	568.8194	-1.400	Na I	16973.368	1.5	34548.766	1.5	KP
568.9783	568.8205	-0.450	Na I	16973.368	1.5	34548.731	2.5	KP
589.1583	588.9951	0.117	Na I	0.000	0.5	16973.368	1.5	NBS
589.7558	589.5924	-0.184	Na I	0.000	0.5	16956.172	0.5	NBS
615.5929	615.4226	-1.560	Na I	16956.172	0.5	33200.675	0.5	KP
616.2452	616.0747	-1.260	Na I	16973.368	1.5	33200.675	0.5	KP

```
Number of lines in output :  7

Parameter set used in search:

Lower wavelength limit:      560.0000 nm
Upper wavelength limit:      660.0000 nm
Lower limit on log gf:   -3.000
Upper limit on log gf:    1.000
Lower limit on energy for lower level:        0.000 cm^-1
Upper limit on energy for lower level:   100000.000 cm^-1
Lower limit on energy for upper level:        0.000 cm^-1
Upper limit on energy for upper level:   100000.000 cm^-1

Element list:

Na I          ( = 11.00 )

Search program by Claas Heise (heise@fz.telekom.de), Jan 17, 1996
```

References

The atomic line data used in this database are taken from Bob Kurucz' CD-ROM 23 of spectroscopic line calculations. The database contains all lines of the file "gfall.dat" with the following items for each line: **Wavelength; loggf; element code; lower level: energy, J, configuration; upper level: energy, J, configuration; gamma r; gamma s; gamma w; reference code.**

You can view the list of references Bob Kurucz and Barbara Bell used in generating the atomic line data. The list is ordered as follows:
Code for (element . ionization stage)
Abbreviation for authors (as in "Ref" column in line list output)
Bibliographic Reference

List of references sorted by *author code.*

List of references sorted by *element code.*

References for Na I

```
11.00  KP    Kurucz, R.L. and Peytremann, E. 1975, SAO Special Report 362.
11.00  NBS   Wiese, W.L., Smith, M.W., and Glennon, B.M. 1966, NSRDS-NBS 4.
```

Back to the Database Interface.

Problems, questions, comments to:
che@pmp.uni-hannover.de

Last changes: Mon. Nov. 13, 1995 (che)

FIGURE 2. Retrieval for Na I for *Atomic Spectral Line Database* from Harvard-Smithsonian Center for Astrophysics.

of lines in ASD is taken from the NIST 68 [3] compilation, and this compilation only contains a selection of the strongest lines in each wavelength region. So if a user requires a more complete list, he would have to use the references to look into it. It is unlikely that we will ever have truly complete wavelength lists for every element in the database. For some rare- earth elements, a complete list for a single ionization stage can run to 20000 or more lines.)

NIST Physics Laboratory | Physical Reference Data

PHYSICAL REFERENCE DATA

Atomic Spectroscopic Data

- **Atomic Spectra Database**

  Version 2.0 NEW! (new interface & expanded data)
  This database contains critically evaluated NIST data for radiative transitions and energy levels in atoms and atomic ions. Data are included for observed transitions of 99 elements and energy levels of 52 elements. ASD contains data on about 950 spectra from about 1 Å to 200 μm, with about 70,000 energy levels and 90,000 lines, 40,000 of which have transition probabilities. The most current NIST-evaluated data associated with each transition are integrated under a single listing.

  Version 1.3 Updated! (earlier version)
  This database includes some of the existing critically evaluated NIST data on atomic energy levels, transition probabilities, and wavelengths that are reasonably up-to-date. This interactive database has energy level data for over 500 spectra, transition probabilities for Sc through Ni, and wavelength data for spectra of several elements.

- **Ground Levels and Ionization Energies for the Neutral Atoms** Updated!
  This table gives the principal ionization energies (in eV) for the neutral atoms from hydrogen (Z=1) through rutherfordium (Z=104). The spectroscopic notations for the electron configurations and term names for the ground levels are also included.

- **Spectrum of Platinum Lamp for Ultraviolet Spectrograph Calibration**
  An atlas of the spectrum of a platinum/neon hollow-cathode reference lamp in the region 1130 Å to 4330 Å is given, with the spectral lines marked and their intensities, wavelengths, and classifications listed. Graphical figures of the spectrum are included.

- **Bibliographic Database on Atomic Transition Probabilities**
  This interactive database contains references on atomic transition probabilities (oscillator strengths, line strengths, and radiative lifetimes). Both theoretical and experimental papers are listed.

- **Bibliographic Database on Atomic Spectral Line Broadening**
  This interactive database contains references on atomic spectral line broadening (line shapes and shifts). Both theoretical and experimental papers are listed.

http://physics.nist.gov/PhysRefData/contents-atomic.html

FIGURE 3. Home page for *Atomic Spectroscopic Data* for NIST Physical Reference Data.

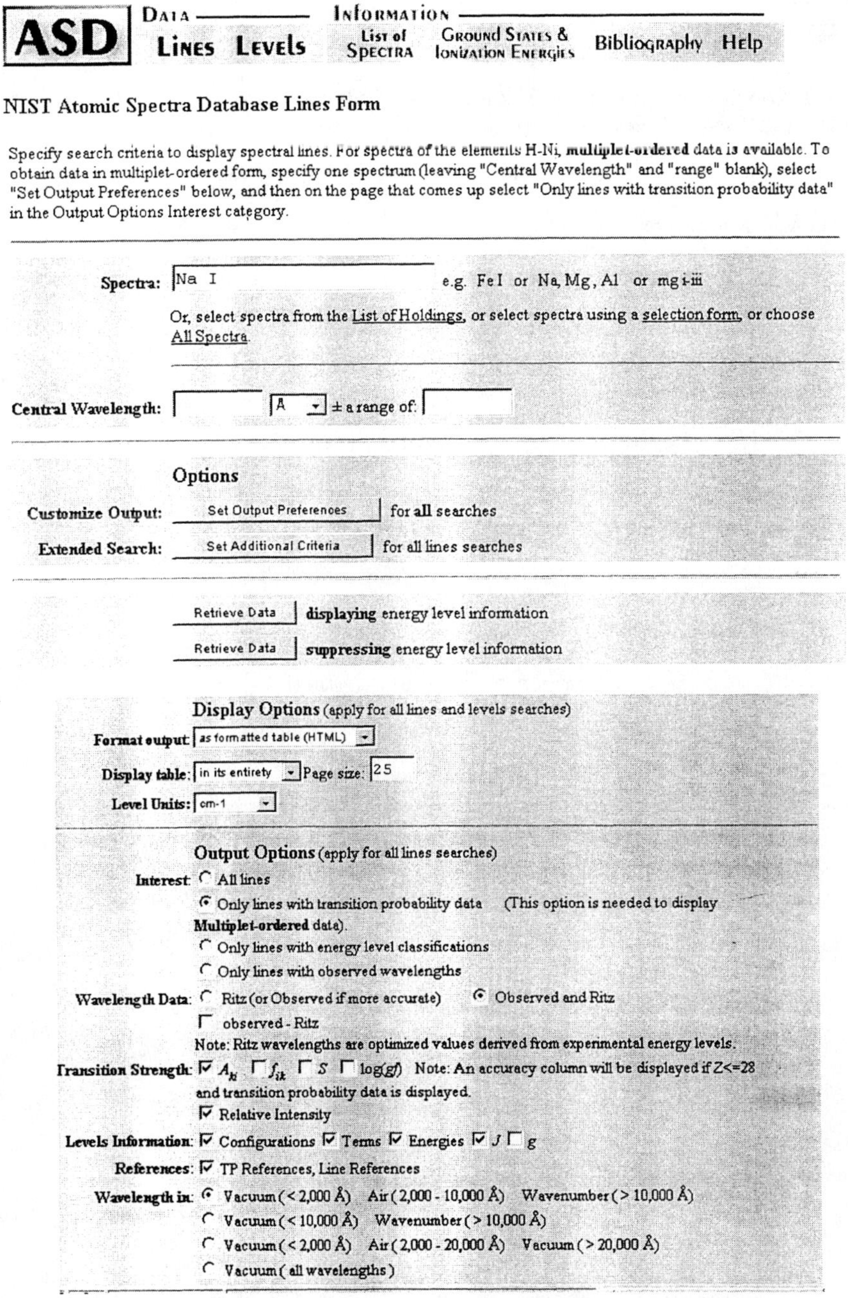

FIGURE 4. Request form for Lines for *NIST Atomic Spectra Database.*

ASD | DATA: **LINES** **LEVELS** | INFORMATION: LIST OF SPECTRA | GROUND STATES & IONIZATION ENERGIES | Bibliography | Help

NIST Atomic Spectra Database Lines Data (Multiplet Ordered)
Na I 466 Lines of Data Found (page 1 of 19)

Wavelength units: vac Å < 2000 Å < air Å < 10,000 Å < vac cm^{-1}

Allowed Transitions

Mult. No.	Configurations	Terms	Observed	Ritz	E_i (cm^{-1})	E_k (cm^{-1})	A_{ki} (10^8 s^{-1})	Acc.	TP Refs.
1	3*s*-3*p*	$^2S-^2P^\circ$		*5 891.94*	0.000 -	*16 967.636*	6.21e-01	A	2,3,4
			5 889.950	5 889.950	0.000 -	16 973.368	6.22e-01	A	CRC
			5 895.924	5 895.924	0.000 -	16 956.172	6.18e-01	A	CRC
2	3*s*-4*p*	$^2S-^2P^\circ$		*3 302.57*	0.000 -	*30 270.72*	2.81e-02	C+	4
			3 302.37	3 302.369	0.000 -	30 272.58	2.81e-02	C+	CRC
			3 302.98	3 302.978	0.000 -	30 266.99	2.81e-02	C+	CRC
3	3*s*-5*p*	$^2S-^2P^\circ$		*2 852.88*	0.000 -	*35 042.03*	6.0e-03	C+	4
			2 852.81	2 852.811	0.000 -	35 042.85	6.0e-03	C+	1s
			2 853.01	2 853.012	0.000 -	35 040.38	6.0e-03	C+	1s
4	3*s*-6*p*	$^2S-^2P^\circ$		*2 680.37*	0.000 -	*37 297.18*	2.26e-03	C+	4
			2 680.34	2 680.341	0.000 -	37 297.61	2.26e-03	C+	4
			2 680.43	2 680.433	0.000 -	37 296.32	2.26e-03	C+	4
5	3*s*-7*p*	$^2S-^2P^\circ$		*2 593.89*	0.000 -	*38 540.68*	1.20e-03	C+	9,10n
			2 593.87	2 593.869	0.000 -	38 540.93	1.20e-03	C+	9,10n
			2 593.92	2 593.919	0.000 -	38 540.18	1.20e-03	C+	9,10n

FIGURE 5. Multiplet-ordered retrieval for Na I for NIST *Atomic Spectra Database.*

"I am a second year chemistry student, and I work in the high temperature science laboratory. One of the areas I am involved in is the study of plasmas and incandescent lamps (combining both practical work and computer modeling). The NIST energy levels and lines compilations have proved to be useful, particularly for the inert gases: He, Ne, Ar, and Kr, present in many incandescent lamps. Any recently compiled data on either Hg I or Hg II will be much appreciated.-SHEFFIELD, ENGLAND" *(Indeed ASD has no energy levels for Hg I or Hg II. The lines are those from NIST 68 [3]. This is a shortcoming that we are in the process of addressing in our new versions. Even though there has not been a recent critical compilation for a particular spectrum, in selected cases the existing material from "Atomic Energy Levels" [4] would suffice for most users, and we intend to use this material where appropriate. In the meantime we have sent this student a preprint of a new NIST investigation of Hg II [5].)*

"I needed the first excited states of the noble gas atoms He, Ne, Ar, Kr, and Xe. By entering He I, Ne I, Ar I, and Kr I, I was able to get information from your database. The method failed for Xe I.-BOSTON, MA, U.S.A." *(This message and the preceding one illustrate the great interest in the spectra and energy levels of the rare gases. As with Hg, ASD has no energy levels for Xe, and we intend to rectify this for our new version. We have also begun a modern compilation for the rare gas atoms that will be included in ASD and also appear as a separate volume. Ver. 2.1 of ASD will have material for a number of additional spectra and have corrections for small data and programming errors that have come to light. It will also have much easier access to the references. A hyperlink at the top of the reference column will produce a separate box with the list of references that appears together with the table being viewed.)*

"I am carrying out experiments with Ar plasmas, and am following the condition of the plasma by monitoring the Ar I resonance lines at 104.8 nm and 106.6 nm. I need the transition probabilities of all the lines that terminate on the upper levels of these resonance lines.-PASADENA,CA, U.S.A." *(This user was guided to the desired values in a telephone conversation. First he used the "lines form" in ASD and requested the lines in Ar I at 105 nm ± 2 nm. This revealed that the upper levels for the two lines were* $3s^23p^5(^2P_{1/2})4s\ ^2[1/2]_1$ *and* $3s^23p^5(^2P_{3/2})4s\ ^2[3/2]_1$*, respectively. By using the "additional criteria" option on the request page, new retrievals were done in which, through the menu of all configurations for Ar I, the lower configuration was successively restricted to each of these two. This produced the retrieval shown in Fig. 6 containing the desired transition probabilities for transitions terminating on* $3s^23p^5(^2P_{1/2})4s\ ^2[1/2]_1$*. Of course, only the values for* $J_i=1$ *were of interest. The data from ASD are also used as input to an online modeling program available through the University of Tennessee [6].)*

ASD | DATA: LINES LEVELS | INFORMATION: LIST OF SPECTRA · GROUND STATES & IONIZATION ENERGIES · Bibliography · Help

NIST Atomic Spectra Database Lines Data (Wavelength Ordered)
Ar I
34 Lines of Data Found (page 1 of 4)

Observed Air (Å)	Ritz Air (Å)	Rel. Int.	A_{ki} ($10^8\ s^{-1}$)	Acc.	E_i (cm^{-1})	E_k (cm^{-1})	Configurations	Terms	J_i-J_k	Type	TP Refs.	Line Refs.
	3 406.180		3.9e-03	D	95 399.8276 -	124 749.805	$3s^2 3p^5 (^2P^\circ 1_{/2})4s - 3s^2 3p^5 (^2P^\circ 1_{/2})7p$	$^2[^1/_2]^\circ - ^2[^1/_2]$	1-0		8n	
	3 563.286		1.2e-03	D	94 553.6652 -	122 609.637	$3s^2 3p^5 (^2P^\circ 1_{/2})4s - 3s^2 3p^5 (^2P^\circ 1_{/2})6p$	$^2[^1/_2]^\circ - ^2[^3/_2]$	0-1		8n	
	3 572.295		5.1e-03	D	95 399.8276 -	123 385.044	$3s^2 3p^5 (^2P^\circ 1_{/2})4s - 3s^2 3p^5 (^2P^\circ 3_{/2})7p$	$^2[^1/_2]^\circ - ^2[^1/_2]$	1-0		8n	
	3 649.832		8.0e-03	D	95 399.8276 -	122 790.547	$3s^2 3p^5 (^2P^\circ 1_{/2})4s - 3s^2 3p^5 (^2P^\circ 1_{/2})6p$	$^2[^1/_2]^\circ - ^2[^1/_2]$	1-0		8n	
	3 670.670		3.1e-04	D	95 399.8276 -	122 635.054	$3s^2 3p^5 (^2P^\circ 1_{/2})4s - 3s^2 3p^5 (^2P^\circ 1_{/2})6p$	$^2[^1/_2]^\circ - ^2[^3/_2]$	1-2		8n	
	3 675.235		4.9e-04	D	95 399.8276 -	122 601.225	$3s^2 3p^5 (^2P^\circ 1_{/2})4s - 3s^2 3p^5 (^2P^\circ 1_{/2})6p$	$^2[^1/_2]^\circ - ^2[^1/_2]$	1-1		8n	
3 770.369	3 770.369	20	7.0e-04	D	94 553.6652 -	121 068.736	$3s^2 3p^5 (^2P^\circ 1_{/2})4s - 3s^2 3p^5 (^2P^\circ 3_{/2})6p$	$^2[^1/_2]^\circ - ^2[^1/_2]$	0-1		8n	219
3 834.679	3 834.678	7	7.5e-03	D	95 399.8276 -	121 470.238	$3s^2 3p^5 (^2P^\circ 1_{/2})4s - 3s^2 3p^5 (^2P^\circ 3_{/2})6p$	$^2[^1/_2]^\circ - ^2[^1/_2]$	1-0		8n	219
	3 894.660		5.7e-04	D	95 399.8276 -	121 068.736	$3s^2 3p^5 (^2P^\circ 1_{/2})4s - 3s^2 3p^5 (^2P^\circ 3_{/2})6p$	$^2[^1/_2]^\circ - ^2[^1/_2]$	1-1		8n	
4 181.884	4 181.884	50	5.61e-03	C	94 553.6652 -	118 459.5975	$3s^2 3p^5 (^2P^\circ 1_{/2})4s - 3s^2 3p^5 (^2P^\circ 1_{/2})5p$	$^2[^1/_2]^\circ - ^2[^1/_2]$	0-1		4n,5n	219
4 191.029	4 191.029	50	5.39e-03	C	94 553.6652 -	118 407.4303	$3s^2 3p^5 (^2P^\circ 1_{/2})4s - 3s^2 3p^5 (^2P^\circ 1_{/2})5p$	$^2[^1/_2]^\circ - ^2[^3/_2]$	0-1		5n	219
4 259.362	4 259.362	200	3.98e-02	C	95 399.8276 -	118 870.9170	$3s^2 3p^5 (^2P^\circ 1_{/2})4s - 3s^2 3p^5 (^2P^\circ 1_{/2})5p$	$^2[^1/_2]^\circ - ^2[^1/_2]$	1-0		4n,5n,6n	219
4 333.561	4 333.561	100	5.68e-03	C	95 399.8276 -	118 469.0508	$3s^2 3p^5 (^2P^\circ 1_{/2})4s - 3s^2 3p^5 (^2P^\circ 1_{/2})5p$	$^2[^1/_2]^\circ - ^2[^3/_2]$	1-2		4n,5n	219
4 335.338	4 335.338	50	3.87e-03	C	95 399.8276 -	118 459.5975	$3s^2 3p^5 (^2P^\circ 1_{/2})4s - 3s^2 3p^5 (^2P^\circ 1_{/2})5p$	$^2[^1/_2]^\circ - ^2[^1/_2]$	1-1		4n,5n	219
4 345.168	4 345.168	25	2.97e-03	C	95 399.8276 -	118 407.4303	$3s^2 3p^5 (^2P^\circ 1_{/2})4s - 3s^2 3p^5 (^2P^\circ 1_{/2})5p$	$^2[^1/_2]^\circ - ^2[^3/_2]$	1-1		4n,5n,6n	219

FIGURE 6. Lines retrieval for Ar I for *NIST Atomic Spectra Database* in which lower level is restricted to levels of the $3s^2 3p^5 (^2P_{1/2})4s$ configuration.

"I'm currently looking for a spectral database on CD-ROM, and I find your site, which is very easy to use and complete. It would be very interesting for us to get this information, but we need to work offline. I'd like to know if Ver. 2.0 of ASD is available on CD-ROM?-ECUBLENS, SWITZERLAND" *(This illustrates the point that, although online accessibility of the data is important, there are still many applications for which stand-alone databases are desirable, such as for identification of unknown species in chemical samples by computer search of standard line-lists. For this it is too time-consuming to download masses of data through the internet. To this end databases on CD-ROM and even floppy disks are still useful. After receiving a number of requests of this type the NIST Standard Reference Data Program has resumed distribution of NIST Standard Reference Database 38 on floppy disk [7]. This disk contains all of the data from NIST 68 as well as ionization energies, nuclear moments, and other atomic data. This database is still widely used for analytical spectroscopy. A related database [8] under development will provide the strongest lines with classifications and transition probabilities, low energy levels of the ground and first excited configurations, and ionization energies for neutral and singly ionized atoms. This will be available in hard copy, on the web, and as an* ***E-book****. E-books, which resemble hand-held palm-type computers, are becoming useful for storage of large amounts of material that need to be transported easily, such as training and maintenance manuals. E-books can contain the equivalent of at least 4000 pages of text.)*

"Our company manufactures precision optical glasses used for commercial applications such as cameras, microlithography, stepper lenses, and precision astronomical telescopes. We are now planning to make equipment to measure refractive indices at 157 nm, and we need a suitable light source. We found data for "Wavelengths and Intensities of a Pt/Ne Hollow Cathode Lamp." Where can we buy this lamp?-SAGAMIHARA CITY, JAPAN." *(This refers to the database for the platinum lamp mentioned in Fig. 3, and of course we referred them to the current manufacturer of the lamp. Moreover, this message illustrates the coming need for accurate wavelength calibration in the vacuum ultraviolet (VUV) in connection with deep ultraviolet microlithography. The lasers presently used in this process are a KrF excimer laser operating at 248 nm and an ArF excimer laser operating at 193 nm. In order to compress even more circuit elements onto integrated circuit (IC) chips, the next generation of microlithography will use a* F_2 *molecular laser operating at 157 nm. The refractive indices of the materials that are used to focus the IC masks on the substrates change rapidly with wavelength in the VUV, and for the optics to be designed properly, the wavelengths of the lasers must be known to high accuracy. Hence the interest in our atlas of the Pt/Ne lamp.)*

"We manufacture ArF lasers used for microlithography at 193 nm. We routinely calibrate the wavelength of the laser by use of a neutral carbon line at 193.0906 nm that appears as a impurity in the laser gas. Our spectrometer system monitors the laser output. A dip in the output is observed when the laser wavelength coincides with that of the carbon line. In order to check our calibration we recently built a *see-through* Pt/Ne hollow cathode lamp and passed the laser light though this lamp. We obtain a dip in the signal when the laser wavelength coincides with one of the strong Pt lines in this region. Unfortunately, using your Pt lamp atlas, there is a discrepancy in the calibration from these two methods of about 0.0015 nm, whereas we require an accuracy of 0.00015 nm. Can the Pt atlas be in error at this level?-CALIFORNIA, U.S.A." (*Since the stated uncertainty of Pt lines in this region of our atlas ranges from 0.00005 nm to 0.00020 nm, an error at the 0.0015 nm level would be very hard for us to accept. The problem was solved by our suggestion that the high gas pressure in the laser had shifted the wavelength of the carbon line to this degree. Calculations of the expected shift then brought the two calibrations into agreement. The company now uses the Pt lamp routinely for their calibrations.)*

"We will be visiting in the U.S. in August 1999 and would appreciate to have with you a discussion about absolute wavelength standards and methods of calibration in the range 193 nm and 157 nm.-GÖTTINGEN, GERMANY." *(This was a manufacturer of lasers for deep uv microlithography that needed to determine the wavelength of its F_2 laser to high accuracy. In discussions with the company it was decided that the best approach would be to ship one of their lasers to NIST to observe with our 10.7-m normal-incidence vacuum spectrograph, using the Pt/Ne lamp for wavelength calibration. This experiment was carried out in the early part of 2000. The laser was coupled to the spectrograph by a beam line constructed in Göttingen. The beam line was designed to avoid shifts in line positions arising from possible differences in illumination of the grating by the two sources. About 100 measurements were made of the lasing F_2 transitions under different operating conditions. Values for the wavelengths were determined with an accuracy of ±0.00010 nm. These values were about 0.001 nm higher than the previously accepted values.)*

BIBLIOGRAPHIC DATABASES

Atomic Transition Probabilities

The NIST Atomic Transition Probability Bibliographic Database [9] has been updated to include about 450 new papers from June 1996 through May 1999. A request form and a typical retrieval are shown in Figs. 7 & 8.

Atomic Energy Levels and Spectra

This database contains approximately 8500 references covering July 1968 through December 1998. It is expected to come online sometime during 2000. The request and retrieval pages will be similar to those for the transition probabilities bibliographic database. Another bibliographic database in this area is the BIBL Atomic Spectral Data Bibliographic Database developed at the Institute of Spectroscopy in Troitsk, Russia. This database came online in 1999. [10] In addition to spectroscopic references, it contains references on excitation and ionization by electrons and photons, multiphoton processes, Auger decay, and ion-electron recombination as well.

NIST Physics Laboratory — Physical Reference Data

Version 6.0 - Disclaimer

Atomic Transition Probability Bibliographic Database

[1]**Data Compilers:**
J. R. Fuhr and H. R. Felrice

[2]**Web Database Developers:**
K. Olsen and J. Hwang

[1]National Institute of Standards and Technology, Physics Laboratory, Atomic Physics Division
[2]NIST, Physics Laboratory, Office of Electronic Commerce in Scientific and Engineering Data

Welcome to NIST's bibliographic database on atomic transition probabilities. The database presently contains 6113 references, dating from 1914 through May 1999.

These papers contain numerical data, comments, and review articles on atomic transition probabilities (oscillator strengths, line strengths, or radiative lifetimes), and are part of the collection of the **Data Center on Atomic Transition Probabilities** at NIST.

I. References with Numerical Data:
☑ Element, ☐ Isoelectronic Sequence, ☐ Stage of Ionization

II. ☐ Publications of Special Interest

III. Additional Criteria:
☐ Word/Pattern in Title, ☐ Author, ☐ Journal, ☑ Year of Publication
☑ Experiment / Theory, ☐ Specific Experimental / Theoretical Method

[Display Search Form] [Clear Form]

Element:

Select one element from either of the following lists:

Order by Atomic Number	Order by Element Name
18. Argon, Ar	89. Actinium, Ac
19. Potassium, K	13. Aluminum, Al
20. Calcium, Ca	95. Americium, Am
21. Scandium, Sc	51. Antimony, Sb
22. Titanium, Ti	18. Argon, Ar
23. Vanadium, V	33. Arsenic, As
24. Chromium, Cr	85. Astatine, At
25. Manganese, Mn	56. Barium, Ba
26. Iron, Fe	97. Berkelium, Bk
27. Cobalt, Co	4. Beryllium, Be

Year of publication: From: 1998 To: 2000

Method: ◉ Experiment ○ Theory

If you want to sort the list of references, check one of these boxes:
◉ by the published year, ○ by the first author's last name.

[Search for References] [Clear Form]

FIGURE 7. Request form for *NIST Atomic Transition Probability Bibliographic Database.*

Publications on Atomic Transition Probabilities

1998 to 2000

Experiment

that include atomic number(s) Z = 26

3 references found:

Selected Fe II lifetimes and f-values suitable for a solar abundance study,
R. Schnabel, M. Kock, H. Holweger,
Astron. Astrophys. **342**, 610 (1999).

A linear Paul trap for radiative lifetime measurements on ions,
M. Schultz-Johanning, R. Schnabel, M. Kock,
Eur. Phys. J. D **5**, 341 (1999).

Solar EUV line identifications from delayed beam-foil spectra,
E. Träbert,
Mon. Not. R. Astron. Soc. **297**, 399 (1998).

Return to main page

FIGURE 8. Fe retrieval for NIST Atomic Transition Probability Bibliographic Database.

Atomic Spectral Line Broadening

The NIST Atomic Spectral Line Broadening Bibliographic Database [11] will be updated with almost 2000 references from July 1978 through March 1992 sometime in 2000.

ACKNOWLEDGMENTS

My thanks to Arlene Robey for assistance with the figures for this manuscript and for the talk at Oxford. This work is supported by the U.S. National Aeronautics and Space Administration and by the Office of Fusion Energy Sciences of the U. S. Department of Energy.

REFERENCES

1. Kelleher, D. E., Mohr, P. J., Martin, W. C., Wiese, W. L., Sugar, J., Fuhr, J. R., Olsen, K., Musgrove, A., Reader, J., Sansonetti, C. J., and Dalton, G. R, "The New NIST Atomic Spectra Database," in *Proceedings of the SPIE: Ultraviolet Atmospheric and Space Sensing: Methods and Instrumentation II*, Vol. **3818**, 170 (1999).
2. Wiese, W. L., "NIST Atomic Spectra Database," in *Proceedings of Atomic and Molecular Data and Their Applications (ICAMDATA) - Second International Conference*, edited by K. Bell and K. Berrington, AIP Conference Proceedings, New York; this volume.
3. Reader, J., Corliss, C. H., Wiese, W. L., and Martin, G. A., "Wavelengths and Transition Probabilities for Atoms and Atomic Ions," Natl. Stand. Ref. Data Ser., Natl. Bur. Stand. (U. S.) **68** (1980).
4. Moore, C. E., *Atomic Energy Levels*, NBS Circular 467 (U.S. Government Printing Office, Washington, D.C.; Vol. I, 1949; Vol. II, 1952; Vol. III, 1958).
5. Sansonetti, C. J., and Reader, J., "Spectrum and Energy Levels of Singly-Ionized Mercury (Hg II)," submitted to Physica Scripta.
6. Spectral Research & Technology, Inc.; URL: http://spectralresearch.com/atom/spectra6/spectra6.html
7. "NIST Spectroscopic Properties of Atoms and Atomic Ions" NIST Standard Reference Database 38, (1992); URL: http://www.nist.gov/srd/nist38.htm
8. Sansonetti, J. E., and Martin, W. C. "Handbook of Most Basic Atomic Data," to be published.
9. Fuhr, J. R., Felrice, H. R., Olsen, K., and Hwang, J.; URL: http://physics.nist.gov/fvalbib
10. Kramida, A. E., and Ryabtsev, A. N.; URL address: http://das101.isan.troitsk.ru/bibl.htm
11. Fuhr, J. R., Felrice, H. R., Olsen, K. and Hwang, J.; URL: http://physics.nist.gov/linebrbib

Plasma Modelling Data

R.E.H. Clark and J.A. Stephens

Atomic and Molecular Data Unit
International Atomic Energy Agency
P.O. Box 100 Wagramer Strasse 5
A–1400 Vienna, Austria

Abstract. The main objective of the Atomic and Molecular Data Unit of the International Atomic Energy Agency is to collect, assess, and make available a variety of data relevant to magnetic fusion energy and other nuclear energy applications. In this paper we present the methods by which the Unit achieves this goal. We also indicate the scope of the resulting databases and present a summary of the method for accessing the databases through the world–wide–web.

INTRODUCTION

The Atomic and Molecular (A+M) Data Unit of the International Atomic Energy (IAEA) is comprised of two professional staff members and one general staff member for support services. The A+M Data Unit is contained in the Nuclear Data Section of the Division of Physics and Chemistry, in the Department of Nuclear Sciences and Applications of the IAEA. The goal of the A+M Data Unit is to provide atomic and molecular physics data for magnetic fusion energy and other nuclear power applications. Once the data have been gathered and assessed they are incorporated into the databases maintained by the Data Unit. The databases are accessed through a world–wide–web interface that is maintained by the Data Unit. There are both numerical and bibliographic databases for fusion energy research maintained by the Unit.

In addition to database development, the Unit publishes the "Atomic and Plasma–Material Interaction Data for Fusion" (APID series) journal as well as the "International Bulletin on Atomic and Molecular data for Fusion" which contains bibliographic information relevant for fusion. It should be noted that the Unit itself does not generate the majority of its database content, rather it collects and assesses the data from Data Centres around the world. In the case of the bibliographic database, for example, the majority of the information is obtained from the Oak Ridge National Laboratory (Controlled Fusion Atomic Data Center) in the US. Further submissions are obtained from the National Institute for Standards and Technology (Physics Laboratory) in the US, the National Institute for Fusion Science in Japan, the Laboratoire de Physique des Gaz et des Plasmas (GAPHYOR) in France, and from the Kurchatov Institute in Russia.

In this paper we will describe the various data collection methods, the resulting databases, and means of accessing these databases and the publications produced by the A+M Data Unit.

CP543, *Atomic and Molecular Data and Their Applications,* edited by K. A. Berrington and K. L. Bell

DATA ACTIVITIES

In this section we describe the various methods used by the A+M Data Unit for collecting and assessing data to be added to the databases. An overview of all of the activities of the Unit relevant to this task is shown in Figure 1.

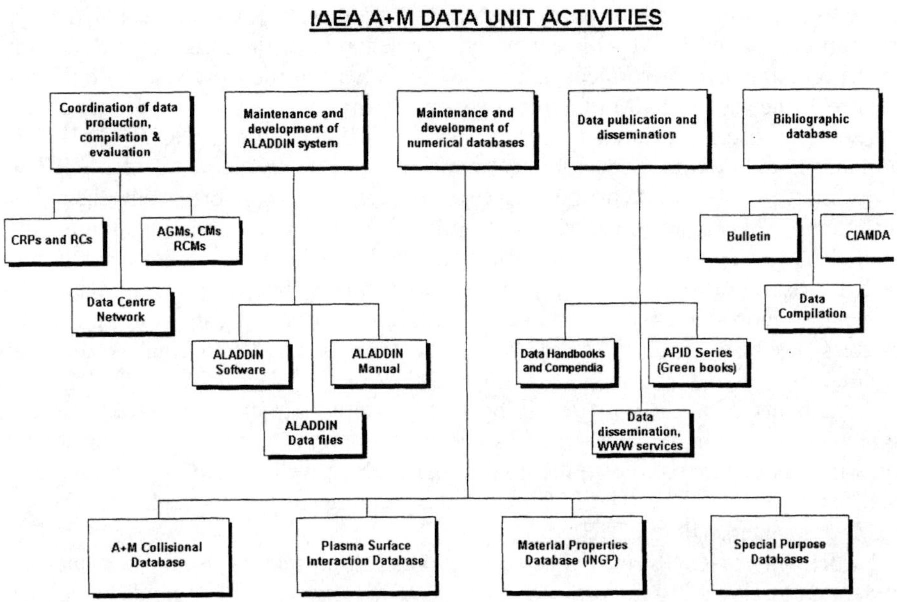

Figure 1. Summary of the activities of the IAEA Atomic and Molecular Unit.

In the Fig. 1 the data gathering activities are in the upper left portion of the chart. These activities include a number of activities identified by their acronyms. Each will be discussed in detail below.

Co–ordinated Research Programs

One of the most effective methods of obtaining data for a specific need is the IAEA Co–ordinated Research Program (CRP). A CRP is a fixed–length program, normally three years in length, with the occasional possibility of a two–year extension. The CRP focuses on a selected, well–defined topic. Approximately ten to twelve researchers in the appropriate field are identified and invited to join an approved CRP. An initial Research Co–ordination Meeting (RCM) is held. At this meeting the individual participants adopt an overall plan for the CRP and agree on a work plan for the program. As work progresses contact among the participants is closely maintained. For a CRP of three–year duration there is a second RCM held during the third year to present results achieved during the execution of the program, and a final reporting along with a publication format is planned. For a five–year program the second RCM will review the first segment of the program and plan tasks for the remainder of the project. A third RCM is held in the final year with the same purpose as the second RCM in a three–year program.

There are often Research Contracts (RC) associated with a CRP. The RC is a mechanism to provide some limited financial support for members of a CRP who have little or non–existent possibilities for financial support from their home institution. This financial support is provided directly to the home institution. If the researcher moves to a different institution the financial support does not transfer. Each RC is reviewed each year to insure that appropriate progress is made. The financial support can only be renewed if a there is progress report showing sufficient progress has been made. At the end of the term of the RC a final report is also required.

Participants of a CRP not receiving a Research Contract are offered a Research Agreement. This agreement specifies that the researcher is participating in the CRP and will make contributions to the goal of the CRP as well as provide material for the final report of the CRP. The IAEA provides funding for the researcher to attend the RCMs associated with the CRP.

Participants of CRPs are expected to produce high quality atomic and molecular physics data relevant to the project. These data are expected to be published either in external journals, or as an IAEA publication, for example in the Data Unit's APID journal. Each participant is expected to contribute a substantial final summary of work performed in the project. It is required that a CRP produce a final report to the IAEA summarising the work done.

The CRP on "Atomic and Plasma–Wall Interaction Data for Fusion Reactor Divertor Modelling" held its second and final RCM on 8–9 March 1999. A Summary Report was issued. Final review papers from the participants have been assembled and will be edited by Dr. R. Janev to appear in the APID series.

During the year 2000 there are two CRPs in their final year. The final RCM for the CRP "Charge–exchange cross section data for fusion plasma studies" will be held on 25–26 September 2000 in Vienna, and the final RCM for the CRP "Plasma–material interaction data for mixed plasma facing materials in fusion reactors" will be held on 16–17 October 2000 in Vienna. At those meetings Summary Reports will be prepared and arrangements for the collection of final reviews and data generated by these CRP will take place. There will be substantial amounts of new data generated by these CRPs which will become available to the community through publication in the APID series and through additions to the A+M Data Unit database available through the world–wide–web.

During the year 2001 two new CRPs will be initiated, one on the topic of plasma diagnostics, the other on molecular processes in fusion plasmas. The scope and participants of these CRPs will be determined through discussions with the IFRC A+M Subcommittee, other outside Consultants, and within the A+M Data Unit. It is planned to have both new CRPs hold their initial RCMs in the year 2001. It is anticipated that these will be three–year CRPs. We are also considering the possibility of starting an additional CRP in the following year, likely on the topic of tritium retention in fusion reactor wall materials.

Advisory Group Meetings

A more specific activity of the A+M Data Unit is the IAEA Advisory Group Meeting (AGM). The AGM is a mechanism of gathering researchers together for a specific topic in fusion energy research and development. Unlike the CRP, the AGM is not normally an ongoing program, rather it is a one–time meeting with a specific topic to be addressed, such as the gathering of information on a single type of physical process of interest to the fusion community. There are no Research Contracts associated with the AGM, since it is not a continuing program, although participants are typically funded by the IAEA to attend.

It is possible to hold more than one AGM on the same topic. The Data Centre Network meetings, discussed below, convene every two years as an AGM. This makes it possible to bring together representatives of A+M Data Centres from around the world to plan activities for the next two–year period.

A possible objective for an AGM is to bring together experts in an area with the potential for initiation of a CRP. The AGM would offer a forum to discuss the possible scope of a CRP and its potential participants. The AGM in itself usually provides a very valuable preliminary assessment of the selected field.

Consultant Meetings

The most condensed of the meeting activities is the Consultant Meeting (CM), which usually involves one to five researchers spending a period of time at the IAEA. The period of time may range from one to several weeks. The researchers are invited to come to the IAEA to provide assistance with a specific project that can be addressed by a small number of experts in the field. Such projects may include, for example, the calculation of a small number of cross sections for a particular class of processes. It can also include the addition of a specific capability to the A+M Data

Unit, such as developing or enhancing its web interface to the database, or provide for the numerical fitting of sets of selected collisional or particle–surface interaction data.

In order to bring in consultants there must be a clearly defined project that can be effectively addressed by the experts. The project must be defined in writing and approved by the IAEA. After the normal approvals have been obtained the CM takes place. Following conclusion of the CM, consultants are required to submit a written summary of the project, its outcome and results.

The A+M Data Unit typically invites several Consultants to the Agency each year on a wide variety of topics. These meetings have resulted in substantial benefit to the Agency in additions to its databases and in development of its data processing capabilities.

Data Centre Network

One of the most important avenues for collection of A+M data is the Data Centre Network (DCN) which currently is comprised of 13 Data Centers from around the world. A diagram of the data centers is shown in Fig. 2.

Representatives of the DCN meet every two years in Vienna as an ongoing Advisory Group Meeting. At this meeting participants assess the needs in the fusion community for A+M data and plan how to best meet those needs. Work among the data centers is co–ordinated through the meeting and through electronic communications. Data generated from the DCN collaboration is collected at the A+M Data Unit and added to the databases at the IAEA.

There is close co–operation among the data centers. Substantial interaction takes place among the data centers and with the A+M Data Unit continually through the world–wide–web. Currently there are preliminary investigations being conducted on the possibility of forming a stronger electronic link of the data center databases. The goal of this linkage is to make it possible for an inquiry to one data center to search for the availability of the requested data in the databases of other data centers. This would greatly increase the ease with which an individual request would be fulfilled.

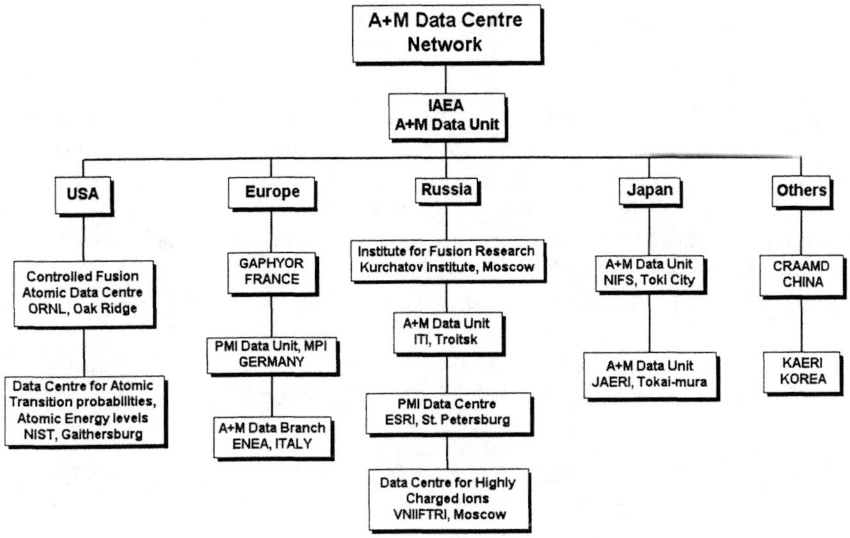

Figure 2. The Atomic and Molecular Data Centre Network.

The International Fusion Research Council Subcommittee

In addition to the data generation activities an important means of advice for the A+M Data Unit is the International Fusion Research Council (IFRC) subcommittee on atomic and molecular data. This subcommittee normally meets every two years to provide advice to the A+M Data Unit on current activities and to plan for the next two–year budget cycle. The final report from the IFRC subcommittee serves as a very useful outline for the Unit in the IAEA budget allocations. The makeup of the IFRC Subcommittee is international and includes approximately ten members, varying somewhat from each two–year cycle.

The Subcommittee meeting lasts two days. During the meeting the activities of the A+M Data Unit over the past two years is reviewed and compared to the recommendations from the previous Subcommittee meeting. The Subcommittee then discusses issues for the next two–year period including possible CRPs, AGMs, and other activities. The Subcommittee also functions as an advisory board for the APID series publication and has the duty to select members of the editorial board for that publication. The Subcommittee chairman prepares a document including meeting conclusions and recommendations as well as an Executive Summary to be distributed to the full IFRC as well as to the appropriate management at the IAEA.

Technical Committee Meetings

Another activity of the A+M Data Unit is to hold Technical Committee Meetings (TCM) on various topics. The TCM can be useful in holding a scientific meeting on a topic not normally covered by other meetings, for example, the IFRC is held as a TCM. One series of TCMs on the topic of atomic and molecular data for fusion reactor technology has been held over a number of years. The last meeting of this series was held in Cadarache, France in 1992. The A+M Data Unit is planning to hold another meeting in the series in 2001 at a location to be determined.

AVAILABLE DATABASES

The Atomic and Molecular Data Unit maintains and develops online bibliographic and numerical databases for atomic and molecular collisional and radiative processes, atomic and molecular structure characteristics, and particle–surface interaction processes for use in fusion energy research and plasma science applications. The critically assessed and recommended numerical data are ALADDIN [1] formatted, and can now be evaluated in tabular and graphical form from a web interface [2]. The numerical databases are established through experts participating in IAEA CRPs, AGMs and Consultants' Meetings as discussed above. We have recently added a comprehensive database for the chemical sputtering of fusion relevant species for Be, C, W, and doped graphite targets and some selected compounds.

The A+M Data Unit bibliographic database is stored in the IAEA Atomic and Molecular Data Information System (AMDIS), and contains references, reaction indices, and collision energy ranges for atomic and molecular processes relevant to fusion and plasma research. This database is developed by contributions from the A+M Data Centre Network, and contributions are regularly edited and published as the International Bulletin on Atomic and Molecular Data for Fusion (annually or biannually), and periodically (every 5–7 years) as a CIAMDA compilation. Presently the bibliographic database is available through a Telnet service with user registration [2].

Planned additions to the numerical databases in the near future include the addition of a comprehensive database for physical sputtering and radiation enhanced sublimation processes for Be, C, W and selected compounds to the particle–surface interaction category, and a database for calculated radiated power losses for fusion–relevant atomic species to the collisional category. Future work also includes the bibliographic database being redesigned to employ a user–friendly web interface similarly to the numerical databases.

PUBLICATIONS

To indicate the types of publications the A+M Data Unit produces and distributes, we provide here some recent examples:

Atomic and Plasma–Material Interaction Data for Fusion Volume 7A, "Particle Induced Erosion of Be, C and W in Fusion Plasmas. Part A: Chemical Erosion", by A.A. Haasz et al. (1998), 277 pgs.

Atomic and Plasma–Material Interaction Data for Fusion Volume 7B, "Particle Induced Erosion of Be, C and W in Fusion Plasmas. Part B: Physical Sputtering and Radiation Enhanced Sublimation", by W. Eckstein et al. (to be submitted in 2000), approximately 250 pgs.

Atomic and Plasma–Material Interaction Data for Fusion Volume 8, "Elastic and Related Transport Cross Sections for Collisions among Isotopomers of $H^+ + H$, $H^+ + H_2$, $H^+ + He$, $H + H$, and $H + H_2$", by P.S. Krstic and D.R. Schultz (1998), 699 pgs.

CIAMDA 98, An Index to the Literature on Atomic and Molecular Collision Data Relevant to Fusion Research (1998), 635 pgs.

International Bulletin on Atomic and Molecular Data for Fusion, Numbers 56–57 (1999), 171 pgs.

IAEA Nuclear Data Section report INDC(NDS)–413, "IAEA International Database on Irradiated Nuclear Graphite Properties", by T.D. Burchell et al., (2000), 35p.

The last two publications listed above are available free of charge, while the APID and CIAMDA series publications may be purchased through the IAEA Sales and Promotion Unit.

REFERENCES

[1] R. A. Hulse, "A Labeled Atomic Data Interface for Fusion Applications", IAEA Report INDC(NDS)–211/GA (1988).

[2] A summary of the A+M Data Unit's programs and access to the ALADDIN database server are located at http://www–amdis.iaea.org/.

Astrophysical and Fusion Data Needs

Atomic Physics Requirements in Fusion Research

Kurt Behringer

Max-Planck-Institut für Plasmaphysik, EURATOM Association,
D85748 Garching, Germany

Abstract. Nuclear fusion research, using magnetically confined plasmas, needs a great amount of atomic data in the fields of plasma diagnostics, plasma spectroscopy and modelling. Conventional spectroscopy and some specialised techniques are being used for analysing plasma behaviour and plasma composition. Hydrogen data are required, as well as coefficients for impurities originating from the vacuum vessel walls and protection tiles. Some noble gases are being used for radiative cooling or transport studies of highly ionised ions. For measuring metal concentrations, diagnostic lines from the plasma interior are recorded, sometimes of highly ionised species like Zn-like tungsten. The ionisation balance and the emitting shells must be calculated on the basis of the relevant atomic data and of the transport coefficients. For high heat-load areas, carbon is mostly used at present, which is fully ionised in the plasma and can only be measured by charge-exchange recombination spectroscopy. Impurity particle influxes at the boundary are derived from lines of neutrals or low ionisation stages, a method, which can also be extended to the band emission of molecules. In this way, chemical erosion of carbon is being investigated. In cold divertor plasmas, hydrogen molecules also play some role, e.g. enhancing hydrogen ion recombination. In addition to the usual coefficients for ionisation, recombination, charge exchange and radiation for all relevant elements, molecular data are required in model calculations of the plasma boundary and of divertors. Some examples of interpretation methods and experimental results in these fields are presented.

INTRODUCTION

In fusion experiments with magnetically confined plasmas, basic atomic data are required in many different areas. Plasma spectroscopy is widely employed for diagnostics of fusion experiments. For its interpretation, as well as for general understanding and prediction of the plasma behaviour, model calculations are being carried out for the core plasmas and for edge and divertor regions. These models require complete sets of atomic data for the relevant elements and ionisation stages. A few cases will be discussed later in detail, but it may be worth-while to present a short overview.

As a simple example, ion temperatures, flow velocities and rotation velocities are derived from Doppler broadening and shifts of hydrogen and impurity lines. Velocities perpendicular to the magnetic field are investigated with respect to macroscopic electrical fields in the plasmas, which are thought to be of high importance for suppressing instabilities and for improving confinement. Measurements of electrical fields are a

CP543, *Atomic and Molecular Data and Their Applications,* edited by K. A. Berrington and K. L. Bell

very important issue at present, and novel diagnostic methods would be most welcome. The motional (vxB) Stark effect of hydrogen atoms in a high energy beam traversing the B-field region is being analysed with respect to magnetic field components, in particular, to the poloidal field and the plasma current density, which is also responsible for stability and confinement. Then, there is the traditional area of impurities and particle transport investigations. As a standard technique, impurity densities are derived from emission spectroscopy in the VUV and X-ray region. Modelling of the ionisation balance including diffusion is required for such an analysis. In the case of light impurities, which are fully ionised in the plasma core, these densities are obtained from charge exchange recombination spectroscopy with neutral hydrogen beams, usually the heating beams. Dedicated diagnostic beams often use helium or lithium (possibly sodium in the future) for measuring electron densities and electron temperatures. In the edge and divertor regions, conventional spectroscopy is still employed for temperature and electron density diagnostics, as well as for neutral density measurements. Here, molecular processes can also become of importance, which means that molecular densities and effective dissociation rate coefficients are required.

While emission spectroscopy always observes line-of-sight integrals, laser-induced fluorescence allows investigation of particle densities including some molecules with spatial resolution, e.g. in edge or divertor plasmas. However, due to the substantial effort, it is seldomly used in fusion devices. Two-photon, Doppler-free LIF may be required for determining the isotope ratio D/T in later burning plasmas. This is again an important task with few alternatives.

Of course, plasma spectroscopy is also employed in the field of plasma-wall-interaction. Particle influxes of atoms, ions and molecules are determined from their line emission. Conclusions are being drawn on chemical erosion and other impurity mechanisms, including those of high-Z materials. At present, carbon is still the most wide-spread wall material in fusion experiments. However, because of some anticipated problems, in particular, chemical erosion and carbon-tritium co-deposition, other materials are being tested as well.

Finally, very many atomic data of medium quality are required as input for plasma modelling, including all hydrogen and impurity reactions and total radiation losses. Selected spectroscopic diagnostics are being used to confirm these models.

The following elements are most important in present fusion experiments: hydrogen and deuterium, carbon, silicon (wall conditioning) and the metals of the vacuum vessels. Tungsten is being investigated as a possible alternative to carbon. Additionally, test gases are frequently added for radiative cooling or for particle transport studies. These are helium, neon, argon, krypton and xenon.

In the following, some examples in the field of fusion plasma spectroscopy will be discussed, namely, the measurement of impurity densities, radiation and transport, the investigation of particle influx densities and studies of chemical erosion. Divertor diagnostics, hydrogen opacity effects and hydrogen molecules will be mentioned briefly and some comments will be made on the importance of divertor modelling. In most of these fields, models for plasma reactions are required as discussed below.

DIVERTORS

Divertor tokamaks shall be discussed here as examples of fusion experiments, e.g. the medium-size ASDEX Upgrade tokamak [1]. These are toroidal devices with external coils for the toroidal field and an internal plasma current producing the poloidal field. The magnetic field lines thus wind around the torus and, in many transits, form magnetic flux surfaces. The plasma ions are essentially tied to these surfaces (actually to the somewhat displaced drift surfaces) and the heat flow along the field lines is much higher than perpendicular to the field.

At the outer edge, the plasmas must eventually be bounded by material walls. In the simplest case, this is a special piece of metal or carbon acting as a "limiter". However, due to the difference in parallel and perpendicular heat conductivities, a very thin "scrape-off layer" develops and the specific heat load becomes very high. This causes material problems and impurities can be set free which can easily enter the main plasma. As an alternative, many present-day tokamaks use a magnetic multipole configuration, which causes the field lines outside the so-called separatrix to bend away from the plasma into the "divertor chamber", usually located at the bottom of the vacuum vessel. In this way, the heat load can be spread over a larger wall area and, in addition, the plasma cools down by radiation from something like 50-100eV in the mid-plane to 10eV or less. Thus, the plasma ions gain much less energy in the sheath potential at the walls and the problem of sputtering is greatly reduced. Impurities are then mainly produced by chemical processes, but they can also be retained, at least to some extent, in the divertor chamber. Wall conditioning by means of boron or silicon deposition reduces the problem of oxygen impurities. Divertor modelling is of high importance to find the optimum geometry and operating conditions of the device. Spectroscopic diagnostics in the visible, the VUV and the X-ray region are provided for the main plasma, but the divertor region is also monitored by many spectroscopic sight-lines along the plasma fan viewing both in poloidal direction and also toroidally, tangential to the magnetic field.

The core plasmas are fuelled by gas valves, by neutral hydrogen beams and by recycling neutral atoms and molecules from the divertor regions. Ions diffuse out of the plasma and stream into the divertor chamber along the magnetic field lines. These flows can be measured by means of the Doppler shift of impurity ion lines in the plasma scrape-off region. The lines also have Doppler profiles due to their kinetic temperature and they are split into Zeeman components by the magnetic field. On ASDEX Upgrade, these flows were investigate e.g. by analysis of visible CIII triplet lines. Depending on viewing line, the Zeeman π and σ components have different intensity, and their convolution with the Doppler profile determines the appearance of the line. Along the magnetic field and in the opposite direction, the lines are shifted up or down in wavelength with respect to the perpendicular view. This is clear proof of plasma flows, and the velocity can be derived from these measurements, typically 10-20km/s, which corresponds to a Mach number of $\approx$0.5. More details can be found in [2].

MODELS

In low-density plasmas, the most important processes are electron impact collisions and spontaneous emission of radiation. This leads to the following simplest models: electron impact dissociation of molecules is balanced by diffusion and ionisation. Ions produced by electron impact ionisation at the plasma edge are lost by flows or diffusion. In the plasma interior, radiative recombination becomes more and more important, until it balances ionisation in steady state, resulting in the so-called corona ionisation equilibrium. Electron impact excitation from the ground state with density n_1 to upper states m with n_{m} is balanced by spontaneous emission to all lower levels with Einstein coefficients A, which is the corona excitation model. In its simplest form it can be written:

$$n_1 n_e X_{1,\mathrm{m}}(T_e) = n_{\mathrm{m}} \sum_{\mathrm{k<m}} A_{\mathrm{m,\,k}} \,. \tag{1}$$

In Equation (1), n_e and T_e are electron density and temperature, X is the electron excitation rate coefficient. The photon emission coefficient for a particular spectral line emitted by transition m→j becomes

$$\dot{N}_{\mathrm{Ph,\,m,\,j}} = n_1 n_e X_{1,m} \frac{A_{\mathrm{m,j}}}{\sum A_{\mathrm{m,k}}} = n_1 n_e X_{1,\mathrm{m}} B_{\mathrm{m,j}}\,, \tag{2}$$

with the branching ratio B. Emission spectroscopy observes line-of-sight integrated photon radiances L_{ph} over co-ordinate ℓ:

$$L_{ph} = \int_0^\infty n_1 n_e X_{1,\mathrm{m}} B_{\mathrm{m,j}} \, d\ell \,. \tag{3}$$

The simple corona model must be complemented by other processes like ionisation, recombination, charge exchange, redistribution of excited states by proton/electron collisions and collisional de-excitation, leading to collisional-radiative models. In most cases, metastable levels must also be considered, and for hydrogen, optical thickness of resonance lines must be taken into account. This leads to effective excitation, ionisation and recombination coefficients as functions of T_e, n_e and neutral density n_0, $X(T_e, n_e, n_0)$, $S(T_e, n_e, n_0)$, $\alpha(T_e, n_e, n_0)$, which can be metastable resolved (metastable states as separate species) or unresolved (population of metastables computed from plasma parameters within the ionisation stage). Collisional-radiative calculations for fusion frequently make use of the international project ADAS (Atomic Data Analysis System) [3], which supplies basic input data and programs for computing effective coefficients. In ADAS, high quality data exist for H, He, Be, C, N, O and are being generated for Ne and Ar. Lower quality data are available for many elements.

In order to derive medium- or high-Z impurity densities from line radiation, the ionisation balance is required or, more precisely, the radiating shell must be known. It can be calculated from the system of continuity equations for ionisation stages Z:

$$\frac{\partial}{\partial t} n_Z + div\,\vec{\Gamma}_Z = n_{Z-1} n_e S_{Z-1} + n_{Z+1} n_e \alpha_{Z+1} - n_Z n_e (S_Z + \alpha_Z), \qquad (4)$$

where the first two terms on the right hand side are ionisation and recombination sources from neighbouring stages, and the third term means losses of the considered ionisation stage due to the same processes.

For the particle flux density Γ, one usually adopts the sum of a diffusive and a convective term:

$$\vec{\Gamma}_Z = -D_Z grad\, n_Z + \vec{v}_Z n_Z. \qquad (5)$$

The transport coefficients are still not very well known from theory, since they are often determined by instabilities, i.e. anomalous. They are usually derived from experiment. Effective ionisation and recombination coefficients, possibly including charge exchange, are required from the atomic physics community. Furthermore, one needs diagnostic line excitation data and coefficients for total radiation, to compare the final results with bolometer measurements. All these calculations are done in various impurity transport codes, e.g. STRAHL [4].

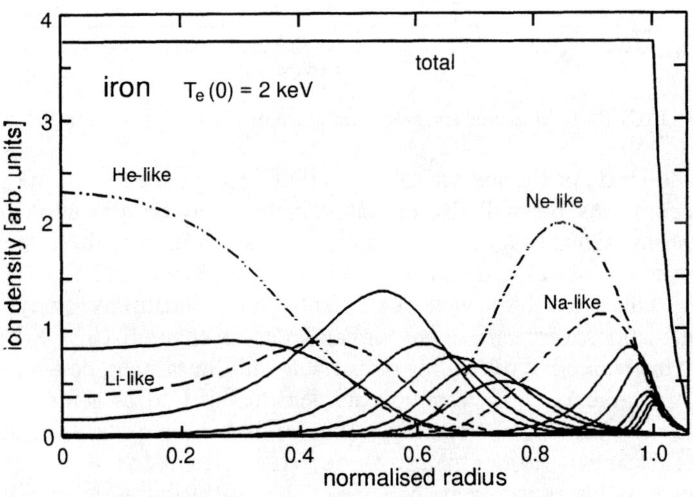

FIGURE 1. Calculations of iron ionisation stages over normalised radius for a plasma discharge with 2keV axis electron temperature. Purely diffusive transport.

An example of such calculations for iron and purely diffusive transport is presented in Fig. 1. The electron temperature profile with an axis value of 2keV results in the shells shown as a function of the normalised plasma radius with the noble gas-like ionisation stages standing out. The total density is flat except for the scrape-off region, where parallel losses come in. In the plasma centre, the ions are in corona ionisation balance, while the edge is affected by transport, as can be recognised from the overlapping shells. An experimental result from JET (Joint European Torus, Culham, England) is shown in Fig. 2 over the outer half of the radius. In this case, the radiation shell of magnesium-like nickel was recorded by means of a rotating mirror and the profile was Abel inverted. A comparison with pertinent calculations shows that, in this case, the shell position and width is determined by diffusion and the corona balance does not fit at all. Thus, the diffusion coefficient can be determined.

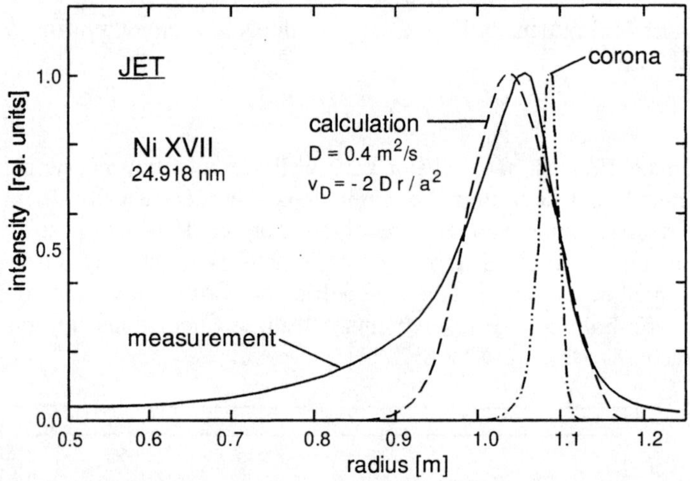

FIGURE 2. Measured and calculated radiation shells of NiXVII in JET.

Metastable levels of the ionisation stages will be populated from the ground state by electron collisions, but will also be linked individually by ionisation and recombination to the ionisation stages below and above. In principle, these states must be treated as separate species and so included in the transport calculations. For that purpose, ADAS contains all relevant coefficients. As a sensitivity study, metastable-resolved calculations were carried out for carbon in the edge of ASDEX Upgrade. Figure 3 shows the respective results in comparison with the simple, non-resolved model. Plotted are the ground and metastable state densities of CIII as a function of plasma radius. As can be concluded from Fig. 3, the difference is quite small and does not warrant the substantial effort of metastable-resolved calculations.

Total radiation losses in corona balance or with diffusion are also being predicted by the transport codes. They demonstrate that impurity radiation, in particular at higher T_e, increases with Z and therefore, the tolerable concentrations of high-Z elements are much smaller than e.g. the one of carbon. The so-called cooling rates are often used to derive impurity concentrations from bolometer measurements of total radiation.

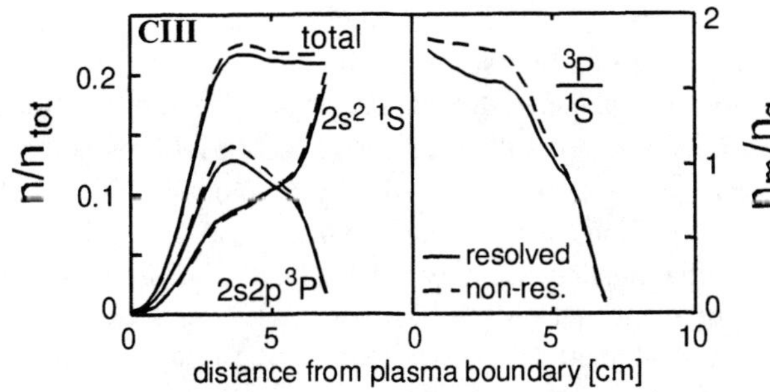

FIGURE 3. Spatial profiles of CIII ground and metastable states, metastable resolved or unresolved.

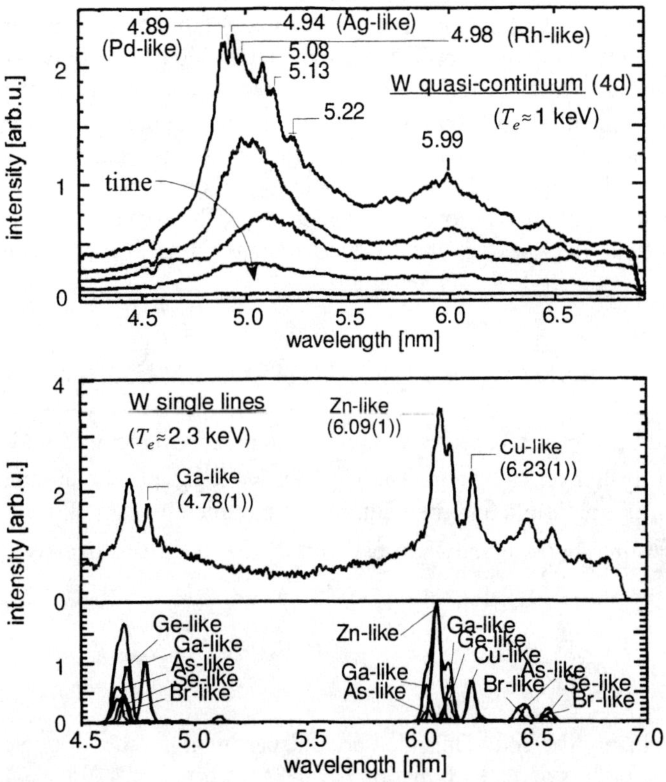

FIGURE 4: Spectra of tungsten, recorded in ASDEX Upgrade after W laser blow-off.

As an alternative to carbon, tungsten is being investigated as wall material in ASDEX Upgrade by covering gradually increasing portions of the walls by tungsten. For diagnostic purposes, tungsten is being injected into the plasma by evaporation of thin films from a glass plate with a laser (laser blow-off). At temperatures of about 1keV, tungsten in the plasma emits very many lines from 4f and 4d shells in the wavelength region around 5 nm, known as the quasi-continuum. Such a spectrum is shown on the top of Fig. 4. The W density is first sustained by repetitive laser blow-off, then it decays with a time constant characteristic for the edge diffusion coefficient. The corresponding concentrations of about 10^{-4} were derived from total radiation. The detection limit for the quasi-continuum limits the sensitivity to about 5×10^{-6} [5]. At higher electron temperatures, W is further ionised and emits spectra with individual lines, which are more suitable for diagnostics, as shown at the bottom of Fig. 4. The pertinent and ionisation stages were identified on the basis of HULAC code predictions [5]. Altogether, the field of high-Z materials poses many open questions and still requires a substantial effort on the atomic physics side.

Atomic data for ionisation, recombination and radiation are also required for modelling divertor plasmas. As an example, it is important to know, under what circumstances hydrogen ions can recombine in the plasma volume before reaching the material walls. As determined by atomic physics, net recombination is only possible, if the electron temperature drops to 1eV or less and the recombination rate coefficient is then some 10^{18}m^3/s. Calculations show that net volume recombination can only reduce the ionisation degree if the flow velocity is 1km/s or less, as compared to the 10 - 20km/s measured above. This means that the plasma in the divertor must not only cool down, but also lose momentum due to elastic collisions with atoms and charge exchange processes. Optical thickness of the hydrogen Lyman lines becomes important and shifts these conditions to lower temperatures.

PARTICLE FLUX MEASUREMENTS

A flux density Γ_k of neutral species k entering the plasmas is ionised very close to the boundary and further on its way in. The particles leave the plasma in ionised states. If recombination is negligible and the neutrals are completely ionised, i.e. $\Gamma_k(\infty)=0$, the integral ionisation rate along the flight path must equal the influx density $\Gamma_k(0)$.

$$\frac{d}{d\ell}\Gamma_k = -n_k n_e S_k \qquad \Gamma_k(0) - \cancel{\Gamma_k(\infty)} = \int_0^\infty n_k n_e S_k d\ell \,. \tag{6}$$

Excitation is closely linked to ionisation and the pertinent number of photons observed along the flight path is given by Equation (3). This leads to

$$\Gamma_k(0) = \int n_k n_e S_k d\ell \frac{L_{Ph}}{\int n_k n_e X_k B \, d\ell} \approx L_{Ph} \boxed{\frac{S_k}{X_k B}}, \tag{7}$$

with an average ratio of ionisation and excitation rate coefficients. The method can be extended to ion lines, where particle losses in lower stages must be taken into account, and also to bands of molecules or radicals.

For the purpose of such particle influx measurements, *S/XB* ratios were calculated with collisional-radiative models for many elements, ionisation stages and lines [6]. Depending on the individual case, they depend little or more strongly on electron temperature and somewhat on electron density. A still open problem is the measurement of tungsten influx from WI lines, since theoretical excitation data are not yet available.

At very low electron temperatures, recombination can become the dominating population mechanism of excited states, and it is then possible to interpret line intensities in terms of recombination processes per photon. For hydrogen lines, opacity of Lyman lines must possibly be taken into account in these calculations.

MOLECULES

In fusion edge plasmas and, in particular, in low-temperature divertors, bands of various molecules are being observed like CH/CD, C_2, BH/BD and others. The appearance of these bands can be modelled on the basis of molecular constants and rotational temperatures T_{rot} can be derived. In technical low-pressure discharges, these temperatures often reflect the heavy-particle kinetic temperatures, but in fusion the molecules are too short-lived for equilibration. The observed T_{rot} are usually of the order of a few 1000K, which is certainly not the distribution with which they come off the walls. Presumably, these results indicate that they were formed in the plasma by the break-up of higher molecules and gained rotational energy in the dissociation process.

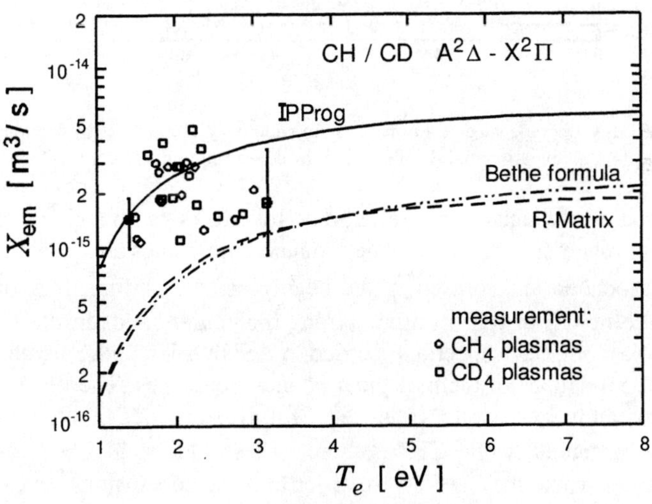

FIGURE 5: CH/CD A-X excitation rate coefficients from several calculations and from measurements.

Similar to atoms, the total intensity of these bands can be used for deriving influx densities. For that purpose, electron impact excitation rate coefficients are required for the molecules or radicals, which are not always readily available. As an example, Fig. 5 shows rate coefficients for the CH radical from several sources, which are equally valid for CD. The Bethe formula and the more elaborate impact parameter method (IPProg) [7] use the optical f-value. There is also a recent calculation based on the R-Matrix formalism [8], which should be of quite high accuracy. Figure 5 also shows measurements obtained by laser-induced fluorescence in low-pressure H_2 and D_2 microwave plasmas [9]. Unfortunately, since the dissociation degree is very small, the latter suffer substantially from the competing contribution of dissociative excitation of CH_4 and higher hydrocarbons, which is causing quite large error limits. As can be seen in Fig. 5, there is certainly need for clarification and improvement.

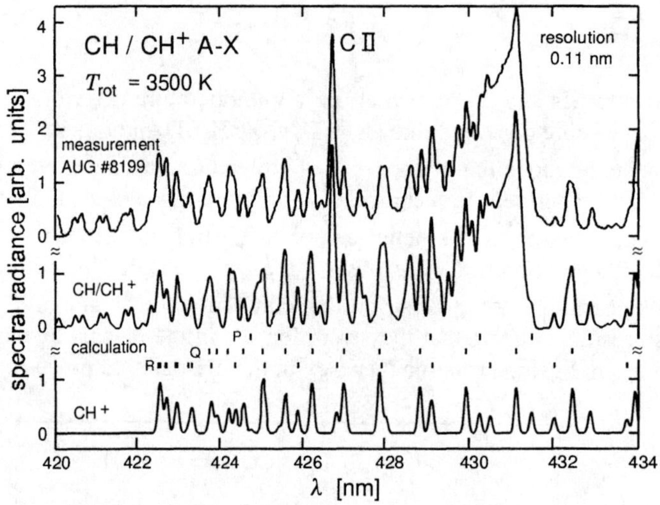

FIGURE 6: ASDEX Upgrade spectra around 430 nm showing CH and CH^+ molecular bands. Simulations explain all experimental details and allow determination of total band intensities.

An important mechanism in present-day tokamaks is the chemical erosion of graphite by hydrogen or its isotopes. The produced molecules are CH_4/CD_4 and others, which are dissociated and ionised in the edge plasmas. During their break-up, they eventually become CH or CH^+ emitting strong molecular bands in the visible spectral range around 430 nm. Such spectra, recorded in ASDEX Upgrade, are shown in Fig. 6, together with simulations of the molecular band appearances. The details of the measured spectrum can be very well reproduced. Still, it is obvious that the intensity of CH can be easily measured, while CH^+ is more or less hidden in this strong band. The situation is rather worse for CD/CD^+. In order to draw conclusions on CH_4, the break-up of this molecule must be calculated, including a great number of reactions and particle diffusion. The pertinent electron collision rates have been compiled some time ago by Ehrhardt and Langer [10], but there are also more recent data. The hydrocarbon dissociation and ionisation has been calculated including diffusion losses of neutrals and of ions to obtain an effective value for the number of dissociation processes per photon in CH [11]. The small intensity of CH^+ indicates a very effective re-deposition of ionised radicals in the plasma. Such calculations have formed the basis of investigations of chemical erosion in ASDEX Upgrade [12]. The results showed a clear H/D isotope dependence and indicated also a decrease in the yield ($\approx$1%) with higher hydrogen fluxes, which would be very beneficial for future fusion machines.

MOLECULAR HYDROGEN

In fusion experiments, molecular hydrogen can be important in edge and divertor plasmas. A high percentage of the hydrogen ions will recycle from the walls in the form of H_2 or D_2 and the ratio of atomic to molecular influxes is an interesting quantity. Hydrogen molecular band radiation also offers a possibility for T_e diagnostics in the very plasma edge. Furthermore, vibrationally excited molecules in higher v states may contribute to hydrogen ion recombination (molecular assisted recombination) by means of two processes:

$$H_2(v) + e \rightarrow H + H^- \qquad \text{dissociative attachment}$$

$$H^- + H^+ \rightarrow H + H \qquad \text{mutual neutralisation}$$

and

$$H_2(v) + H^+ \rightarrow H + H_2^+ \qquad \text{charge exchange}$$

$$H_2^+ + e \rightarrow H + H \qquad \text{dissociative recombination.}$$

Studies of H_2 and D_2 radiation were carried out in low-pressure laboratory discharges, in order to develop various diagnostic methods, in particular, to derive the ground state vibrational population. A collisional-radiative code including atomic and molecular hydrogen [13] was employed for the calculation of excitation, ionisation and dissociation rates. However, it must be kept in mind that molecular hydrogen is quite different from molecular deuterium.

The idea of measuring the hydrogen ground state vibrational population is the following: during electronic excitation, this distribution will be projected up into excited states by electron collisions according to the Franck-Condon principle. The distribution in the upper state can be derived from hydrogen molecular band emission, e.g. the Fulcher bands. Investigations in hydrogen and deuterium confirmed this idea [14], such that similar investigations could be carried out in the ASDEX Upgrade divertor. In contrast to laboratory plasmas, where heavy particle collisions and wall contact influence vibrational distribution, the population in divertors should be dominated by electron collisions and therefore only depend on T_e. Figure 7 shows such calculated results (equilibrium vibrational population) in the electron temperature range between 1eV and 12eV in hydrogen. Applying the above diagnostic method to ASDEX Upgrade, precisely the shown variation of the v=1/v=0 population was observed during a density ramp, leading to low electron temperatures and divertor detachment. This was interpreted as a variation in T_e from its usual value above 10eV, as measured by electrical probes, down to 1eV, a value confirmed by other spectroscopic diagnostics. At the same time, quite reliable information was available for the vibrational temperatures in these plasmas.

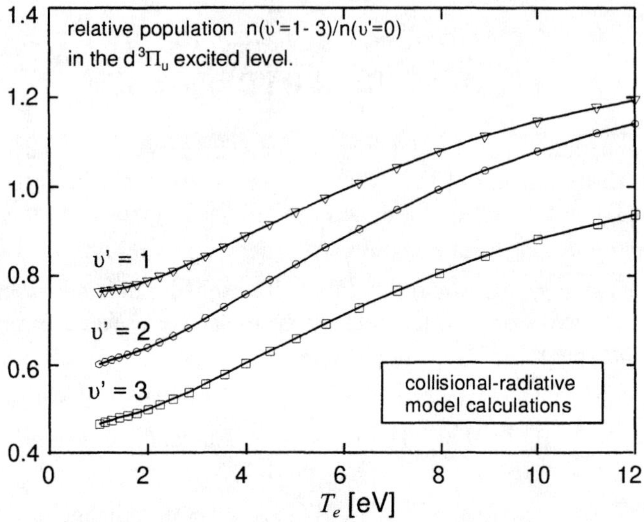

FIGURE 7: Calculated variation in relative vibrational distribution of the H_2 ground state as a function of electron temperature.

On the basis of these results, vibrational excitation and its influence could be taken into account in model calculations for the ASDEX Upgrade divertor plasmas. It turned out that vibrational population leads to a very important increase in dissociation and ionisation rate coefficients. Molecular assisted recombination does play some role in the outmost layer of the plasma fan, where it even becomes larger than the usual ion recombination. But the molecular density drops very rapidly towards the plasma interior, and the global numerical solution for the plasma is little changed. All these investigations must be repeated for deuterium, and a comprehensive collection of rate coefficients and models for deuterium still remains to be compiled. Also, the question is still open, whether the molecules come off the walls in vibrationally excited states.

SUMMARY

In nuclear fusion research, a great number of atomic data is required for the interpretation of plasma spectroscopy, as well as for modelling bulk and divertor plasmas. Modelling needs parameterised, effective rate coefficients for a great number of processes like ionisation, recombination, excitation, charge exchange and total radiation.

The range of elements and ionisation stages extends from highly ionised high-Z ions like tungsten, which may exist in the plasma core of future reactors and are being investigated now, to neutrals and molecules in edge plasmas and divertors. Some noble gases are being used for radiative cooling or for transport studies of highly ionised ions. For measuring metal concentrations, diagnostic lines from the plasma interior are used, sometimes of highly ionised species like Zn-like tungsten. The ionisation balance and the emitting shells must be calculated on the basis of the relevant atomic data and of the transport coefficients. Spectrally integrated radiation from bolometers or soft X-ray diodes is a very valuable diagnostic tool, to be compared with code predictions for the relevant impurity species. For high heat-load areas, carbon is mostly used at present, which is fully ionised in the plasma and can only be measured by charge-exchange recombination spectroscopy. Impurity particle influxes at the boundary are derived from lines of neutrals or low ionisation stages, a method, which can also be extended to the band emission of molecules. However, electron excitation rate coefficients for molecules are still relatively scarce. Chemical erosion of carbon has been investigated by means of CH/CD band emission. In cold divertor plasmas, hydrogen molecules also play some role, e.g. enhancing ion recombination. Relevant diagnostics and rate coefficients are being established. For the plasma codes, data with moderate accuracy are often sufficient, coefficients for diagnostic lines, used for density measurements or code verification, must be known with higher precision. There is still substantial need for more atomic data, but an interface is often required between producers of basic data and fusion researchers. The ADAS system and the ADAS group have e.g. taken over such a role.

REFERENCES

1. Gruber, O., many authors, *Nucl. Fusion* **39**, 1321 (1999)
2. Gafert, J., Behringer, K., Coster, D., Dorn, C., Kallenbach, A., Schneider, R., Schumacher, U., and the ASDEX Upgrade Team, *J. Nucl. Mater.* **266-269**, 365 (1999).
3. Summers H P 1999: ADAS User Manual Version 2.1 [http://patiala.phys.strath.ac.uk/adas] University of Strathclyde, Glasgow
4. Behringer K., Description of the Impurity Transport Code "STRAHL", Report JET-R(87)08, 1987
5. Asmussen, K., Fournier, K.B., Laming, J.M., Neu, R., Seely, J.F., Dux, R., Engelhardt, W., Fuchs, J.C., ASDEX Upgrade Team, *J. Nucl. Mater.* **38**, 967 (1998)
6. Behringer K., Summers H.P., Denne B., Forrest M. and Stamp M., *Plasma Phys. Contr. Fusion* **31**, 2059 (1989)
7. Burgess A. and Summers H. P., *Mon. Not. R. Astr. Soc.* **174**, 345 (1976)
8. Berrington K.A., Burke P.G., Darlington W.G.V. et al.: Linking Belfast and ADAS Databases, Radiative Recombination Data for JET Impurity Species and Vibronic Excitation of CD, JET Order No: JP5/10061, 1997
9. Schütte Th., Thesis Universität Stuttgart, Germany, 1995
10. Ehrhardt A.B. and Langer W.D., Collisional Processes of Hydrocarbons in Hydrogen Plasmas, Report PPPL-2477, Sept. 1987
11. Behringer K., *J. Nucl. Mater.* **176 & 177**, 606 (1990)
12. Kallenbach A., Bard A., Coster D. et al., *J. Nucl. Mater.* **266-269**, 343 (1999)
13. Sawada K. and Fujimoto T., *J. Appl. Phys.* **78**, 2913 (1995)
14. Fantz U. and Heger B., *Plasma Phys. Contr. Fusion* **40**, 2023 (1998)

Atomic Data in X-ray Astrophysics: the Chandra Emission Line Project

N. S. Brickhouse
Smithsonian Astrophysical Observatory
60 Garden Street
Cambridge, MA 02138

Abstract

With the launches of the Chandra X-ray Observatory (CXO) and the X-ray Multimirror Mission (XMM), high resolution X-ray spectroscopy of cosmic sources has begun. Early, deep observations of three stellar coronal sources will provide not only invaluable calibration data, but will also give us benchmarks for the atomic data under collisional equilibrium conditions. Analysis of the Chandra X-ray Observatory data, and data from other telescopes taken simultaneously, for these stars is ongoing as part of the Emission Line Project. Goals of the Emission Line Project are (1) to determine and verify accurate and robust diagnostics and (2) to identify and prioritize issues in fundamental spectroscopy which will require further theoretical and/or laboratory work. The Astrophysical Plasma Emission Database will be described in some detail, as it is introducing standardization and flexibility into X-ray spectral modeling.

Spectral models of X-ray astrophysical plasmas can be generally classified as dominated by either collisional ionization or by X-ray photoionization. While the atomic data needs for spectral models under these two types of ionization are significantly different, there are overlapping data needs, as I will describe. Early results from the Emission Line Project benchmarks are providing an invaluable starting place, but continuing work to improve the accuracy and completeness of atomic data is needed. Addtionally, we consider the possibility that some sources will require that both collisional ionization and photoionization be taken into account, or that time-dependent ionization be considered. Thus plasma spectral models of general use need to be computed over a wide range of physical conditions.

CP543, *Atomic and Molecular Data and Their Applications,* edited by K. A. Berrington and K. L. Bell

1 Introduction

The Emission Line Project (ELP) is a collaborative effort, organized by the Chandra X-ray Observatory Center, to improve the plasma spectral models used to analyze and fit X-ray spectral observations. During the first phase of the Emission Line Project we are analyzing the high quality spectra of three stellar coronal targets that are being obtained for the purposes of calibrating the Chandra transmission gratings. For select ELP observations by Chandra, overlapping measurements have been obtained by the Extreme Ultraviolet Explorer (EUVE), the HST Space Telescope Imaging Spectrograph, Beppo-SAX, and the NSF's Very Large Array radio telescope. Broadening the wavelength coverage provides additional important tests for the atomic models.

Stellar coronal spectra are rich in emission lines from high ionization states of cosmically abundant elements. The line source calibration targets — Procyon, Capella, and HR 1099 — were specially chosen because they exhibit emission corresponding to a wide range of plasma temperatures, dominated by $T_e \sim 2 \times 10^6$, 6×10^6, and 2×10^7 K, respectively (Fig. 1). The three sources emit over a continuous range of temperatures, such that most elements and ionization states accessible to the Chandra gratings will be exhibited in these spectra. Current spectral models of collisionally ionized plasmas will be tested against the observations to assess the problem areas and to help set priorities for fundamental theoretical and experimental spectroscopy.

2 Assumptions of Collisional Plasma Emission Models

Given the electron temperature T_e, electron density N_e, and elemental abundances, models of collisionally ionized plasma calculate the power radiated from continuum and line emission processes. These models are applicable to approximately half the X-ray sources studied. Other sources have strong X-ray continuum emission which photoionizes the plasma, producing line emission through a process of radiative recombination followed by cascades. The dominant lines are significantly different in the two cases; however, there are overlapping and complementary atomic data needs.

A number of assumptions are generally made in collisionally ionized plasma models, particularly that the plasma is optically thin and non-relativistic. Continuum emission from bremsstrahlung, radiative recombina-

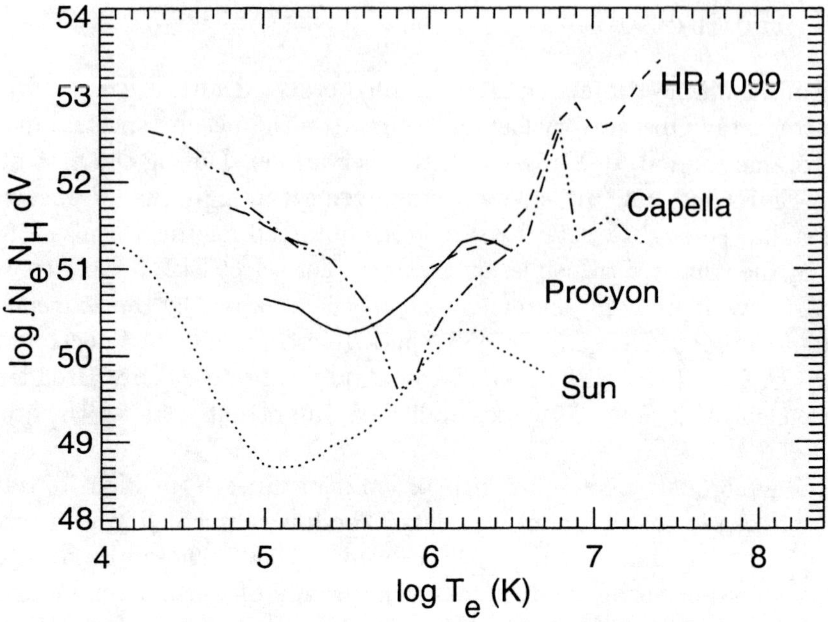

Figure 1: Emission measure distributions for Procyon (Drake, Laming, & Widing 1995), Capella (Dupree et al. 1993; Brickhouse 1996), and HR 1099 (Griffiths & Jordan 1998) compared with that of the quiescent Sun (Raymond & Doyle 1981). Procyon is shown as the solid line, Capella is dash-dotted, HR 1099 is dashed, and the Sun is dotted. The spectral resolution of the EUVE spectrometers allows the measurement of isolated emission lines which have been used to construct the stellar emission measure distributions. ASCA spectra set limits on the highest temperatures in the plasma.

tion, and 2-photon processes are included. Collisional impact cross sections are integrated over Maxwellian velocity distributions, and it is assumed that the radiation is not affected by electric and magnetic fields. Photoionization and photoexcitation are generally not included.

Often the emission is assumed to be time-independent; however, non-equilibrium ionization models may be constructed using a physical model. For emission lines which may be populated by recombination cascades, it is important to take the ionization balance into account, and thus one must link the ionization and level population calculations. Another common assumption of the "coronal approximation" is that the electron density is low. In that case the population of energy levels is dominated by transitions from the ground state. Generally today, the level populations are calculated using a rate matrix to account for all transitions, such that density effects on the level populations can be properly treated (up to densities $\sim 10^{14}$ cm^{-3}). Density effects on the ionization balance are still treated crudely, when treated at all.

The physical assumptions of ionization equilibrium and negligible optical depth appear to be good to a first order approximation. Since the coronal emission is weighted by N_e^2 for collisionally excited lines, the observed spectrum (time-averaged over several hours to several days) is naturally biased toward plasma with the shortest equilibration times. The good agreement among lines emitted from the same ionization state with the theoretical predictions, and particularly for branching ratios, can place tight constraints on the optical depths involved. To observe the breakdown of these assumptions, if applicable, will require high spectral resolution, such as provided by the XMM Reflection Grating Spectrometer (RGS) and Chandra Low Energy Transmission Gratings (LETG) and High Energy Transmission Gratings (HETG).

It is interesting to note that most of the simplifying assumptions break down somewhere in the solar corona! Measured particle distributions in the solar wind are non-Maxwellian (Olgivie & Scudder 1978), the strongest X-ray line in the quiescent corona, Fe XVII λ15.013, appears optically thick in active region loops (Schmelz et al. 1997), and the He II λ303.8 is influenced by recombination following photoionization of He II by the EUV corona (Athay 1988). Nevertheless, the assumptions work well to first order for most emission, thus allowing us to proceed.

In the absence of a significant radiation field, the line intensity I_{line} is the product of the population of the upper level N_k and the atomic transition probability A_{kj} from level k to j.

$$I_{line} = N_k A_{kj} \tag{1}$$

For a two-level atom collisions excite the level and thus I_{line} is proportional to the collisional excitation rate C_{jk} from the lower level j.

$$I_{line} = N_j N_e C_{jk} \tag{2}$$

More generally, the level populations are calculated, such that

$$I_{line} = \frac{1}{4\pi R^2} \varepsilon_{At} EM \tag{3}$$

where ε_{At} is the atomic emissivity, and EM is the emission measure

$$EM(\Delta T_e) = \int_{V(\Delta T_e)} N_e N_H dV \tag{4}$$

In principle, if the plasma is isothermal and isochoric, line ratio diagnostics provide the cleanest measurements of the physical conditions of the emitting region. In practice the astrophysical plasma is neither, and one must construct multi-component models. A high quality X-ray spectrum (defined as having sufficient bandpass, spectral resolution, signal-to-noise, and calibration) provides a wealth of information to infer the distribution of temperature and possibly density. Emission measure distributions are typically constructed by an inversion of Equation 3, such that a good fit is provided to the spectrum, either globally or to specified emission lines.

The temperatures, densities, emission measures, and abundances derived from such methods are subject to numerous uncertainties beyond the statistical errors associated with the observations, including instrument calibration, line blending, the atomic data, and the simplifying assumptions themselves. The ASCA spectrum of Capella could not be fit by the standard models, and eventually required the examination of all of these issues. The biggest improvement in the fit came with the addition of emission lines from $n > 5$ for Fe XVII to XIX, which had not previously been calculated (Brickhouse et al. 2000; D. A. Liedahl & N. S. Brickhouse 2000, in preparation).

3 Stellar Coronae as Tests of Models

Questions raised over the past several years about the validity of results from moderately resolved spectra from ASCA and EUVE, such as the elemental abundances of elliptical galaxies (Arimoto et al. 1997) and of stellar coronae

(Jordan et al. 1998) have led many of us to the conclusion that the models need to be tested and validated; however, given the number of assumptions upon which the spectral models are based, and our limited understanding of stellar coronae, how can we be so confident that the ELP data will provide useful benchmarks?

Before answering that question it is important to emphasize that astrophysical spectra are no substitute for laboratory atomic physics experiments and well calibrated spectroscopic measurements under controlled conditions, and we fully support efforts such as the ones reported in this proceedings; however, the numbers of atomic rates for potentially important emission lines far exceed any conceivable laboratory resources. Moreover, the wide range of temperatures and densities observed in stellar coronae include conditions which are difficult to replicate under laboratory conditions. Thus, high quality astrophysical spectra provide a complementary data set for assessing the quality of the spectral models.

The answer, then, to the question of how useful the ELP data are as benchmarks, is twofold. (1) We begin with the underlying premise that we understand the emission processes in these sources reasonably well. These sources have been well observed with ASCA and EUVE, such that we do not expect gross departures from equilibrium. Since the emission from the strong collisionally excited lines is naturally weighted toward N_e^2, thermalization and equilibration are most likely to apply for the emission lines that we are most able to observe. Furthermore, EUV branching ratios place tight constraints on the optical depth effects for Capella (Brickhouse, Raymond, & Smith 1995). (2) The breakdown of the classical assumptions would have observable consequences. The Chandra observations (and other coordinated measurements) provide spectra of such high quality that consistency checks are built in. The broad bandpass provides multiple diagnostic lines, such that issues resulting from line blends can be identified. Multiple ionization states of different elements provide a cross-check on the ionization balance models, which underlie the derived shape of the emission measure distributions. While a few lines may have optical depths of order unity, these will easily be identified by line ratio measurements.

4 Preliminary Emission Line Project Data Analysis

The first light papers for the LETG (Brinkman et al. 2000) and for the HETG (C. Canizares et al. 2000) support the argument above that these data are useful for testing the plasma spectral emission models. Since the instrument calibration and data reduction and processing are still at an early stage, the results reported here are preliminary, and are intended to give an understanding of how the data can be used. We present two examples:

- Optical depth. Fe XVII λ15.014 is the strongest emission line in the X-ray spectrum of Capella. By comparing its ratio with the optically thin Fe XVII λ15.265 one can test the optical depth of λ15.014, and thus place constraints on the degree of resonance scattering in the atmosphere. The Capella line ratio appears to be consistent with the assumption of negligible net optical depth (Brinkman et al. 2000; C. Canizares et al., in preparation). While the theoretical predictions tend to be somewhat higher than the laboratory measurements (see Brown et al. 1998), and the jury is still somewhat out on the best data, the Capella measurements provide additional support for the idea that we can neglect optical depth effects in our analysis of these spectra.

- The identification of spectral regions with "missing lines." Figure 2 shows a small section of the Capella spectrum, which had presented modeling difficulties (Brickhouse et al. 2000) for the ASCA analysis. We can clearly identify lines of Fe XVII based on the work of Brown et al. (1998), which are not yet included in the plasma spectral models.

Other tests are planned once the data are felt to be reliable. Collisional ionization equilibrium can be tested by comparing $\Delta n = 0$ transitions in the EUV to $\Delta n = 1$ transitions in the X-ray, exploiting the broad bandpass of the LETG. For example, the ratio of the Fe XVIII λ93.92 intensity to the Fe XVIII λ14.22 intensity is temperature-sensitive due primarily to the Boltzmann factor in the collision rate. By comparing the line ratio temperature to the temperature derived from the ratios among adjacent ions, one tests the local temperature relative to the degree of ionization. Of course, the underlying collision strengths and the ionization balance models themselves are also in need of testing for these lines. Ultimately the comparison among several lines from the same ion will help to validate the collision strengths, while comparing lines from different elements will be useful in testing the ionization balance.

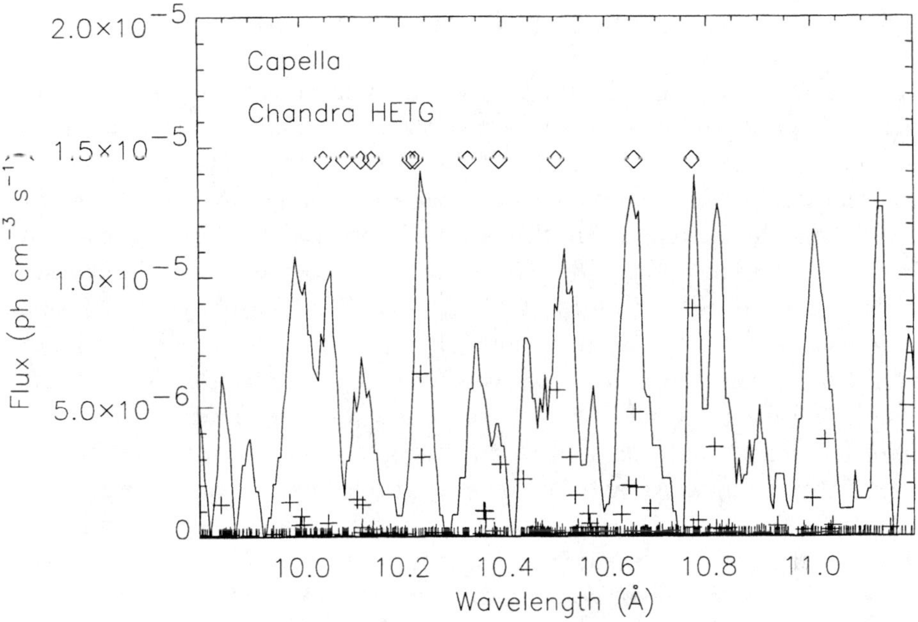

Figure 2: Region of the observed Chandra HETG spectrum of Capella compared with model emission lines, showing the "missing lines." The smoothed spectrum shown is the Observation Identification 1099, Medium Energy Grating spectrum, minus first order. Plus symbols mark the observed lines calculated using an APEC model (Smith et al. 1999) scaled to the strongest line shown, while diamonds mark the wavelengths of measured Fe XVII lines (Brown et al. 1998) from energy levels with principal quantum numbers $n > 5$. Such lines were estimated to account for nearly 30% of the flux in the waveband between 10 and 11 Å in the ASCA spectrum of Capella (Brickhouse et al. 2000).

5 The Astrophysical Plasma Emission Database

Several plasma emission codes for collisionally ionized astrophysical sources exist in the public domain, notably the Raymond-Smith code (Raymond 1998, and reference therein) and the MEKAL code (Kaastra, Mewe, & Nieuwenhuijzen 1996, and references, therein). Our group is modernizing and updating the Raymond-Smith code to improve flexibility and user interface, as well as to update the atomic data and to use level-to-level computations when possible (Smith et al. 1998). The Astrophysical Plasma Emission Code (APEC), as it is now called, makes use of a general format for atomic data through a separate database, the Astrophysical Plasma Emission Database, or APED. This database provides access to multiple formats for atomic data as well as to multiple datasets, for flexibility in changing atomic data. An additional value of this flexibility is the ease with which comparisons can be made among the atomic rates — by calculating level populations with different inputs, the impact on the observable spectral features can be determined, not just differences among rates themselves.

The database also provides, when possible, error estimates, which are of great interest to astrophysical plasma modelers. Laboratory wavelengths are listed along with the theoretical lines to improve spectral models, of importance for line identification and for line deblending. The ELP tests will make use of the existing first version of APED (now publicly available), while improving the atomic data through benchmarks and new rate calculations.

The interest in the X-ray astronomy community for using this type of format is growing, especially as spectral modeling software packages become available to the Guest Observers of the satellite missions, including the extension to X-ray photoionized plasmas. As an example, the inclusion of accurate wavelengths even for weaker lines under collisional ionization conditions, will vastly improve the line identification for lines observed under the more extreme X-ray photoionized conditions. As the idea of a shared public database gains momentum, our ability to utilize accurate atomic data and to respond quickly to improvements in data will grow. One of the major advantages is that the burden of compiling and benchmarking the plasma codes will be shared among a more diverse group of users.

6 Conclusions

Our understanding of the coronal spectrum is sufficiently mature that the Emission Line Project will help to identify and solve problems with the plasma spectral emission models. Problems with the models affect both high spectral resolution data, for which the primary analysis will involve the measurement of specific diagnostic lines, and low to moderate resolution data, for which global spectral models are needed. This project has widespread implications for other types of astrophysical sources, including, galaxies, supernova remnants, clusters of galaxies, and stellar coronae.

The Chandra calibration data for Procyon, Capella, and HR 1099 have now been obtained, with many of the data sets available through the Chandra public archives. Preliminary results support the usefulness of these data for testing models. Meanwhile, calibration and processing improvements continue. The ELP website (http://asc.harvard.edu/elp/ELP.html) gives information on the distribution, collection, and organization of data as well as links to useful software and plasma models. We are in the process of producing an emission line catalog consisting of measured wavelengths, fluxes, line widths, and identifications. Improvements will be available as quickly as possible through the Astrophysical Plasma Emission Database.

Broad support in the X-ray astronomy community has made the Emission Line Project possible.

Acknowledgements

We acknowledge the support of Harvey Tananbaum and the Chandra X-ray Observatory Center NAS8-39073. We also appreciate the contributions of the Emission Line Project Core Working Group: Jeremy Drake, Dan Dewey, Richard Edgar, Terry Gaetz, John Houck, David Huenemoerder, Vinay Kashyap, Duane Liedahl, Randall Smith, Doug Swartz, and Brad Wargelin. Some of this material has been presented elsewhere (Astrophysical Plasmas: Codes, Models, and Observations, Mexico City, 1999, to be published in the Revista Mexicana de Astronomia y Astrofisica, Serie de Conferencias, ed. S. J. Arthur, N. Brickhouse and J. Franco).

References

[1] Arimoto, N., Matsushita, K., Ishimaru, Y., Ohashi, T., & Renzini, A. 1997, ApJ, 477, 128

[2] Athay, R. G. 1988, ApJ, 329, 482

[3] Brickhouse, N. S. 1996, in IAU Colloq. 152, Astrophysics in the Extreme Ultraviolet, ed. S. Bowyer & R. F. Malina (Dordrecht: Kluwer), 105

[4] Brickhouse, N. S., Dupree, A. K., Edgar, R. J., Liedahl, D. A., Drake, S. A., White, N. E., & Singh, K. P. 2000, ApJ, 530, 387

[5] Brickhouse, N. S., Raymond, J. C., & Smith, B. W. 1995, ApJS, 97, 551

[6] Brinkman, et al. (23 authors) 2000, ApJ, 530, L111

[7] Brown, G. V., Beiersdorfer, P., Liedahl, D. A., Widmann, K. & Kahn, S. M. 1998, ApJ, 502, 1015

[8] Canizares, C. R. 2000, ApJ, accepted

[9] Drake, J. J., Laming, J. M., Widing, K. G., 1995, ApJ, 443, 393

[10] Dupree, A. K., Brickhouse, N. S., Doschek, G. A., Green, J. C., & Raymond, J. C. 1993, ApJ, 418, L37

[11] Griffiths, N. W., & Jordan, C. 1998, ApJ, 497, 883

[12] Jordan, C., Doschek, G. A., Drake, J. J., Galvin, A. B., & Raymond, J. C. 1998, in ASP Conf. Ser. 154, Cool Stars, Stellar Systems, and the Sun, Tenth Cambridge Workshop, ed. R. A. Donahue & J. A. Bookbinder (San Francisco: ASP), 263

[13] Kaastra, J. S., Mewe, R., & Nieuwenhuijzen, H. 1996, in UV and X-ray Spectroscopy of Astrophysical and Laboratory Plasmas, K. Yamashita and T. Watanabe (eds.) (Tokyo: Universal Academy Press, Inc.), 411

[14] Ogilvie, K. W., & Scudder, J. D. 1978, JGR, 83, 3776

[15] Raymond, J. C. 1998, in Hot Thin Plasmas in Astrophysics, ed. R. Pallavicini (Dordrecht:Kluwer), 3

[16] Raymond, J. C., & Doyle, J. G. 1981, ApJ, 247, 686

[17] Schmelz, J. T., Saba, J. L. R., Chauvin, J. C., & Strong, K. T. 1997, ApJ, 477, 509

[18] Smith, R. K., Brickhouse, N. S., Raymond, J. C., & Liedahl, D. A. 1998, in Proceedings of the First XMM Workshop on "Science with XMM", ed. M. Dahlem (Noordwijk, The Netherlands)

Atomic Data Production

Theory of Electron Scattering by Atoms, Ions and Molecules

P. G. Burke

Department of Applied Mathematics and Theoretical Physics, The Queen's University of Belfast, Belfast BT7 1NN, Northern Ireland, UK

Abstract. In this review, the theory of low energy electron scattering by light atoms and ions is first presented. The theory is then extended to treat electron scattering by heavy atoms and ions where relativistic effects are important. Electron scattering by atoms and ions at intermediate and high energies and the threshold behaviour of the ionization cross section are then reviewed. Finally the theory of scattering of electrons by molecules is discussed, where the possibility of exciting degrees of freedom associated with the nuclear motion is considered. In all of these areas, computer program packages which enable accurate scattering data to be calculated are referenced.

ELECTRON SCATTERING BY ATOMS AND IONS

Scattering at Low Energies

Wave equation, scattering amplitude and cross section

In this section we consider the process

$$e^- + A_i \rightarrow A_j + e^- \tag{1}$$

where A_i and A_j are the initial and final bound states of the target atom or ion and where the velocities of the incident and scattered electrons are of the same order or less than that of the target electrons actively involved in the the collision. We also assume that all relativistic effects can be neglected which restricts the present treatment to atoms and ions with small nuclear charge Z. These restrictions will be relaxed later.

The time-independent Schrödinger equation which describes the scattering of an electron by an atom or ion containing N electrons and with nuclear charge Z is

$$H_{N+1}\Psi = E\Psi \tag{2}$$

where E is the total energy of the $(N+1)$-electron system and H_{N+1} is the non-relativistic Hamiltonian, which is given in atomic units by

CP543, *Atomic and Molecular Data and Their Applications*, edited by K. A. Berrington and K. L. Bell

$$H_{N+1} = \sum_{i=1}^{N+1} \left(-\frac{1}{2}\nabla^2 - \frac{Z}{r_i} \right) + \sum_{i>j=1}^{N+1} \frac{1}{r_{ij}}. \tag{3}$$

In this equation, $r_{ij} = |\mathbf{r_i} - \mathbf{r_j}|$ where $\mathbf{r_i}$ and $\mathbf{r_j}$ are the vector coordinates of the ith and jth electron relative to the origin of coordinates taken to be the target nucleus, which is assumed to have infinite mass.

We also introduce the target eigenstates and possibly pseudostates Φ_i and the corresponding eigenenergies e_i by the equation

$$\langle \Phi_i | H_N | \Phi_j \rangle = e_i \delta_{ij} \tag{4}$$

where H_N is the N-electron target Hamiltonian defined by Eq. (3) with $N+1$ replaced by N. We will assume in the following discussion that accurate target states can be calculated, this being a necessary criterion to obtain accurate scattering cross sections. The solution of Eq. (2) corresponding to the process defined by Eq. (1), then has the asymptotic form

$$\Psi_i \underset{r \to \infty}{\sim} \Phi_i \chi_{\frac{1}{2}m_i} e^{ik_i z} + \sum_j \Phi_j \chi_{\frac{1}{2}m_j} f_{ji}(\theta, \phi) \frac{e^{ik_j r}}{r} \tag{5}$$

where $\chi_{\frac{1}{2}m_i}$ and $\chi_{\frac{1}{2}m_j}$ are the spin eigenfunctions of the incident and scattered electrons, where the direction of spin quantization is usually taken to be the incident electron beam direction, and $f_{ji}(\theta, \phi)$ is the scattering amplitude, the spherical polar coordinates of the scattered electron being denoted by r, θ and ϕ. Also the wave numbers k_i and k_j of the incident and scattered electron in Eq. (5) are related to the total energy E of the electron atom system by

$$E = e_i + \frac{1}{2}k_i^2 = e_j + \frac{1}{2}k_j^2. \tag{6}$$

When the target is an ion, then logarithmic phase factors caused by the long-range Coulomb potential must be included in the exponents in Eq. (5). We will discuss the effects of this long-range Coulomb potential below.

The differential cross section for a transition from an initial state $|i\rangle \equiv |\mathbf{k}_i, \Phi_i, \chi_{\frac{1}{2}m_i}\rangle$ to a final state $|j\rangle \equiv |\mathbf{k}_j, \Phi_j, \chi_{\frac{1}{2}m_j}\rangle$ can now be obtained by considering the incident and scattered fluxes in Eq. (5). We obtain

$$\frac{d\sigma_{ji}}{d\Omega} = \frac{k_j}{k_i} |f_{ji}(\theta, \phi)|^2. \tag{7}$$

The total cross section is obtained by averaging over the initial spin states, summing over the final spin states and integrating over all scattering angles.

In order to solve the Schrödinger equation (2) to determine the scattering amplitude and and cross section at low energies, most approximation methods start from a partial wave expansion of the total wave function with the form

$$\Psi_i^\Gamma(\mathbf{X}_{N+1}) = \mathcal{A}\sum_{j=1}^{n}\tilde{\Phi}_j^\Gamma(\mathbf{x}_1,\ldots,\mathbf{x}_N;\hat{\mathbf{r}}_{N+1}\sigma_{N+1})r_{N+1}^{-1}F_{ji}^\Gamma(r_{N+1}) + \sum_{j=1}^{m}\chi_j^\Gamma(\mathbf{X}_{N+1})b_{ji}^\Gamma \quad (8)$$

where $\mathbf{X}_{N+1} \equiv \mathbf{x}_1,\ldots,\mathbf{x}_{N+1}$ represents the space and spin coordinates of all $N+1$ electrons, $\mathbf{x}_i \equiv \mathbf{r}_i\sigma_i$ represents the space and spin coordinates of the ith electron and $\mathcal{A}$ is the operator that antisymmetrizes the first expansion with respect to exchange of all pairs of electrons in accordance with the Pauli exclusion principle. The channel functions $\tilde{\Phi}_j^\Gamma$ in Eq. (8) are obtained by coupling the orbital and spin angular momenta of the target states Φ_j with those of the scattered electron to form eigenstates of the total orbital and spin angular momenta L and S, their z components M_L and M_S and the parity π, where the quantum numbers

$$\Gamma \equiv LSM_LM_S\pi \quad (9)$$

are conserved in the collision, since relativistic effects can be neglected. The χ_j^Γ in the second expansion in Eq. (8), are square integrable correlation functions which allow for additional correlation effects not adequately represented by the first expansion which can in practise include only a limited number of target states and possibly pseudostates. These correlation functions are constructed from the same orbital basis that is used to construct the target states Φ_j. The second expansion also includes terms that ensure completeness if, as is usual, the functions F_{ji}^Γ which represent the scattered electron are orthogonalized to the orbitals used to construct the Φ_j.

We now substitute expansion (8) into the Schrödinger equation (2), project onto the channel functions $\tilde{\Phi}_j^\Gamma$ and onto the square integrable functions χ_j^Γ and eliminate the expansion coefficients b_{ji}^Γ between these equations. We then obtain n coupled integrodifferential equations satisfied by the radial functions F_{ji}^Γ, with the form

$$\left(\frac{d^2}{dr^2} - \frac{\ell_j(\ell_j+1)}{r^2} + \frac{2Z}{r} + k_j^2\right)F_{ji}^\Gamma(r) = 2\sum_{l=1}^{n}\{V_{jl}^\Gamma(r)F_{li}^\Gamma(r)$$
$$+\int_0^\infty \left[W_{jl}^\Gamma(r,r') + X_{jl}^\Gamma(r,r')\right]F_{li}^\Gamma(r')dr'\}, \quad j=1,\ldots,n, \quad (10)$$

where ℓ_j is the orbital angular momentum of the scattered electron and V_{jl}^Γ ,W_{jl}^Γ and X_{jl}^Γ are the local direct, non-local exchange and non-local correlation potentials. If the correlation functions in expansion (8) are not included, Eqs. (10) are called the close coupling equations. The scattering amplitude and cross sections can be obtained by solving Eqs. (10), for each set of conserved quantum numbers Γ, subject to the following K-matrix asymptotic boundary conditions

$$F_{ji}^\Gamma(r) \underset{r\to\infty}{\sim} k_j^{-1/2}(\sin\theta_j\delta_{ji} + \cos\theta_j K_{ji}^\Gamma), \quad (k_j^2 \geq 0)$$
$$F_{ji}^\Gamma(r) \underset{r\to\infty}{\sim} 0, \quad (k_j^2 < 0) \quad (11)$$

where

$$\theta_j = k_j r - \frac{1}{2}\ell_j \pi + \frac{z}{k_j} \ln 2k_j r + \sigma_j \tag{12}$$

and where $z = Z - N$ is the residual charge on the ion and $\sigma_j = \arg\Gamma(\ell_j + 1 - iz/k_j)$ is the Coulomb phase shift. The S-matrix $\mathbf{S}^\Gamma$ and the T-matrix $\mathbf{T}^\Gamma$ are related to the K-matrix $\mathbf{K}^\Gamma$, defined by Eqs. (11), by the matrix equations

$$\mathbf{S}^\Gamma = \frac{\mathbf{I} + \mathbf{i}\mathbf{K}^\Gamma}{\mathbf{I} - \mathbf{i}\mathbf{K}^\Gamma}, \quad \mathbf{T}^\Gamma = \mathbf{S}^\Gamma - \mathbf{I} = \frac{2\mathbf{i}\mathbf{K}^\Gamma}{\mathbf{I} - \mathbf{i}\mathbf{K}^\Gamma} \tag{13}$$

where the dimensions of the matrices in these equations are $n_a \times n_a$ where n_a is the number of open channels $(k_j^2 \geq 0)$ at the energy E under consideration, for the given Γ. The hermiticity and time reversal invariance of the Hamiltonian ensures that $\mathbf{K}^\Gamma$ is real and symmetric and hence from Eq. (13), that $\mathbf{S}^\Gamma$ is unitary and symmetric. The total cross section is then given by

$$\sigma_{\mathrm{Tot}}(i \to j) = \frac{\pi}{2k_i^2(2L_i+1)(2S_i+1)} \sum_{LS\pi\ell_i\ell_j} (2L+1)(2S+1)|T_{ji}^\Gamma|^2 \tag{14}$$

in units of a_0^2, which describes a transition from an initial target state $\alpha_i L_i S_i$ to a final target state $\alpha_j L_j S_j$. It is also useful to define a collision strength by

$$\Omega(i,j) = k_i^2(2L_i+1)(2S_i+1)\sigma_{\mathrm{Tot}}(i \to j) \tag{15}$$

which is dimensionless and symmetric with respect to interchange of the initial and final states denoted by i and j.

The coupled integrodifferential equations (10) are solved using general computer program packages. These include packages based on the convergent close coupling method, the R-matrix method, the J-matrix method, the linear algebraic equations method, the noniterative integral equations method and the Kohn and the Schwinger variational methods. These methods and the associated computer program packages have been reviewed in several publications [1–4]

During the 1980's and 1990's there have been many calculations of low energy electron ion colision strengths. An accurate knowledge of the corresponding effective collision strengths, obtained by averaging the collision strength over a Maxwell distribution of electron velocities, is required in analyses of observations of atomic processes in many laboratory and astrophysical plasmas. As an example of these calculations, we show in Figure 1, the collision strengths for three forbidden transitions from the ground state in e^-–Fe^+ scattering, calculated using the R-matrix method [5], compared with distorted wave calculations [6]. In the R-matrix calculation, 38 states in LS coupling were included in expansion (8), corresponding to all $3d^7$, $3d^6 4s$ and $3d^6 4p$ quartet and sextet spin states, yielding collision strengths for 703 transitions. The results shown in Figure 1, which correspond to the forbidden transitions $a^6D \to a^4F$, $a^6D \to a^4D$ and $a^6D \to a^4P$, illustrate the importance of resonances in low energy electron scattering.

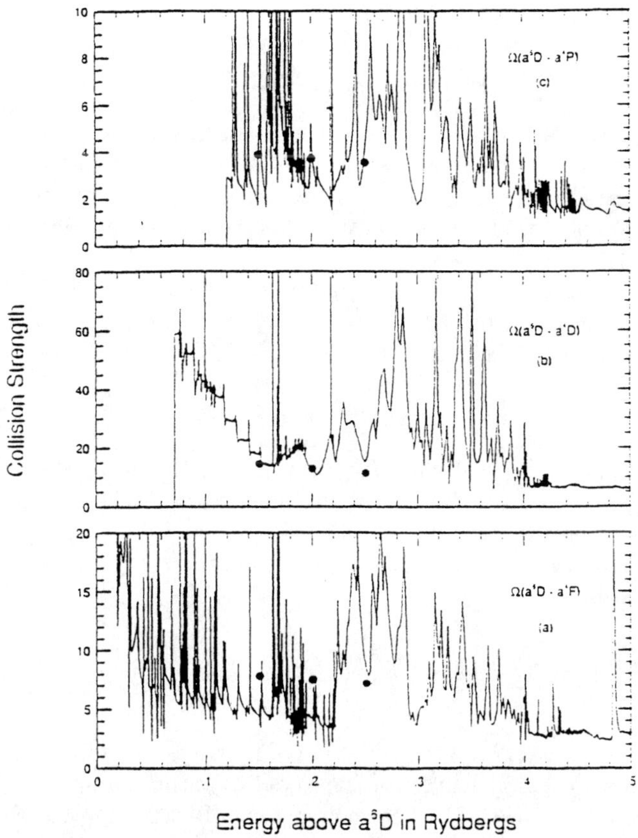

FIGURE 1. e^-–Fe^+ collision strengths for excitation of the a^6D ground state to the metastable states: (a) a^4F, (b) a^4D and (c) a^4P. Full curve, 38-state R-matrix calculation [5]; full circles, distorted wave calculation [6].

Relativistic effects for heavy atoms and ions

As the nuclear charge Z of the target increases, relativistic effects become important even for low energy electron scattering. There are two ways in which relativistic effects play a role. First, there is a direct effect corresponding to the relativistic distortion of the wave function describing the scattered electron by the strong Coulomb potential near the nucleus. Second, there is an indirect effect caused by the change in the charge distribution of the target due to the use of relativistic target wave functions.

For atoms and ions with small Z, the K-matrices can first be calculated in LS coupling, neglecting relativistic effects. The K-matrices are then recoupled to yield transitions between the fine-structure levels of the target. This is best achieved by introducing a pair-coupling scheme

$$\mathbf{L_i} + \mathbf{S_i} = \mathbf{J_i}, \quad \mathbf{J_i} + \boldsymbol{\ell}_i = \mathbf{K_i}, \quad \mathbf{K_i} + \mathbf{s_i} = \mathbf{J}, \tag{16}$$

where $\mathbf{L_i}$ and $\mathbf{S_i}$ are the orbital and spin angular momenta of the target, $\mathbf{J_i}$ is the total angular momentum of the target, $\boldsymbol{\ell}_i$ and $\mathbf{s_i}$ are the orbital and spin angular momenta of the scattered electron and $\mathbf{J}$ is the total angular momentum which with the parity π is conserved. This transformation has been implemented in a computer program [7]

For atoms and ions with higher Z values relativistic effects must be included by retaining relativistic terms in both the target and scattering Hamiltonians. One widely used approach for achieving this is to use the Breit-Pauli Hamiltonian [8]

$$H_{N+1}^{\rm BP} = H_{N+1}^{\rm nr} + H_{N+1}^{\rm rel}, \tag{17}$$

where $H_{N+1}^{\rm nr}$ is defined by Eq. (3) and $H_{N+1}^{\rm rel}$ consists of the one-body relativistic terms. Relativistic effects using this approach have been incorporated in an R-matrix program for a general atom [9].

For high Z atoms and ions the most satisfactory approach is to adopt the Dirac Hamiltonian defined by

$$H_{N+1}^{\rm D} = \sum_{i=1}^{N+1} \left(c\boldsymbol{\alpha}.\mathbf{p} + \beta' c^2 - \frac{Z}{r_i} \right) + \sum_{i>j=1}^{N+1} \frac{1}{r_{ij}}, \tag{18}$$

where $\beta' = \beta - 1$ and $\boldsymbol{\alpha}$ and β are the usual Dirac matrices. The expansion of the total wave function for a particular set of conserved quantum numbers J, M_J and π takes the general form of Eq. (8) where now j-j coupling is adopted [10].

As an example of electron scattering by heavy atoms, we show in Figure 2 R-matrix calculations [11] compared with measurements of the Stokes' parameter η_1 for light emitted at 90° to the incident beam direction in excitation of the 6s6p $^3\mathrm{P}_1^o$ state of Hg by polarized electrons [12]. The calculations, which were carried out using the Breit-Pauli Hamiltonian, included the five target states $6\mathrm{s}^2\ ^1\mathrm{S}_0$, 6s6p $^3\mathrm{P}_{0,1,2}^o$ and 6s6p $^1\mathrm{P}_1^o$ in the expansion of the total wave function. The scattering process at these low energies is dominated by resonances which have been assigned to be a $6\mathrm{s}6\mathrm{p}^2\ ^4\mathrm{P}_{5/2}^e$ resonance at 4.9 eV, a $6\mathrm{s}6\mathrm{p}^2\ ^4\mathrm{P}_{3/2}^e$ resonance at 5.0 eV, a $6\mathrm{s}6\mathrm{p}^2\ ^2\mathrm{D}_{3/2}^e$ resonance at 5.5 eV and a $6\mathrm{s}6\mathrm{p}^2\ ^2\mathrm{D}_{5/2}^e$ resonance at 5.5 eV [11].

Resonances and quantum defect theory

We have seen in the previous section that resonances play an important role in low energy electron scattering by atoms and ions. General resonance theories have been developed by Feshbach [13,14] and Fano [15] and their role in electron scattering by atoms and molecules has been reviewed by Burke [16] and by Schulz [17].

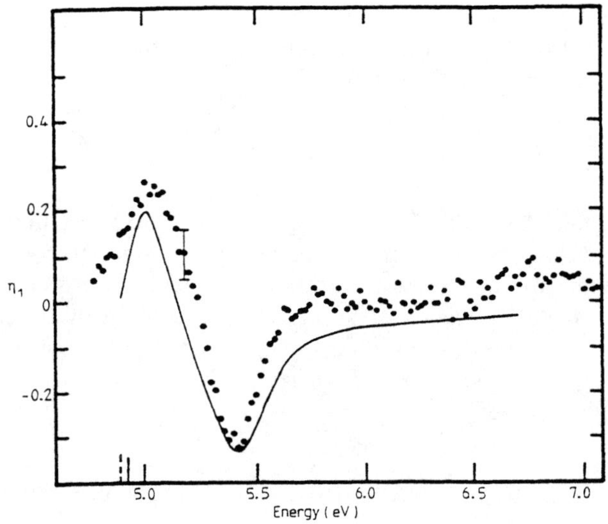

FIGURE 2. Stokes' parameter η_1 for the 6s6p $^3P_1^o$–6s^2 $^1S_0$ transition in Hg excited by polarized electrons. Full curve, Breit-Pauli R-matrix calculation [11]; full circles, experimental measurements [12].

It is useful to distinguish two types of resonances, namely open-channel or shape resonances and closed channel or Feshbach resonances. Shape resonances occur when the effective potential between the electron and the target has a characteristic shape with an inner attractive well and an outer repulsive barrier, the latter usually caused by the centrifugal repulsion. Feshbach resonances occur for scattered electron energies just below excitation thresholds where the scattered electron is captured temporarily into a bound state with the target in an excited state.

Multichannel quantum defect theory (MQDT) developed by Seaton [18,19] enables whole series of resonances to be predicted in electron-ion scattering rather than individual resonances as in the above mentioned theories. MQDT uses general analytic properties of the Coulomb wave functions to relate the S-matrices and the K-matrices above and below the threshold to which the resonance series converge. In particular, the K-matrix defined by Eq. (11) can be written below this threshold as

$$\mathbf{K}^{\Gamma}_{\mathrm{op}} = \mathbf{K}^{\Gamma}_{\mathrm{oo}} - \mathbf{K}^{\Gamma}_{\mathrm{oc}} \left(\mathbf{K}^{\Gamma}_{\mathrm{cc}} + \tan \pi \boldsymbol{\nu}_c\right)^{-1} \mathbf{K}^{\Gamma}_{\mathrm{co}}, \tag{19}$$

where the elements of the sub-matrices $\mathbf{K}^{\Gamma}_{\mathrm{oo}}$, $\mathbf{K}^{\Gamma}_{\mathrm{oc}}$, $\mathbf{K}^{\Gamma}_{\mathrm{co}}$ and $\mathbf{K}^{\Gamma}_{\mathrm{cc}}$ are defined by partitioning the K-matrix into channels which are open and channels which are closed below this threshold as follows:

$$\mathbf{K}^{\Gamma} = \begin{bmatrix} \mathbf{K}^{\Gamma}_{\mathrm{oo}} & \mathbf{K}^{\Gamma}_{\mathrm{oc}} \\ \mathbf{K}^{\Gamma}_{\mathrm{co}} & \mathbf{K}^{\Gamma}_{\mathrm{cc}} \end{bmatrix}. \tag{20}$$

The K-matrix elements on the right-hand-side of Eq. (19) are determined by analytical continuation from the K-matrix elements calculated above this threshold. Also in Eq. (19), ν_i is defined in the closed channels by

$$k_i^2 = -\frac{z^2}{\nu_i^2}. \tag{21}$$

The factor $\tan \pi \boldsymbol{\nu}_c$ causes the last term in Eq. (19) to contain an infinity of poles converging to the threshold of interest from below, giving rise to the observed resonance series which we have seen, for example in Figure 1. Equation (19) thus enables cross sections and resonance structure to be predicted in an energy region where some channels are closed, by calculating the K-matrix at a few energies where these channels are open and then extrapolating the K-matrix to the energy region where the channels are closed.

Scattering at Intermediate and High Energies

We now consider electron atom and electron ion scattering at electron impact energies greater than the ionization threshold of the target. In this case, an infinite number of channels are open, so they cannot all be included explicitly in the expansion of the total wave function. Several approaches have been developed to treat scattering at these energies. These include extensions of low energy methods based on retaining pseudostates in expansion (8), time-dependent methods, optical potential methods, distorted wave methods and extensions of the first Born approximation by including higher-order terms in the Born series. We discuss these approaches in turn below.

Pseudostate methods

In this approach, the target eigenstates retained in expansion (8) are augmented by the inclusion of a set of square-integrable pseudostates $\overline{\Phi}_i$ which are orthogonal to the retained eigenstates. These pseudostates are usually defined by diagonalizing the target Hamiltonian H_N in a square-integrable basis, yielding an equation analogous to Eq. (4)

$$\langle \overline{\Phi}_i | H_N | \overline{\Phi}_j \rangle = \overline{e}_i \delta_{ij}, \tag{22}$$

where the energies $\overline{e}_i$ lie just below or above the ionization threshold. The pseudostates $\overline{\Phi}_i$ then represent in an average way the higher lying Rydberg states and the continuum states of the target. Hence, accurate excitation and ionization cross sections can be calculated at "intermediate energies" defined to lie between the ionization threshold and several times this threshold.

A proposal to include pseudostates in the e^-–H scattering expansion was first made in a study of low energy e^-–H elastic scattering [20], and cross sections at intermediate energies were first calculated for e^-–H 1s → 2s and 1s → 2p excitation using this approach [21]. Recently, the convergent close-coupling (CCC) method for e^-–H scattering has been developed [22] in which both the target eigenstates and pseudostates are expanded in a Laguerre basis, giving accurate results for excitation and ionization cross sections at intermediate energies. This approach has been further extended to e^-–He scattering at intermediate energies [23] and has been used to obtain accurate electron alkali atom scattering cross sections at intermediate energies [24]. We also mention that the R-matrix method has been extended to include pseudostates [25,26]. The latter R-matrix with pseudostates (RMPS) approach [26], has enabled accurate intermediate energy electron scattering cross sections to be calculated for a number of light atoms and ions including H, He, Be, Be^+, B, B^{2+}, C^{3+} and O^{5+}.

The role of the continuum states in the scattering process can be determined directly by studying the ionization cross section. In this case the sum of the excitation cross sections to pseudostates whose energies $\bar{e}_i$ in Eq. (22) lie above the ionization threshold, provides an estimate of the ionization cross section. A more accurate determination of the scattering amplitude can be obtained by determining the overlap of the pseudostates with the discrete and continuum spectrum of the target atom. We show in Figure 3, the e^-–H ionization cross section and spin asymmetry calculated using the CCC approach [27] compared with total cross section measurements [28] and spin asymmetry measurements [29,30]. This work showed that including pseudostates in expansion (8) was capable of giving accurate ionization cross sections as well as excitation cross sections at intermediate energies.

In addition to the direct ionization process shown in the case of e^-–H scattering in Figure 3, indirect ionization processes play an increasingly important role for complex atoms and ions and can even dominate the ionization cross section for heavy ions. This was first demonstrated in the ionization cross section of Ba^+ where a three fold increase was observed [31] due to excitation of inner-shell electrons which led to excited resonance states which subsequently autoionized yielding Ba^{2+}.

Time-dependent methods

In these methods [32] the time-dependent Schrödinger equation is solved directly to yield electron impact excitation and ionization cross sections. In the case of electron scattering by atomic hydrogen or by an atom or ion with one electron outside of a closed inert shell the total wave function for each conserved $LS\pi$ symmetry is expanded as follows

$$\Psi^{LS\pi}(\mathbf{r}_1, \mathbf{r}_2, t) = \sum_{\ell_1\ell_2} (r_1 r_2)^{-1} P_{\ell_1\ell_2}^{LS}(r_1, r_2, t) \mathcal{Y}_{\ell_1\ell_2}^{LS}(\hat{\mathbf{r}}_1, \hat{\mathbf{r}}_2) \tag{23}$$

where

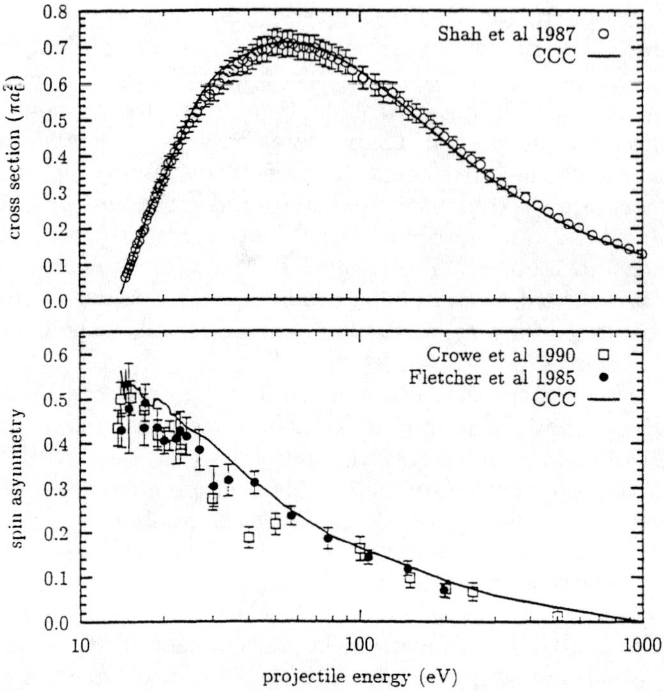

FIGURE 3. Total ionization cross section and spin asymmetry for e^-–H scattering. Full curves, CCC calculations [27]; open circles, total ionization cross section measurements [28]; open squares [29] and full circles [30], spin asymmetry measurements.

$$\mathcal{Y}^{LS}_{\ell_1\ell_2}(\hat{\mathbf{r}}_1, \hat{\mathbf{r}}_2) = \sum_{m_{\ell_1} m_{\ell_2}} (\ell_1 m_{\ell_1} \ell_2 m_{\ell_2} | L M_L) Y_{\ell_1 m_{\ell_1}}(\theta_1, \phi_1) Y_{\ell_2 m_{\ell_2}}(\theta_2, \phi_2). \tag{24}$$

Substituting expansion (23) into the time-dependent Schrödinger equation and projecting onto the coupled angular functions $\mathcal{Y}^{LS}_{\ell_1\ell_2}$ then yields a set of coupled differential equations for $P^{LS}_{\ell_1\ell_2}(r_1, r_2, t)$.

As an example of the application of this time-dependent approach we show in Figure 4 the total electron impact ionization cross section for Mg^+ calculated using the time-dependent, RMPS and CCC methods compared with experimental measurements [33]. The good agreement of the three calculations using completely independent methods and computer programs is very encouraging and shows that ionization cross sections for atoms and ions at intermediate energies can now be reliably calculated at least for simple targets.

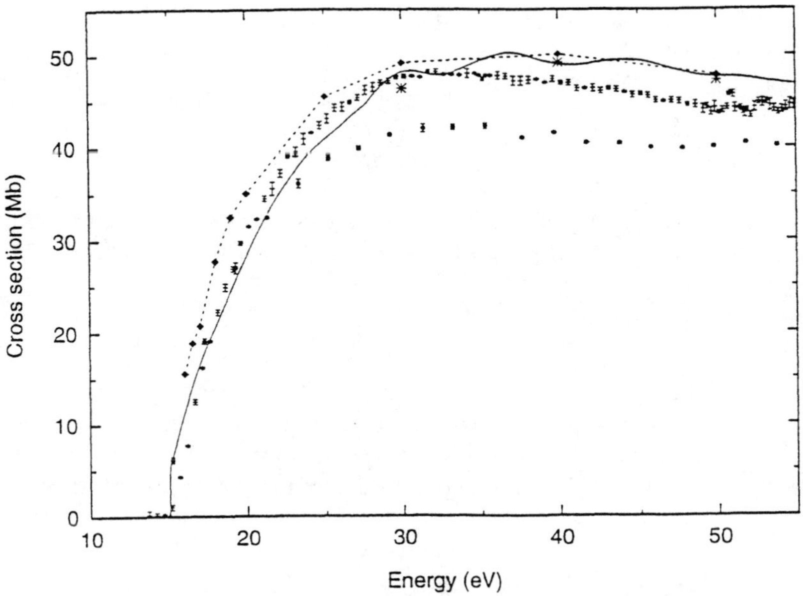

FIGURE 4. Total ionization cross section for e$^-$–Mg$^+$ scattering. Full curve, RMPS calculation [33]; dashed curve, CCC calculation [33]; stars, time-dependent calculation [33]; crosses with error bars, experimental measurements [34]; full circles with error bars, experimental measurements [35].

Optical potential methods

These methods are also capable of yielding accurate results at intermediate energies for electron atom scattering. The optical potential can be expressed in terms of Feshbach projection operators P and Q [13,14] where P projects onto those low energy channels that can be treated explicitly, for example by solving the coupled equations (10), and Q projects onto the remaining infinity of coupled channels including the continuum. At sufficiently high energies the optical potential can be expanded in a perturbation series, where the second-order term is given by

$$V_{\rm opt}^{(2)} = -PVQ\frac{1}{T_e + H_N - E - i\epsilon}QVP, \qquad (25)$$

In this equation V is the electron atom (ion) interaction potential, T_e is the kinetic energy of the scattered electron and H_N is as before the target Hamiltonian. The lowest-order terms of perturbation theory can be converted into an *ab-initio* local complex potential that has been used to describe the elastic scattering of electrons and positrons from a number of atoms [36].

The optical potential calculated in second-order has been used to describe e$^-$–H scattering at intermediate energies [37]. Also an optical potential approximation, called the coupled-channel optical (CCO) method, that goes beyond second-order and makes allowance for electron exchange has been introduced [38]. This method has been applied with considerable success to treat electron scattering by atomic hydrogen and by sodium atoms [39].

Distorted wave methods

The distorted wave series for the T-matrix can be derived from the Lippman-Schwinger integral equation [40]. The first-order distorted wave amplitude can be written as [41]

$$T_{ji}^{(1)} = 2\langle F_j^-(\mathbf{x}_{N+1})\Phi_j(\mathbf{X}_N)\,|(V - U_j)\mathcal{A}|\,\Phi_i(\mathbf{X}_N)F_i^+(\mathbf{x}_{N+1})\rangle \tag{26}$$

where Φ_i and Φ_j are the initial and final states of the target atom and V is the interaction potential between the scattered electron and the atom. The distorting potential U_j is the final state spherically symmetric approximation for V, and the final state distorted wave F_j^- describing the scattered electron is a solution of the equation

$$(T_e + U_j - \frac{1}{2}k_j^2)F_j^- = 0 \tag{27}$$

corresponding to plane wave plus ingoing wave boundary conditions. The initial state distorted wave F_i^+ is obtained similarly using the initial state distorting potential U_i and satisfies plane wave plus outgoing wave boundary conditions. Finally in Eq. (26) $\mathcal{A}$ is the usual antisymmetrization operator.

The first-order distorted wave approximation gives accurate results efficiently at high energies. However at intermediate energies, higher-order terms in the perturbation expansion must be evaluated if quantitative agreement with experiment is to be obtained at all scattering angles and for all transitions.

Finally we mention a procedure which goes beyond the first-order distorted wave approximation [42,43]. This describes the ionization of a complex atom or ion by a fast electron, where the interaction of an ejected slow electron with the residual ion is treated using a close coupling expansion analogous to Eq. (8). This gives an accurate representation of resonances in the final state.

Born series methods

In the high energy domain, which typically extends from several times the ionization threshold upwards, methods based on the Born series converge rapidly and can be used to obtain accurate electron scattering results. Ignoring electron exchange for the moment, the Born series for the direct scattering amplitude can be written as [44]

$$f = \sum_{n=1}^{\infty} f_{\mathrm{B}n} \tag{28}$$

where the nth term $f_{\mathrm{B}n}$ contains the interaction V between the scattered electron and the target atom or ion n times and the Green's function

$$G_0^+ = (H_N + T_e - E - i\epsilon)^{-1} \tag{29}$$

$(n-1)$ times where H_N and T_e have been defined following Eq. (25). It is important to retain consistently all terms in the Born series with similar energy and momentum transfer $\Delta = |\mathbf{k}_i - \mathbf{k}_f|$ dependencies. For elastic scattering, the scattering amplitude converges to the first Born approximation for all Δ, but at lower energies it is necessary to include the second Born terms as well as $\Re e f_{\mathrm{B3}}$ to obtain the cross section correct to k^{-2}. In the forward direction, convergence to the first Born approximation is slow because the contribution from $\Im m f_{\mathrm{B2}}$ which, from the optical theorem, corresponds to loss of flux into all open channels. For inelastic scattering, the first Born approximation does not yield the correct high energy limit for large Δ. Instead this comes from $\Im m f_{\mathrm{B2}}$.

Since the third Born amplitude is difficult to calculate, the scattering amplitude can be calculated to third-order using the eikonal-Born-series (EBS) approximation [45]:

$$f_{\mathrm{EBS}} = f_{\mathrm{B1}} + f_{\mathrm{B2}} + f_{\mathrm{G3}} + g_{\mathrm{och}} \tag{30}$$

where f_{G3} is the third-order term in the expansion of the Glauber amplitude [46] in powers of V, which can be more easily calculated than f_{B3}, and g_{och} is the Ochker approximation [47] to the electron exchange amplitude. The EBS approximation has been used successfully when perturbation theory converges rapidly, namely at higher energies, at small and intermediate scattering angles and for light atoms.

Threshold Behaviour of the Ionization Cross Section

We conclude this section on electron scattering by atoms and ions by considering the behaviour of the ionization cross section close to threshold. It was pointed out in a fundamental paper by Wigner that the derivation of threshold laws does not require a detailed knowledge of the collision dynamics in the "reaction zone" where the scattering particles are close together. Instead they are determined by the long-range interactions between the particles which determine the asymptotic form of the wave function [48]. This result can then be used to derive threshold laws describing the interaction of two particles resulting in the excitation of one or both of these particles.

Wannier extended Wigner's theory to the single ionization of atoms or ions by electrons, where the final state consists of three charged particles interacting through long-range Coulomb potentials [49]. In analogy with Wigner's analysis, Wannier pointed out that it is not necessary to know the detailed behaviour of the electrons in the reaction zone, but only to assume that the distribution in phase space of the two outgoing electrons is uniform (i.e. quasiergodic) as they enter the zone where they interact only through the long-range Coulomb potentials. Introducing the hyperspherical coordinates

$$R = (r_1^2 + r_2^2)^{\frac{1}{2}}, \quad \tan\alpha = \frac{r_2}{r_1} \tag{31}$$

where r_1 and r_2 are the radial coordinates of the two outgoing electrons, Wannier noted that for large enough R, the Coulomb potential varies sufficiently slowly for classical mechanics to be valid even when the total energy of the electrons $E = E_1 + E_2$ tends to zero. He then showed that the ionization cross section for hydrogen-like ions is dominated by the region in the neighbourhood of the saddle point in the potential energy surface at

$$\alpha = \frac{\pi}{4}, \quad \theta_{12} = \pi \tag{32}$$

where θ_{12} is the angle between the two outgoing electron vectors $\hat{\mathbf{r}}_1$ and $\hat{\mathbf{r}}_2$. By expanding the potential about this saddle point he showed that for total angular momentum $L = 0$ the threshold law for the total ionization cross section is given by

$$\sigma_{\text{ion}} \sim E^m \tag{33}$$

where

$$m = -\frac{1}{4} + \frac{1}{4}\left(\frac{100z - 9}{4z - 1}\right)^{\frac{1}{2}} \tag{34}$$

where z is the charge on the residual ion. When $z = 1$ then $m \simeq 1.127$ and as $z \to \infty$ then $m \to 1$. Also, the width of the angular distribution in θ_{12} is

$$\Delta\theta_{12} \sim E^{\frac{1}{4}} \tag{35}$$

with a maximum at $\theta_{12} = \pi$ so that the angular distribution becomes more concentrated as $E \to 0$.

Experiments on the threshold ionization of helium confirmed both the Wannier threshold law Eq. (33) and the behaviour of the angular correlation function given by Eq. (35) [50]. Subsequently many (e, 2e) studies of the angular correlation of the outgoing electrons near threshold have been made, allowing both the electron energies and their directions to be measured [51].

In spite of the success of Wannier's threshold theory of ionization it has proved difficult to develop a theory which yields the absolute magnitude of the threshold ionization cross section and predicts the the corrections to the theory as the energy increases from threshold. Such a theory requires an accurate treatment of the scattering process in the reaction zone, discussed above, which links up at some suitably chosen large value of R to the asymptotic form of the wave function corresponding to two outgoing electrons in the field of the residual ion. Recently a number of advances in this direction have been made. These include a semi-classical approach to threshold ionization which has yielded absolute total and differential cross sections for e^- – He scattering [52], an application of the hyperspherical close coupling method which has yielded results in good agreement with Wannier's theory [53] and the direct solution of the Schrödinger equation on a lattice which has yielded excellent agreement with experiment for the e^- – H triple differential ionization at 4 eV above threshold [54,55] . These approaches, together with ongoing developments of pseudostate methods discussed earlier should enable accurate ionization cross sections to be calculated close to threshold for simple atomic targets.

ELECTRON SCATTERING BY MOLECULES

Basic Processes

The processes that occur in electron scattering by molecules are considerably more varied than those that arise in electron scattering by atoms and atomic ions because of the possibility of exciting degrees of freedom associated with the motion of the nuclei. Thus as well as electronic excitation and ionization, which also occur in electron atom scattering, additional processes that can now take place include rotational and vibrational excitation

$$e^- + AB_{vj} \to AB_{v'j'} + e^-, \tag{36}$$

dissociation and dissociative ionization

$$\begin{aligned} e^- + AB_{vj} &\to A + B + e^- \\ e^- + AB_{vj} &\to A + B^+ + e^- + e^- \end{aligned} \tag{37}$$

and dissociative attachment and recombination

$$\begin{aligned} e^- + AB_{vj} &\to A + B^- \\ e^- + AB^+_{vj} &\to A + B, \end{aligned} \tag{38}$$

where in these equations AB_{vj} means that the molecule is in the (vj) vibrational and rotational state, and where for simplicity we have restricted our attention to diatomic molecules, although the same processes will occur for polyatomic molecules.

A further complexity which arises in electron molecule scattering is the multicentre and nonspherical nature of the scattering problem, resulting in the need to solve very much larger sets of coupled equations. The multicentre integrals that arise in the evaluation of the Hamiltonian matrix elements are also more difficult to calculate than the integrals that occur in electron atom scattering.

Laboratory Frame Representaton

We consider first the derivation of coupled integrodifferential equations describing low energy electron molecule scattering in the laboratory frame of reference. Our approach closely parallels that given earlier for low energy electron atom scattering. The time-independent Schrödinger equation describing the electron molecule system is

$$(H_m + T_e + V)\Psi = E\Psi \tag{39}$$

where H_m is the target molecule Hamiltonian, which includes both rotational and vibrational kinetic energy terms, T_e is the kinetic energy operator of the scattered electron and V is the electron molecule interaction potential.

$$V(\mathbf{R}, \mathbf{r_m}, \mathbf{r}) = \sum_i \frac{1}{|\mathbf{r} - \mathbf{r}_i|} - \sum_i \frac{Z_i}{|\mathbf{r} - \mathbf{R}_i|}. \tag{40}$$

In Eq. (40), $\mathbf{R}$ represents the coordinates $\mathbf{R_i}$ of all the nuclei with charges Z_i, $\mathbf{r}_m$ represents the coordinates $\mathbf{r}_i$ of the target electrons and $\mathbf{r}$ represents the coordinates of the scattered electron. The total energy E in Eq. (39) refers to the frame of reference where the centre of mass of the whole system is at rest.

As in the case of electron atom scattering we introduce target eigenstates and possibly pseudostates Φ_i by the equation

$$\langle \Phi_i | H_m | \Phi_j \rangle = e_i \delta_{ij} \tag{41}$$

where these states depend on both rotational and vibrational coordinates. We then expand the total wave function Ψ, in analogy with Eq. (8), in the form

$$\Psi_i = \mathcal{A} \sum_j \Phi_j(\mathbf{R}, \mathbf{r_m}) F_{ji}(\mathbf{r}) + \sum_j \chi_j(\mathbf{R}, \mathbf{r_m}, \mathbf{r}) b_{ji} \tag{42}$$

where the spin variables have been suppressed and we have not carried out a partial-wave decomposition of the wave function $F_{ji}(\mathbf{r})$ representing the scattered electron, for notational simplicity. The subscripts i and j now represent the ro-vibrational as well as the electronic state of the target molecule.

Coupled integrodifferential equations for the functions F_{ji} can be obtained by substituting expansion (42) into Eq. (39) and projecting onto the target states Φ_j and onto the square integrable functions χ_j. After eliminating the coefficients b_{ji} we find that the functions F_{ji} satisfy the coupled equations

$$(\nabla^2 + k_j^2) F_{ji}(\mathbf{r}) = 2 \sum_l (V_{jl} + W_{jl} + X_{jl}) F_{li}(\mathbf{r}), \tag{43}$$

where $k_j^2 = 2(E - e_i)$ and where V_{jl}, W_{jl} and X_{jl} are the local direct, non-local exchange and non-local correlation potentials. By expanding F_{ji} in partial waves we obtain a set of coupled integrodifferential equations analogous to Eqs. (10) for electron scattering by atoms and ions.

These equations have been solved by a number of workers to obtain accurate rotational excitation cross sections for simple diatomic molecules such as H_2, N_2 and CO. However, if vibrational and electronic excitation cross sections are required in addition to rotational excitation cross sections, major computational problems arise due to the very large number of channels which are then coupled. This difficulty can best be overcome by adopting the molecular frame representation discussed below.

Molecular Frame Representation

In this approach a Born-Oppenheimer separation of the electronic and nuclear motion is made. The electronic motion is first determined with the nuclei held fixed. This is referred to as the fixed-nuclei approximation. The molecular rotational and vibrational motion is then included in a second step of the calculation. This procedure owes its validity to the large ratio of the nuclear mass to the electronic mass and can be adopted when the collision time is much shorter than the periods of molecular rotation and vibration. Thus it is expected to be valid when the scattered electron energy is not close to threshold or does not coincide with that of a narrow resonance. In such situations further developments described below are needed to obtain reliable cross sections. The fixed-nuclei approximation is not new, having been used since the 1930s to describe the scattering of slow electrons by diatomic molecules [56,57]. However in recent years it has been widely used as the basis of *ab-initio* computational methods that are yielding the most accurate cross sections for electron scattering by diatomic and polyatomic molecules.

In order to formulate the scattering process in the fixed-nuclei approximation, we adopt a frame of reference that is rigidly attached to the molecule, where the centre of gravity of the molecule is chosen as the origin of the coordinates and where, for a diatomic molecule, the z axis is chosen to lie along the internuclear axis. We then look for solutions of the time-independent Schrödinger equation

$$(H_{\mathrm{el}} + T_e + V)\psi = E\psi \tag{44}$$

where H_{el} is the electronic part of the target Hamiltonian obtained by assuming that the target nuclei have fixed coordinates, denoted collectively by $\mathbf{R}$. It follows that H_{el} is related to H_m in Eq. (39) by

$$H_m = H_{\mathrm{el}} + T_R \tag{45}$$

where T_R is the kinetic energy operator for the rotational and vibrational motion of the nuclei.

The solution of Eq. (44) proceeds in an analogous way to the solution of Eq. (2) for low energy electron scattering by atoms and ions. We adopt an expansion which is formally equivalent to Eq. (8), where the bound orbitals and continuum orbitals are expanded in terms of symmetry adapted angular functions that transform as the appropriate irreducible representation (IRR) of the molecular point group [58]. Substituting this expansion into Eq. (44) and projecting onto the corresponding channel functions and onto the square integrable functions, yields a set of coupled integrodifferential equations satisfied by the radial functions describing the motion of the scattered electrons. These equations have the same form as Eqs. (10), where the channel indices i, j and ℓ now represent the component of the IRR as well as the electronic state of the target. Also the conserved quantum numbers represented by Γ in Eqs. (10) now include the IRR and for light molecules the total spin.

The final step is to solve the coupled integrodifferential equations for each set of nuclear coordinates $\mathbf{R}$ of importance in the scattering process, subject to K-matrix boundary conditions defined by Eq. (11). The S-matrix and hence the cross sections for each fixed $\mathbf{R}$ can then be obtained.

In the last few years three major *ab initio* computational procedures have been developed to solve the fixed-nuclei equations for electron scattering by diatomic and polyatomic molecules allowing electronic excitation processes to be calculated. These are the complex Kohn variational method [59], the Schwinger variational method [60] and the R-matrix method [61,62]. All of these approaches use standard multi-centre quantum chemistry structure packages that have been appropriately modified for scattering. In addition to these *ab initio* methods, in the case of low energy electron scattering by polyatomic molecules involving only the ground electronic state, a number of methods and computer programs have been developed which replace the non-local exchange and correlation potentials by equivalent local potentials and which adopt single-centre expansions for the bound and continuum orbitals. This makes the solution of the coupled equations tractable for larger molecular systems like benzene and SF_6 [63].

As an example of recent calculations of electron scattering by polyatomic molecules, we show in Figure 5 R-matrix and measured total cross sections for electron scattering by CO_2 [64,65]. Two R-matrix calculations using the fixed-nuclei approximation were undertaken. The first included the electronic ground state in a "static exchange plus polarization" (SEP) model. The second used an expansion that included the electronic ground state and the seven linear electronically excited states $K1\pi_g^3 5\sigma_g\ ^{1,3}\Pi_g$ and $K1\pi_g^3 2\pi_u\ ^3\Sigma_u^+$, $^{1,3}\Delta_u$ and $^{1,3}\Sigma_u^-$ in the expansion of the total wave function, where K denotes the inner-shell electrons. All of these linear excited states are located in the range 7.5 – 8.3 eV above the CO_2 ground state are possible parents for a Feshbach resonance. Both theory and experiment show a threshold peak due to a $^2\Sigma_g^+$ virtual state and a broad $^2\Pi_u$ shape resonance in the energy range 4–6 eV. The inclusion of the electronically excited states in the R-matrix expansion brings the peak position of the $^2\Pi_u$ resonance into better agreement with experiment, but it still lies about 1 eV too high probably due both to incomplete allowance for the polarization of the target by the scattered electron and to the omission of vibrational motion in the theoretical model.

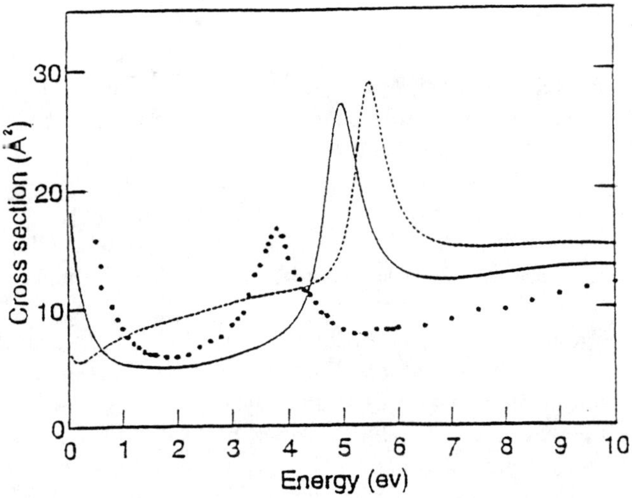

FIGURE 5. Total cross section for $e^- - CO_2$ scattering. Full curve, multistate R-matrix calculation [64]; dashed line, SEP calculation [64]; dots, total cross section measurements [65].

Nuclear motion effects

In this section we discuss briefly how observables involving the nuclear motion such as rotational and vibrational excitation and dissociative attachment can be obtained from the solutions of the fixed-nuclei equations.

The most widely used approach for including the nuclear motion is the adiabatic nuclei approximation first used in studies of nuclear rotational excitation by neutron scattering [66]. In the case of diatomic molecules in a $^1\Sigma$ state the scattering amplitude for a transition between electronic, vibrational and rotational states defined by $ivjm_j$ and $i'v'j'm'_j$ is given by

$$f_{i'v'j'm'_j,ivjm_j}(\mathbf{k}.\hat{\mathbf{r}}) = \langle \chi_{i'v'}(R) Y_{j'm'_j}(\hat{\mathbf{R}}) | f_{i'i}(\hat{\mathbf{k}.\mathbf{r}};\mathbf{R}) | \chi_{iv}(R) Y_{jm_j}(\hat{\mathbf{R}}) \rangle \tag{46}$$

where $f_{i'i}(\hat{\mathbf{k}.\mathbf{r}};\mathbf{R})$ is the fixed-nuclei scattering amplitude, which depends parametrically on the nuclear coordinates $\mathbf{R}$, and χ_{iv} and Y_{jm_j} are the molecular vibrational and rotational eigenfunctions respectively. This approximation is valid provided that the collision time is short compared with the vibration and/or the rotation time. Hence it can be accurately applied in non resonant regions or in the neighbourhood of broad resonances.

The adiabatic-nuclei approximation breaks down in the neighbourhood of narrow resonances or close to thresholds [67]. A straightforward way of including nonadiabatic effects that arise in vibrational excitation is to retain the vibrational terms in the Hamiltonian but to still treat the rotational motion adiabatically. Hence instead of solving Eq. (44), we solve the following equation

$$(H_{\rm el} + T_{\rm vib} + T_e + V)\tilde{\psi} = E\tilde{\psi} \tag{47}$$

where $T_{\rm vib}$ is the kinetic energy operator for the vibrational motion and where the other quantities have the same meaning as in Eq. (44). Adopting a frame of reference in which the molecule has fixed spacial orientation and separating out the the angular variables of the scattered electron we can then derive integrodifferential equations coupling the target vibrational states as well as the electronic states. This approach has been adopted with success by several workers [68]. However it is computationally demanding since the number of coupled channels can become very large.

We have already remarked that vibrational excitation and dissociative attachment are particularly important in resonance regions. As a result, a number of approaches have been developed for describing these processes based on resonance theories [69–71]. The basic idea is that a series of fixed-nuclei resonance states $\psi_n^{(r)}$ are introduced for a range of values of the internuclear coordinate R, either by imposing Siegert outgoing wave boundary conditions [72] or by introducing Feshbach projection operators [13,14]. The amplitude for a transition from an initial electronic-vibrational state iv to a final state $i'v'$ is then given by

$$T_{i'v',iv} = \sum_n \langle \chi_{i'v'}(R')\zeta_{ni'}(R')|G_n^{(r)}(R',R)|\zeta_{ni}(R)\chi_{iv}(R)\rangle \tag{48}$$

where $\chi_{iv}(R)$ are the vibrational eigenfunctions, ζ_{ni} are the "entry amplitudes" from the initial or final electronic states into the resonance states $\psi_n^{(r)}$ and $G_n^{(r)}$ are the Green's functions that describe the propagation in the intemediate resonance states $\psi_n^{(r)}$. Dissociative attachment can be described by a straightforward extension of this theory. The fixed-nuclei R-matrix method has also been extended to treat vibrational excitation and dissociative attachment [73].

A number of other approaches have been introduced to obtain accurate vibrational excitation cross sections in low energy electron molecule scattering from fixed-nuclei calculations. These include the energy-modified adiabatic phase matrix method, which retains the vibrational kinetic energy in the fixed-nuclei scattering matrix and approximates the continuum energy, and the first-order non-degenerate adiabatic approximation, which evaluates the fixed-nuclei scattering matrix off the energy momentum shell in order to ensure strict conservation of energy. These methods have recently been reviewed [74].

ACKNOWLEDGEMENT

This paper was written while the author was a visitor at ITAMP and was partially supported by the National Science Foundation through a grant for the Institute for Theoretical Atomic and Molecular Physics at Harvard University and Smithsonian Astrophysical Observatory.

REFERENCES

1. Burke, P.G., and Eissner, W.B., in *Atoms in Astrophysics*, edited by P.G. Burke *et al*, Plenum Press, New York and London, 1983, pp. 1–54.
2. Henry, R.J.W., *Rep. Prog. Phys.*, **56**, 327–362(1993).
3. Burke, P.G., and Berrington, K.A., *Atomic and Molecular Processes - An R-matrix Approach*, IOP Publishing, Bristol and Philadelphia, 1993.
4. Bartschat, K., editor *Computational Atomic Physics. Electron and Positron Collisions with Atoms and Ions*, Springer,Berlin, Heidelberg and New York, 1996.
5. Pradhan, A.K., and Berrington, K.A., *J. Phys. B:At. Mol. Opt. Phys.*, **26**, 157–172(1993).
6. Nussbaumer, H., and Storey, P.J., *Astron. Astrophys.*, **89**, 308–313(1980)
7. Saraph, H.E., *Comput. Phys. Commun.*, **3**, 256–268 (1972).
8. Bethe, H.A., and Salpeter, E.E., *Quantum Mechanics of One- and Two-Electron Atoms*, Springer-Verlag, Berlin, Göttingen and Heidelberg,1957, pp.170–183.
9. Berrington, K.A., Eissner, W.B., and Norrington, P.H., *Comput. Phys. Commun.*, **92**, 290–420 (1995).
10. Norrington, P.H., and Grant, I.P., *Comput. Phys. Commun.*, *to be published*(2000).
11. Bartschat, K., Scott, N.S., Blum, K., and Burke, P.G., *J. Phys. B:At. Mol. Phys.*, **17**, 269–277 (1984).
12. Wolcke, A., Bartschat, K., Blum, K., Borgmann, H., Hanna, G.F., and Kessler, J., *J. Phys. B:At. Mol. Phys.*, **16**, 639–655(1983).
13. Feshbach, H., *Ann. Phys. (N.Y.)*, **5**, 357–390(1958).
14. Feshbach, H., *Ann. Phys. (N.Y.)*, **19**, 287–313(1962).
15. Fano, U., *Phys. Rev.*, **124**, 1866–1878(1961).
16. Burke, P.G., *Adv. At. Mol. Phys.*, **4**, 173–219(1968).
17. Schulz, G.J., *Rev. Mod. Phys.*, **45**, 378–422, 423–486(1973).
18. Seaton, M.J., *Mon. Not. Roy. Astron. Soc.*, **118**, 504–518(1958).
19. Seaton, M.J., *Rep. Prog. Phys.*, **46**, 167–257(1983).
20. Burke, P.G., and Schey, H.M., *Phys. Rev.*, **126**, 147–162(1962).
21. Burke, P.G., and Webb, T.G., *J. Phys. B:At. Mol. Phys.*, **3**, L131–L134(1970).
22. Bray, I., and Stelbovics, A.T., *Phys. Rev. A*, **46**, 6995–7011(1992).
23. Fursa, D.V., and Bray, I., *Phys. Rev. A*, **52**, 1279–1298(1995).
24. Bray, I., *Phys. Rev. A*, **49**, R1–R4(1994).
25. Burke, P.G., Noble, C.J., and Scott, P., *Proc. Roy. Soc. A*, **410**, 287–310(1987).
26. Bartschat, K., Hudson, E.T., Scott, M.P., Burke, P.G., and Burke, V.M., *J. Phys. B:At. Mol. Opt. Phys.*, **29**, 115–123(1996).
27. Bray, I., and Stelbovics, A.T., *Phys. Rev. Letters*, **70**, 746–749(1993).
28. Shah, M.B., Elliott, D.S., and Gilbody, H.B., *J. Phys. B:At. Mol. Phys.*, **20**, 3501–3514(1987).
29. Crowe, D.M., Guo, X.Q., Lubell, M.S., Slevin, J., and Eminyan, M., *J. Phys. B:At. Mol. Opt. Phys.*, **23**, L325–L331(1990).
30. Fletcher, G.D., Alguard, M.J., Gay, T.J., Wainwright, P.F., Lubell, M. S., Raith, W., and Hughes, V.W., *Phys. Rev. A*, **31**, 2854–2883(1985).
31. Peart, B., and Dolder, K.T., *J. Phys. B:At. Mol. Phys.*, **1**, 872–878(1968).

32. Pindzola, M.S., and Schultz, D.R., *Phys. Rev. A*, **53**, 1525–1536(1996).
33. Badnell, N.R., Pindzola, M.S., Bray, I., and Griffin, D.C., *J. Phys. B:At. Mol. Opt. Phys.*, **31**, 911–924(1998).
34. Peart, B., Thomason, J.W.G., and Dolder, K., *J. Phys. B:At. Mol. Opt. Phys.* **24**, 4453–4461(1991).
35. Crandall, D.H., Phaneuf, R.A., Falk, R.A., Belić, D.S., and Dunn, G.H., *Phys. Rev. A*, **25**, 143–153(1982).
36. Byron, F.W. Jr., and Joachain, C.J., *J. Phys. B:At. Mol. Phys.*, **14**, 2429–2448(1981).
37. Bransden, B.H., Scott, T., Shingal, R., and Raychoudhury, J., *J. Phys. B:At. Mol. Phys.*, **24**, 4605–4616(1982).
38. McCarthy, I.E., and Stelbovics, A.T., *Phys. Rev. A*, **28**, 2693–2707(1983).
39. Bray, I., and McCarthy, I.E., *Phys. Rev. A*, **47**, 317–326(1993).
40. Goldberger, M.L., and Watson, K.M., *Collision Theory*, John Wiley, New York, London and Sydney, 1994, pp. 197–202.
41. Madison, D.H., in *Many-Body Atomic Physics*, edited by J.J. Boyle and M.S. Pindzola, Cambridge University Press, Cambridge, 1998, pp. 237–260.
42. Jakubowicz, H., and Moores, D.L., *J. Phys. B:At. Mol. Phys.*, **14**, 3733–3760(1981).
43. Bartschat, K., and Burke, P.G., *J. Phys. B:At. Mol. Phys.*, **20**, 3191–3200(1987).
44. Joachain, C.J., in *The Physics of Electronic and Atomic Collisions*, edited by A. Dalgarno *et al*, Am. Inst. Phys., New York, 1990 , pp. 68–81.
45. Byron, F.W., Jr. and Joachain, C.J., *J. Phys. B:At. Mol. Phys.*, **8**, L284–L288(1975).
46. Glauber, R.J., in *Lectures in Theoretical Physics*, edited by W.E. Brittin, Interscience, New York, 1959, pp. 315.
47. Ochker, V.I., *Zh. Eksp. Teor. Fiz.*, **45**, 734–741(1963).
48. Wigner, E.P., *Phys. Rev.*, **73**, 1002–1009(1948).
49. Wannier, G.H., *Phys. Rev.*, **90**, 817-825(1953).
50. Cvejanović, S., and Read, F.H., *J. Phys. B:At. Mol. Phys.* **7**, 1842–1852(1974).
51. Lehmam-Bennani, A., *J. Phys. B:At. Mol. Opt. Phys.*, **24**, 2401–2442(1991).
52. Crothers, D.S.F. *J. Phys. B:At. Mol. Phys.*, **19**, 463–483(1986).
53. Kato, D., and Watanabe, S., *Phys. Rev. A*, **56**, 3687–3700 (1997).
54. McCurdy, C.W., Rescigno, T.N., and Byrum, D., *Phys. Rev. A*, **56**, 1958–1969(1997).
55. Rescigno, T.N., Baertschy, M., Isaacs, W.A., and McCurdy, C.W., *Science* , **286**, 2474–2479(1999).
56. Stier, H.C., *Zeits. f. Physik*, **76**, 439(1932).
57. Fisk, J.P., *Phys. Rev.* **49**, 167–173(1936).
58. Burke, P.G., Chandra, N., and Gianturco, F.A., *J. Phys. B:At. Mol. Phys.*, **5**, 2212–2224(1972).
59. Rescigno, T.N., McCurdy, C.W., Orel, A.E., and Lengsfield, B.H., in *Computational Methods for Electron-Molecule Collisions*, edited by W.M. Huo and F.A. Gianturco, Plenum, New York, 1995, Ch. 1.
60. Huo, W.M., 1995 in *Computational Methods for Electron-Molecule Collisions*, edited by W.M. Huo and F.A. Gianturco, Plenum, New York, 1995, Ch. 15.
61. Pfingst, K., Nestmann, B.M., and Peyerimhoff, S.D., *J. Phys. B:At. Mol. Opt. Phys.*,

27, 2283–2296(1994).
62. Tennyson, J., and Morgan, L.A., *Phil. Trans. R. Soc. A*, **357**, 1161–1173(1999).
63. Huo, W.M., and Gianturco, F.A., Editors *Computational Methods for Electron-Molecule Collisions*, Plenum Press, New York, 1995.
64. Morgan, L.A., *Phys. Rev. Letters*, **80**, 1873–1875(1998).
65. Szmytkowski, C., Zecca, A., Karwasz, G., Oss, S., Maciag, K., Marinković, B., Brusa,R.S., and Grisenti,R., *J. Phys. B:At. Mol. Phys.* **20**, 5817–5825(1987).
66. Chase, D.M., *Phys. Rev.*, **104**, 838–842(1956),
67. Morrison, M.A., *Adv. At. Mol. Phys.*, **24**, 51–156(1987).
68. Chandra, N., and A. Temkin, A., *Phys. Rev. A*, **13**, 188–203, **14**, 507–511(1976).
69. Herzenberg, A., and Mandl, F., *Proc. Roy. Soc. A*, **270**, 48–71(1962).
70. Bardsley, J.N., *J. Phys. B:At. Mol. Phys.*, **1**, 349–364(1968).
71. W. Domcke, W., *Phys. Rep.*, **208**, 97–188(1991),
72. A.F.J. Siegert, A.F.J., *Phys. Rev.*, **56**, 750–752(1939).
73. Schneider, B.I., LeDourneuf, M. and Burke, P.G. *J. Phys. B:At. Mol. Phys.*, **12**, L365–L369(1979).
74. Mazevet, S., Morrison, M.A., Boydstun, O., and Nesbet, R.K., *J. Phys. B:At. Mol. Opt. Phys.*, **32**, 1269–1294(1999).

Applications of collision theory

Igor Bray

Electronic Structure of Materials Centre, The Flinders University of South Australia, G.P.O. Box 2100, Adelaide 5001, Australia

Abstract. The previous decade has seen immense progress in the field of electron-atom scattering. The availability of accurate computed data has revitalised numerical modelling of plasmas in astrophysics, fusion, lighting, and elsewhere. Here we summarise the contribution that the Convergent Close-Coupling (CCC) method has made to the generation of such data.

I INTRODUCTION

I thank the organisers of the ICAMDATA II conference for inviting me to the meeting. It was a very enjoyable, stimulating and productive period for me. The following summarises the talk, of same title, that I gave at the conference.

The simplest electron-atom scattering problem, that of e-H scattering, has avoided complete solution for many decades. Only very recently Rescigno *et al.* [1] have claimed to have solved this three-body problem utilising immense computational resources involving complex $10^6 \times 10^6$ matrices. Whether this is really the case is yet to be determined by extensive comparison with experiment, but indicates that the e-H scattering problem is not a simple one. The core problem is the difficulty of treating both discrete excitation and ionisation of the target. Whereas initially the incoming electron sees a neutral target, subsequent to ionisation the three charged particles are formally interacting out to infinite distances. If the e-H problem is so hard, what about electron-atom scattering for general targets? It is clear that such problems have all the difficulties of the e-H system and more.

Though electron-atom calculations may be difficult the corresponding cross section data is of importance in many scientific and industrial applications. Astrophysical, fusion, lighting, and industrial plasmas all have interactions between electrons and various atomic and ionic species. The knowledge of these allows for an understanding of the plasma yielding information of scientific or industrial importance. For example, knowledge of electron density is important in fusion plasmas, the lighting industry is interested in the excitation cross sections that contribute to the

CP543, *Atomic and Molecular Data and Their Applications,* edited by K. A. Berrington and K. L. Bell

quality of the light, and astrophysicists determine the presence of atomic and ionic species and their abundances.

Thus, there is plenty of motivation for obtaining accurate electron-atom scattering data. The key to the development of a reliable theory is not so much to solve the full three or more body problem, but find a way to extract the largest and most valuable integral cross sections. The Convergent Close-Coupling (CCC) method was developed by Bray and Stelbovics [2] with this aim in mind.

II THE CONVERGENT CLOSE-COUPLING METHOD

We begin by considering atomic hydrogen as the target. The idea of the method is to treat ionisation channels like discrete excitation. In this way consistency between the treatment of the target discrete and the continuous spectrum is restored. Discretisation of the complete target spectrum is imposed by diagonalising the target Hamiltonian H_T, for each target-space orbital angular momentum l, in an orthogonal truncated Laguerre basis

$$\xi_{kl}(r) = \left(\frac{\lambda_l(k-1)!}{(2l+1+k)!}\right)^{1/2} (\lambda_l r)^{l+1} \exp(-\lambda_l r/2) L_{k-1}^{2l+2}(\lambda_l r), \tag{1}$$

where the $L_{k-1}^{2l+2}(\lambda_l r)$ are the associated Laguerre polynomials, and k ranges from 1 to the basis size N_l. The constant λ_l is arbitrary and is chosen so that the lowest energy states are essentially the exact hydrogen eigenstates. Upon diagonalisation the targets states satisfy

$$\langle \phi_{fl}^{(N_l)} | H_T | \phi_{il}^{(N_l)} \rangle = \epsilon_{fl}^{(N_l)} \delta_{fi}, \tag{2}$$

where the ϕ are linear combinations of the ξ. In Fig. 1 the energy levels are given for the $l = 0$ case of diagonalising the atomic hydrogen Hamiltonian using N_l states. We see that with increasing N_l the negative energy states converge to the true discrete spectrum, given in the column labelled by $N_l = \infty$, while the positive energies give an increasingly dense discretisation of the continuum.

The total number $N = \sum_{l=0}^{l_{\max}}$ of states ϕ are used to expand the total wavefunction of the system

$$|\Psi_i^{(+)}\rangle \approx |\Psi_i^{(N+)}\rangle = \sum_{n=1}^{N} |\phi_n^{(N)}\rangle\langle\phi_n^{(N)}|\Psi_i^{(+)}\rangle = \sum_{n=1}^{N} |\phi_n^{(N)} f_{ni}^{(N+)}\rangle, \tag{3}$$

where the unknown one-electron functions f are to be found. Eq. 3 is the approximation made to evaluating the non-relativistic three-body e-H Hamiltonian. Having made this approximation the close-coupling equations may be derived and solved by many numerical schemes. In our case given a total energy E we use the momentum space approach [2]

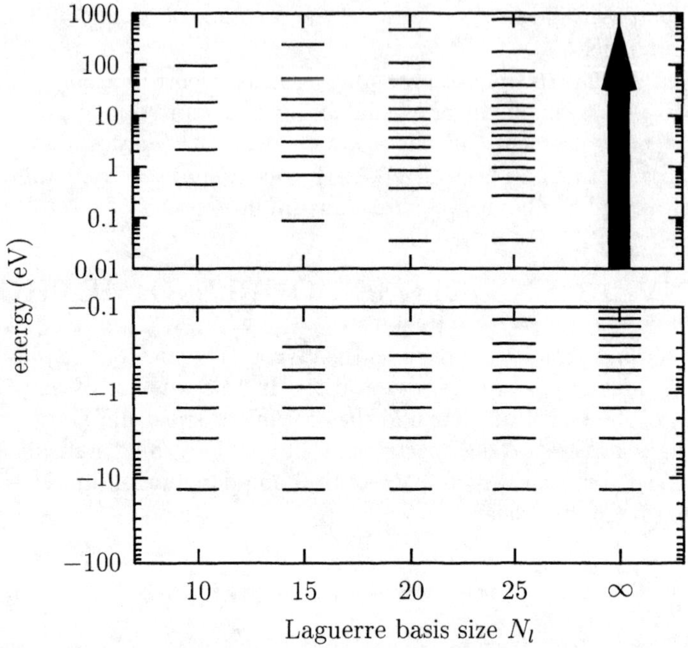

FIGURE 1. The energy levels obtained by diagonalising the hydrogen Hamiltonian in the Laguerre basis 1 for $l = 0$, $\lambda_l = 1$ and indicated values of N_l.

$$\langle \boldsymbol{k}_f \phi_f^{(N)} | T | \phi_i^{(N)} \boldsymbol{k}_i \rangle = \langle \boldsymbol{k}_f \phi_f^{(N)} | V | \phi_i^{(N)} \boldsymbol{k}_i \rangle + \sum_{n=1}^{N} \int d^3k \frac{\langle \boldsymbol{k}_f \phi_f^{(N)} | V | \phi_n^{(N)} \boldsymbol{k} \rangle \langle \boldsymbol{k} \phi_n^{(N)} | T | \phi_i^{(N)} \boldsymbol{k}_i \rangle}{E + i0 - \epsilon_n^{(N)} - k^2/2}, \qquad (4)$$

to solve directly for the T matrix. Variational and R-matrix approaches should yield identical results. The integrated cross sections are then obtained by integrating over the solid angle of the outgoing electron

$$\sigma_{fi}^{(N)} = \int d\Omega |\langle \boldsymbol{k}_f \phi_f^{(N)} | T | \phi_i^{(N)} \boldsymbol{k}_i \rangle|^2. \qquad (5)$$

The calculations are repeated with increasing N until convergence to a desired accuracy is obtained.

III PROOFS OF PRINCIPLE FOR MODEL AND FULL E-H SCATTERING

The first major step in establishing the CCC method was to show that convergence did indeed occur and that this convergence was to the correct answer. It is rare to have known answers in electron-atom scattering systems, but one such case is provided by the Temkin-Poet [3,4] model of electron-hydrogen scattering. In this model the orbital angular momenta are restricted to zero, allowing for a very accurate solution of the problem. While this is not physical the complexity of the treatment of the target continuum is not lost. An example of how the CCC theory works, with increasing size of the calculations, is given in Figs. 2 and 3.

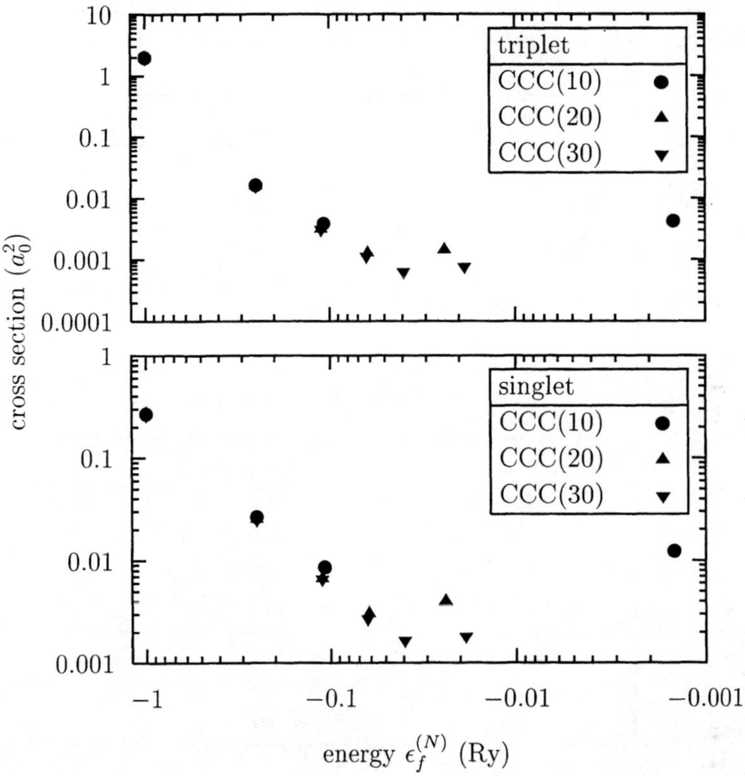

FIGURE 2. The singlet and triplet discrete excitations in the Temkin-Poet model of 3 Ry e-H scattering, as calculated by the CCC method using the indicated number of states.

We consider the case of the incident electron being 3 Ry (40.8 eV) and observe what happens to excitation of both the discrete and the continuous target spectrum. Beginning with the discrete spectrum we see that the $N_l = 10, 20, 30$ calculations all yield much the same answer for both elastic and 2s excitation. However, for the 3s excitation the smallest calculation gives the biggest result with the other two having converged. The last available discrete state in the CCC(10) calculation is unphysical and perhaps surprisingly gives a result that is even higher than for the 3s case. Looking at excitation of the 4s state we again see that the smaller calculation yields the larger cross section with the next available state yielding an even greater cross section. What is happening is that that last available negative-energy state for a given N_l attempts to take into account all the remaining true discrete states (and possibly some of the continuum).

Having considered discrete excitation we turn to ionisation. In this case the unity normalisation of the states $\phi_f^{(N)}$ is inappropriate. Careful analysis of the underlying equations [5] shows that the ionisation amplitude is obtained from $\langle\chi_f^{(-)}|\phi_f^{(N)}\rangle\langle \boldsymbol{k}_f\phi_f^{(N)}|T|\phi_i^{(N)}\boldsymbol{k}_i\rangle$, where χ_f is the Coulomb wave of energy $\epsilon_f = \epsilon_f^{(N)}$. Apart from restoring the continuum normalisation the Coulomb function brings in the Coulomb phase, of importance in ionisation angular distributions.

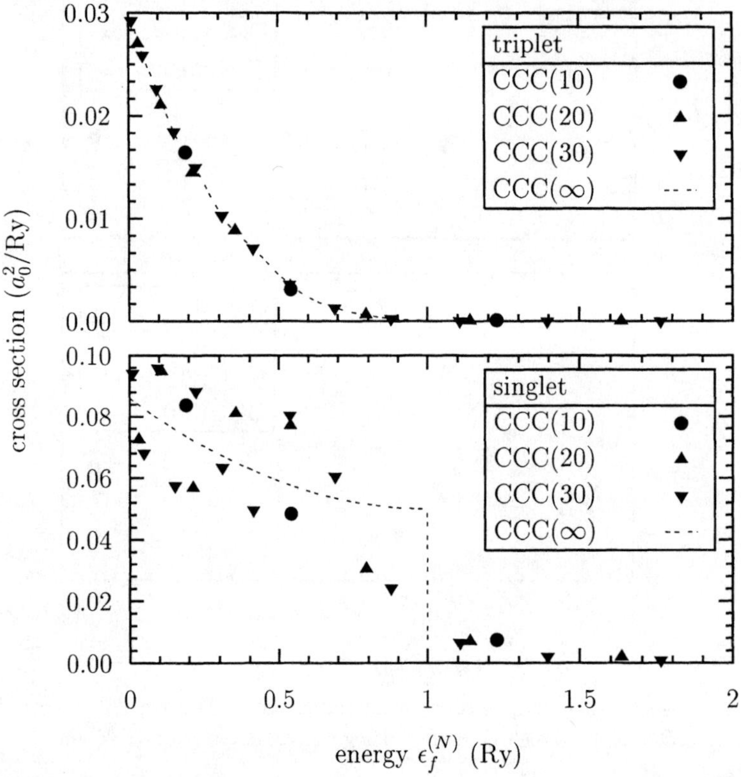

FIGURE 3. The singlet and triplet ionisation in the Temkin-Poet model of 3 Ry e-H scattering, as calculated by the CCC method using the indicated number of states. The curve denoted by CCC(∞) is an estimate, see text.

In Fig 3 we present the ionisation singly differential cross section (SDCS) presented in the form that integration over the energy yields the total ionisation cross section. The triplet case shows good convergence, but the lack of symmetry about $E/2$, where the total energy $E = 2$ Ry, requires explanation. Owing to the Green's function containing E in (4) the secondary on-shell energy integration ranges from 0 to E, and not to $E/2$ as might be expected in formal theory. Since the close-coupling theory is unitary and cannot have double-counting of the ionisation cross section the CCC formalism gives the physical SDCS(e) only for $e \leq E/2$, and near zero cross section for $e > E/2$. The accuracy of the CCC-calculated triplet SDCS has been verified by Jones and Stelbovics [6].

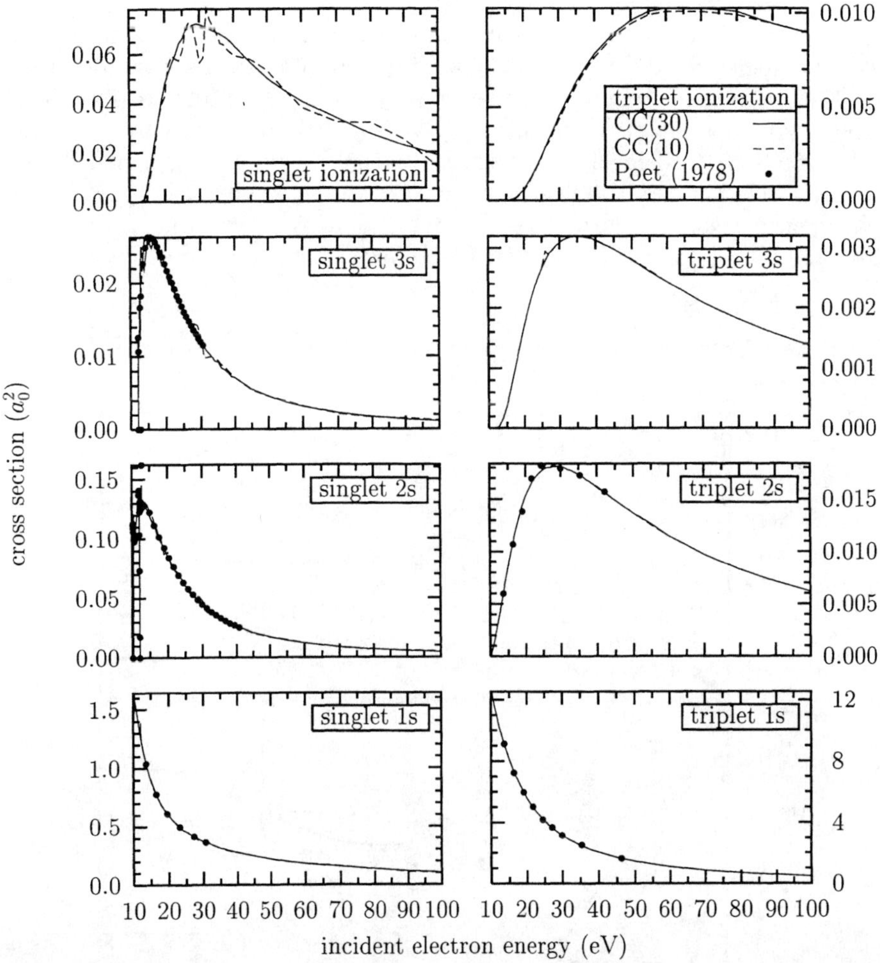

FIGURE 4. The singlet and triplet cross sections in the Temkin-Poet model of e-H scattering, as calculated by the CCC method using the indicated number of states. The near-exact results are due to Poet [4].

We now turn to the singlet case given in Fig. 3. Here we see substantial oscillation in the CCC results. The estimate given is based on an integral preserving quadratic with the SDCS($E/2$) being four times the interpolated estimate. The reason for oscillation is that the theory should get identically zero cross section for $e > E/2$, but unlike the triplet case the cross section at $E/2$ is non-zero. Hence here the finite calculations are attempting to fit a step function of substantial step size [7].

Stelbovics [8] essentially proved this and furthermore showed that at the step the underlying amplitudes converges to half the step size, hence the cross section is a factor of four too low. Note that, though the underlying SDCS is rather oscillatory the integral that yields the total ionisation cross section is not [7].

The problems with the ionisation SDCS as calculated by the CCC method are of little consequence for discrete excitations, including the total ionisation cross sections. Having examined a specific case we present, in Fig. 4, the cross sections as a function of incident electron energy. We see little difference between the CCC(30) and CCC(10) calculations, with the former being smoother and hence likely to be most accurate. Excellent agreement with the results of Poet [4] shows that convergence is to the correct result. This conclusion was one of the first successes of the CCC method [15], and has allowed the immense extension and

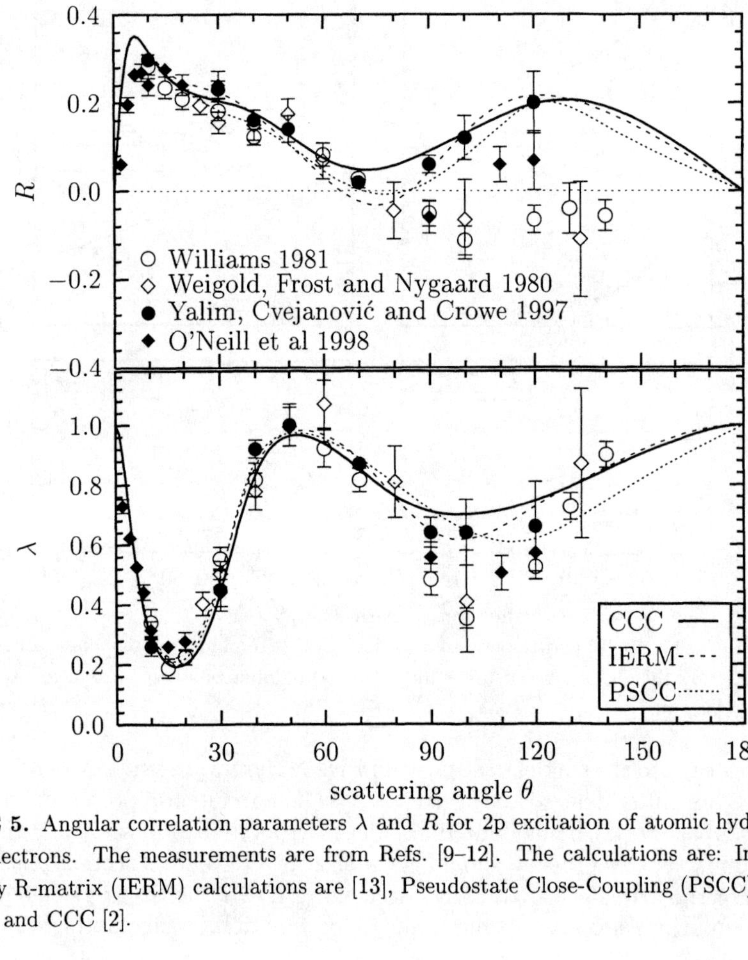

FIGURE 5. Angular correlation parameters λ and R for 2p excitation of atomic hydrogen by 54.4 eV electrons. The measurements are from Refs. [9–12]. The calculations are: Intermediate Energy R-matrix (IERM) calculations are [13], Pseudostate Close-Coupling (PSCC) calculations [14], and CCC [2].

application of the method to realistic problems.

One of the major motivations for developing the CCC method was the inability of the best available theories to describe the angular correlation measurements of the 2p excitation of hydrogen by 54.4 eV electrons [9,10]. Therefore the first application of the CCC method was to this problem [2]. However, the CCC theory yielded much the same result as the earlier theories and so discrepancy with experiment remained, see Fig. 5. Fortunately, subsequent experimental analysis of the problem has suggested much greater consistency with theory [11,12].

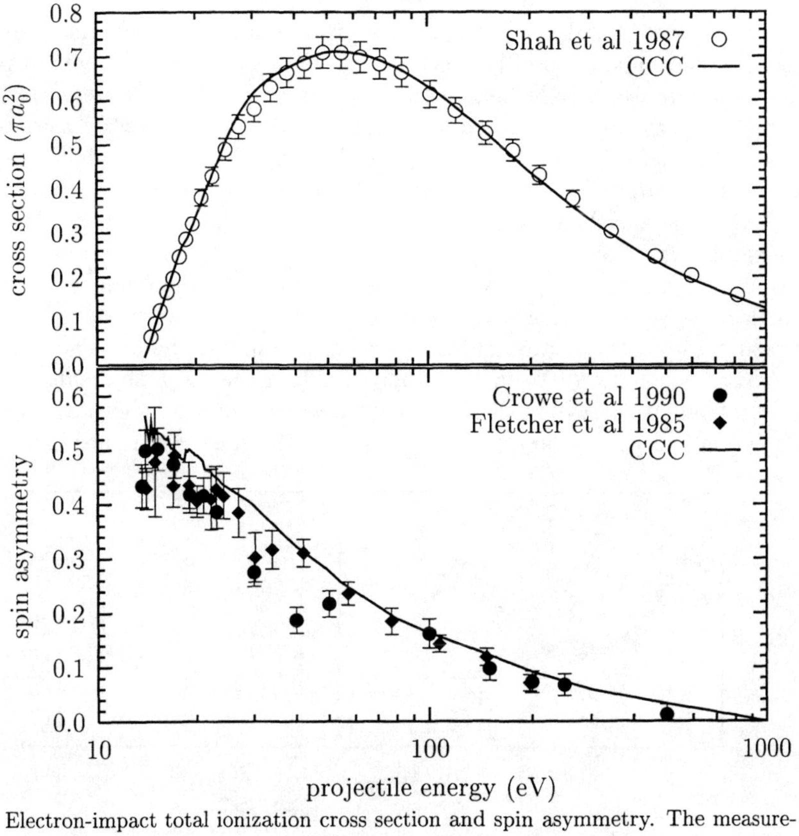

FIGURE 6. Electron-impact total ionization cross section and spin asymmetry. The measurements are from Refs. [16–18], and the CCC calculations from Ref. [19].

The first indication of the forthcoming success of the CCC method was given by extracting the total ionisation cross section from the calculations. This is essentially the sum of the integrated cross sections for the excitation of the positive-energy states. Comparison with experiment, see Fig. 6, yielded excellent agreement, unmatched at the time by any other ab initio theory.

IV APPLICATIONS OF THE CCC THEORY

The previous section gave the foundations of the CCC method and some key examples which indicate that the method is able to yield accurate electron-atom cross section data for discrete transitions at all energies. We now turn to uses of the CCC theory in practical applications. The approach taken is to give a general overview with references to where more detailed information can be found.

A Electron-alkali metal scattering

The knowledge and understanding of plasmas in fusion reactors is an important part of ensuring success of the technology. Various diagnostic tools in probing the plasma are in use. One of these is lithium beam spectroscopy. Here a lithium beam is injected into the plasma and subsequent radiation is analysed to determine the plasma characteristics, such as electron density. The success of such a tool relies on the knowledge of accurate electron-lithium scattering cross sections.

Having convinced ourselves of the utility of the CCC method to electron-hydrogen collisions, extension to alkali metals was performed [20]. This is relatively straightforward since such atoms may be accurately treated by the model of a single valence electron above a frozen Hartree-Fock core. For the heavier alkalis a phenomenological core-polarisation potential is added to improve the accuracy of the atomic structure calculation. With this approximation the electron-alkali metal scattering problem becomes just a three-body problem, almost identical to that of e-H scattering except that more sophisticated potential operators are involved.

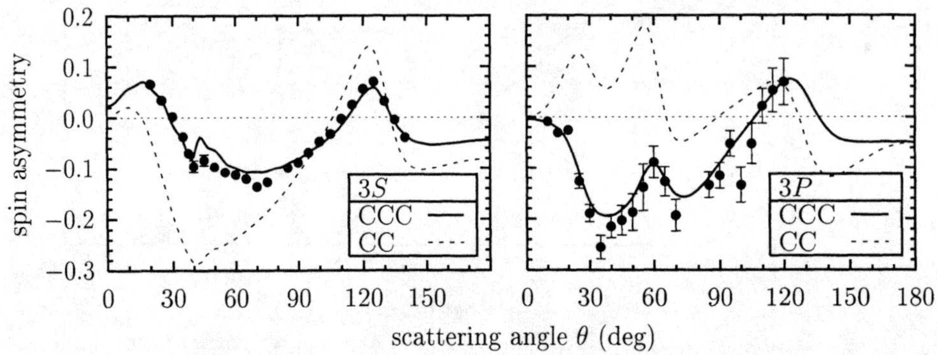

FIGURE 7. Spin asymmetries for elastic scattering and 3P excitation of the ground state of sodium by 20 eV electrons. The CCC calculations treat the full target spectrum, whereas those denoted by CC treat only the sodium discrete spectrum [20]. The experimental data are from Ref. [21].

The testing of the CCC formalism is made particularly stringent by the spin-resolved measurements of electron-sodium elastic and 3P excitation transitions [21–23] and references therein. In Fig. 7 we give the example of spin asymmetries when 20 eV electrons scatter on sodium. The spin asymmetry measures the relative importance of singlet and triplet differential cross sections. Two calculations are given. The one denoted by CC is standard close-coupling which is convergent in the treatment of just the discrete target subspace. The CCC, on the other hand, is convergent in the treatment of both the discrete and the continuous subspaces. What is particularly remarkable about the example given is that the treatment of the target continuum is not only necessary, but has an enormous effect on the calculated spin asymmetries. This very large effect brings about excellent agreement with experiment.

There are many other examples of excellent agreement of the CCC theory with electron scattering on other alkalis, e.g. lithium [24] and potassium [25,26]. Together, these give us great confidence in the accuracy of the CCC method for such scattering systems. As a consequence Schweinzer *et al.* [27] put together a database of cross section data of excitation of lithium by electron (and proton) impact. These data should be valuable to anyone interested in using lithium beam spectroscopy of plasmas.

B Stark broadening

Spectral lines emanating from atoms and ions of a plasma may be broadened due to the interactions between the plasma particles. The resulting line profiles may then be analysed to determine the plasma properties. The underlying theory of such process has been thoroughly developed [28], and requires the knowledge of cross sections for all transitions involving the initial state and all transitions involving the final state. This implies that calculations are particularly difficult and hence initially only various semi-empirical and semi-classical approaches were available. However, with the advance of computational technology and development of sophisticated close-coupling theories fully ab initio quantum mechanical calculations are possible.

To test the accuracy of such calculations we require some accurate benchmark measurements of Stark broadening. Here we run into a problem. Unfortunately such measurements are particularly difficult and presently there are discrepancies in the measured and fully quantum mechanically predicted widths by as much as a factor of two in the case of the B^{2+} ion [29]. The situation is a hotly debated topic [30,31] and is still unresolved. How can we be confident that all of the CCC calculated transitions are accurate? Our confidence stems from comparison with experimental e-B^{2+} cross section data and other calculations.

The extension of the CCC theory to ionic lithium-like targets has also been completed together with atomic alkali targets [20]. The only difference to atomic targets is that the asymptotic projectile wavefunctions are Coulomb rather than plane waves. Thus the demonstrated accuracy for atomic lithium suggests the same for lithiumlike ions. The first application of the CCC method to e-B^{2+} scattering showed that the then available total ionization cross section was around 15%

lower than the calculations [32]. However, most recent experiment and calculations presented by Woitke *et al.* [33] support the CCC calculations.

C The lighting industry

Gas discharges are a valuable source of light extensively utilised by industry. A typical fluorescent lamp consist of argon as the buffer gas and a small amount of mercury which generates UV radiation upon electron impact, and which is subsequently converted to the visible spectrum by a phosphor. Though each light contains little mercury, a neurotoxin, the collective amount of mercury used is substantial given that fluorescent lights dominate all artificial forms of lighting. There is some pressure to replace mercury with other metals such as barium in order to reduce the toxicity and improve the quality of the artificial light. Before such an expensive undertaking is considered it is important to have a complete understanding of the underlying physics of the plasmas. Though the e-Hg lighting system has been in use for many decades, its complete understanding can only be possible if the underlying cross sections are known. Once this system is understood a similar study of e-Ba scattering would allow for a comparison of the two systems to ascertain the relative advantages of either. To this end Osram-Sylvania inc. have contracted us to provide detailed calculations of e-Hg and e-Ba scattering.

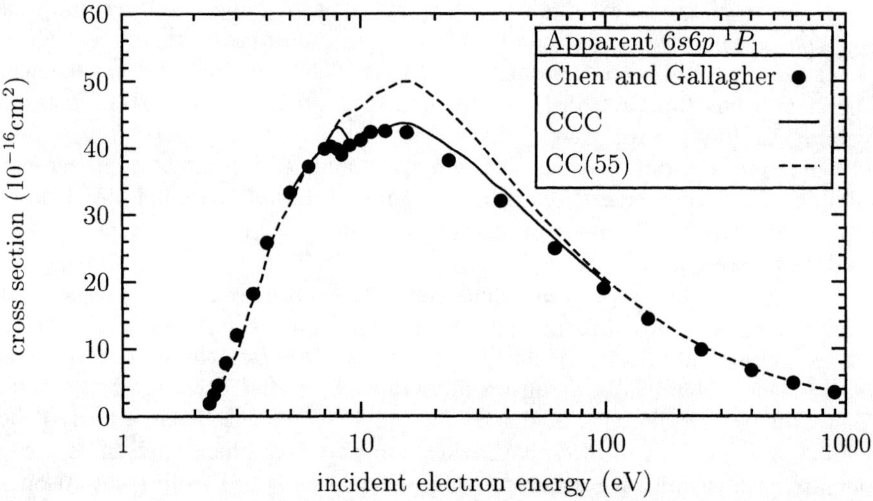

FIGURE 8. Apparent 6^1P^o excitation integrated cross sections for electron scattering from the Ba ground state. The CCC calculation treats the target continuum in addition to the 55 negative-energy states of the CC(55) calculation. The measurements are due to Chen and Gallagher [34] and the calculations are due to Fursa and Bray [37].

Both mercury and barium are very heavy metals whose inner electrons require a relativistic treatment for accurate description. Through self-consistency it can be argued that this extends to the outer electrons also. As yet the CCC method has not been implemented within the relativistic framework. Instead, the highly-successful non-relativistic implementation of the CCC method to e-He scattering [35] has been extended to group two elements within the non-relativistic formalism relying on one- and two-electron core polarisation potentials and the fine-structure approximation to take some account of the relativistic effects [36,37]. The results have been very encouraging, see Fig. 8 for example, but we feel that a full relativistic formalism may be required, particular in the resonance region, where accurate knowledge of target state energies is likely to be very important. Implementation of a relativistic CCC formalism has only just began.

V CONCLUSIONS AND FUTURE DIRECTIONS

We believe that the CCC method is able to yield accurate cross section data for electron impact excitation of hydrogen, helium, and the lighter atoms and ions of groups one and two electronic configuration, at all energies. This includes the total ionisation cross section. As such the method has been utilised in a number of applications with some success, but also some controversies remain. Presently, the extension of the method to incorporate the relativistic formalism is underway. In addition, extension of the method to more complicated atoms such as neon is also being investigated.

ACKNOWLEDGMENTS

The author is indebted to his colleagues Andris Stelbovics, Dmitry Fursa, Alisher Kadyrov and Yuri Ralchenko for their many contributions to the development and application of the CCC theory. Support of the Australian Research Council, the Flinders University of South Australia, and the South Australian Centre for High Performance Computing and Communications is acknowledged.

REFERENCES

1. T. N. Rescigno, M. Baertschy, W. A. Isaacs, and C. W. McCurdy, Science **286**, 2474 (1999).
2. I. Bray and A. T. Stelbovics, Phys. Rev. A **46**, 6995 (1992).
3. A. Temkin, Phys. Rev. **126**, 130 (1962).
4. R. Poet, J. Phys. B **11**, 3081 (1978).
5. I. Bray and D. V. Fursa, Phys. Rev. A **54**, 2991 (1996).
6. S. Jones and A. T. Stelbovics, Phys. Rev. Lett. **84**, 1878 (2000).
7. I. Bray, Phys. Rev. Lett. **78**, 4721 (1997).
8. A. T. Stelbovics, Phys. Rev. Lett. **83**, 1570 (1999).
9. E. Weigold, L. Frost, and K. J. Nygaard, Phys. Rev. A **21**, 1950 (1980).

10. J. F. Williams, J. Phys. B **14**, 1197 (1981).
11. H. Yalim, D. Cvejanovic, and A. Crowe, Phys. Rev. Lett. **79**, 2951 (1997).
12. R. W. O'Neill, P. J. M. van der Burgt, D. Dziczek, P. Bowe, S. Chwirot, and J. A. Slevin, Phys. Rev. Lett. **80**, 1630 (1998).
13. T. T. Scholz, H. R. J. Walters, P. G. Burke, and M. P. Scott, J. Phys. B **24**, 2097 (1991).
14. W. L. van Wyngaarden and H. R. J. Walters, J. Phys. B **19**, 929 (1986).
15. I. Bray and A. T. Stelbovics, Phys. Rev. Lett. **69**, 53 (1992).
16. M. B. Shah, D. S. Elliot, and H. B. Gilbody, J. Phys. B **20**, 3501 (1987).
17. D. M. Crowe, X. Q. Guo, M. S. Lubell, J. Slevin, and M. Eminyan, J. Phys. B **23**, L325 (1990).
18. G. D. Fletcher, M. J. Alguard, T. J. Gay, P. F. Wainwright, M. S. Lubell, W. Raith, and V. W. Hughes, Phys. Rev. A **31**, 2854 (1985).
19. I. Bray and A. T. Stelbovics, Phys. Rev. Lett. **70**, 746 (1993).
20. I. Bray, Phys. Rev. A **49**, 1066 (1994).
21. M. H. Kelley, J. J. McClelland, S. R. Lorentz, R. E. Scholten, and R. J. Celotta, in *Correlations and Polarization in Electronic and Atomic Collisions and (e,2e) Reactions*, No. 122 in *Institute of Physics Conference Series*, edited by P. J. O. Teubner and E. Weigold (IOP, London, 1992), pp. 23–32.
22. S. R. Lorentz, R. E. Scholten, J. J. McClelland, M. H. Kelley, and R. J. Celotta, Phys. Rev. Lett. **67**, 3761 (1991).
23. R. E. Scholten, S. R. Lorentz, J. J. McClelland, M. H. Kelley, and R. J. Celotta, J. Phys. B **24**, L653 (1991).
24. V. Karaganov, I. Bray, and P. J. O. Teubner, J. Phys. B **31**, L187 (1998).
25. K. A. Stockman, V. Karaganov, I. Bray, and P. J. O. Teubner, J. Phys. B **31**, L867 (1998).
26. K. A. Stockman, V. Karaganov, I. Bray, and P. J. O. Teubner, J. Phys. B **32**, 3003 (1999).
27. J. Schweinzer, R. Brandenburg, I. Bray, R. Hoekstra, F. Aumayr, R. K. Janev, and H. P. Winter, Atomic Data and Nuclear Data Tables **72**, 239 (1999).
28. H. R. Griem, *Spectral Line Broadening by Plasmas* (Academic Press, New York, 1974).
29. H. R. Griem, Y. V. Ralchenko, and I. Bray, Phys. Rev. E **56**, 7186 (1997).
30. S. Alexiou, S. Glenzer, and R. W. Lee, Phys. Rev. E **60**, 6238 (1999).
31. H. R. Griem, Y. V. Ralchenko, and I. Bray, Phys. Rev. E **60**, 6241 (1999).
32. I. Bray, J. Phys. B **28**, L247 (1995).
33. O. Woitke, N. Djuric, G. Dunn, M. Bannister, A. Smith, B. Wallbank, N. Badnell, and M. Pindzola, Phys. Rev. A **58**, 4512 (1998).
34. S. T. Chen and A. Gallagher, Phys. Rev. A **14**, 593 (1976).
35. D. V. Fursa and I. Bray, Phys. Rev. A **52**, 1279 (1995).
36. D. V. Fursa and I. Bray, J. Phys. B **30**, 5895 (1997).
37. D. V. Fursa and I. Bray, Phys. Rev. A **59**, 282 (1999).

Kinetics and energetics of electron impact ionization of molecules: ionization cross section, appearance energies and kinetic energy release.

S. Matt, T. Fiegele, G. Hanel, D. Muigg, G. Denifl, K. Becker[a], H. Deutsch[b], O. Echt[c], N. Mason[d], A.Stamatovic[e], P. Scheier and T.D.Märk

Institut für Ionenphysik, Leopold Franzens Universität, Technikerstr.25, A–6020 Innsbruck, Austria

[a] *Dept. Physics, Stevens Institute Technology, Hoboken, NJ 07030, USA*
[b] *Institut für Physik, Universität Greifswald, D–17487 Greifswald, Germany*
[c] *Dept. Physics, University New Hampshire, Durham, NH 03824–3568, USA*
[d] *Dept. Physics, UCL, Gower Street, London WC1E 6BT, United Kingdom*
[e] *Faculty of Physics Beograd, P.O. Box 368, 11001 Beograd, Yugoslavia*

Abstract. Much effort has been recently devoted to the experimental determination of absolute partial and total electron impact ionization cross sections of molecules and radicals due to the ever increasing importance of these cross sections in many applications such as low– and high– temperature plasma physics and chemistry, atmospheric physics and mass spectrometry. In addition supporting calculations have been recently developed in order to allow the analytic quantitative description of these cross section functions for modeling codes. This review summarizes recent experimental and theoretical progress in this area of electron impact ionization including also a short description of (i) recent advances in the field of electron/ion collisions (i.e., kinetic energy release distributions in dissociative excitation reactions of molecular ions) and (ii) a novel approach to the determination of electron induced appearance energies.

INTRODUCTION

The rapidly growing technological importance of (i) low–temperature plasmas in areas such as the plasma–assisted etching of microstructures and the deposition of high quality films [1] and (ii) high–temperature plasmas such as fusion devices [2], has created an urgent need for a variety of molecular electron–induced cross sections (excitation, dissociation, ionization, attachment, etc.). A quantitative description of the fundamental collision processes leading to ionic species in terms of absolute production cross sections, appearance energies, kinetic energy release distribution etc. is required for a detailed understanding of the properties of these plasma reactors as well as for the modeling of these plasmas.

Ionization and attachment cross sections, the appearance energy and the kinetic energy release play an important role in this context, since they determine the ionization balance in the plasma and thus influence the plasma parameters in a crucial way. Unfortunately, the database on collision cross sections for particles of relevance

CP543, *Atomic and Molecular Data and Their Applications,* edited by K. A. Berrington and K. L. Bell

to plasma chemistry and other plasma applications is still limited [1,2]. Due to the urgent need for such data in the past few years we have developed a research program to provide authoritative data on electron interactions with plasma molecules, in particular we have performed extensive experimental and theoretical studies on electron impact ionization and attachment of selected molecules. The measurements include ionization cross sections (as a function of electron energy in the range from threshold up to about 1000 eV), attachment cross sections (as a function of energy from about 0 eV up to several tens of eV), the derivation of corresponding ionization energies for various *neutral* atoms and molecules, and the study of electron impact ionization and dissociation of mass selected *ionized* molecules. Moreover, we have also studied the kinetic energy release distributions of fragment ions formed by electron impact with molecules and mass selected molecular ions. In this review we will only discuss the electron impact ionization of neutral and ionized molecules (for electron impact ionization of atoms see [3] and for electron attachment work see as an example the selected references [4–7]).

IONIZATION CROSS SECTIONS OF MOLECULES

The general approach to the quantitative study of electron impact ionization and dissociative ionization of molecules is that of implementing an electron beam colliding either with a stagnant gas target or with a molecular beam. Assuming single collision conditions (i.e., $N_t L\sigma \ll 1$, where N_t is the target density, L the collision interaction length and σ the cross section) the total positive ion current i_t produced along the interaction volume (defined by L) is

$$i_t = i_e N_t L \sigma_t \qquad (1)$$

where i_e is the electron current and σ_t is the *total* ionization cross section. If the product ions are analyzed with respect to their mass to charge ratio m/ze the measured individual ion currents are given by

$$i_t = i_e N_t L z \sigma_p \qquad (2)$$

where σ_p is the *partial* ionization cross section, i.e., the cross section for the production of a specific ion irrespective of its charge ze. The total ionization cross section is then related to the partial cross sections via the weighted sum

$$\sigma_t = \Sigma z \sigma_p \qquad (3)$$

Experimental determination of the partial or the total cross section therefore requires the accurate measurement of the four quantities in equations (1) and (2). At first sight this appears to be an easy task, however, as it has been discussed in detail in [8,9,10] there are a number of experimental conditions which have to be fulfilled in order to arrive at reliable and accurate cross sections. In the case of absolute total cross sections the most difficult parameter to be measured accurately is the target density N_t, whereas for partial cross sections the proper measurement of the individual ion currents of the fragment ions using mass spectrometry is the limiting factor.

Progress has been very slow considering that (i) such cross sections have been measured since about 1930 [11] and (ii) accurate knowledge of such cross sections is of great importance for mass spectrometry, plasma physics, etc [1,2,12]. Nevertheless in the seventies and early eighties, there were some advances in experimental techniques, the corresponding results of which have been summarized in [12–16] (see also earlier work referenced in these reviews).

Despite these experimental improvements, large discrepancies existed at this time for total cross sections of even simple molecular targets such as NH_3 [17] (Figure 1).

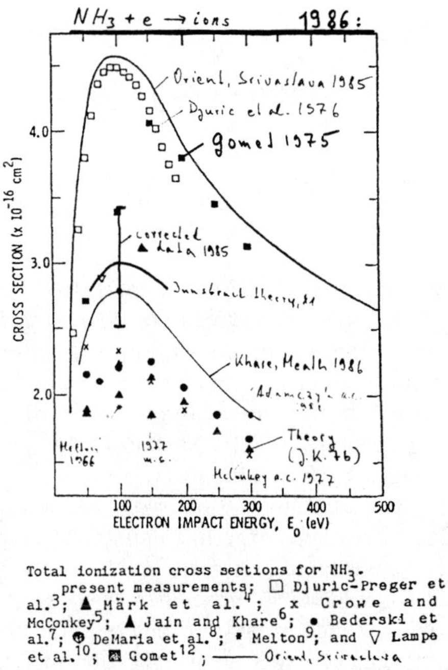

FIGURE 1. Total electron impact ionization cross section for NH_3 as a function of electron energy as reported in a 1986 lecture summarizing the experimental and theoretical results available at that time. This is a facsimile of a transparency used in [17].

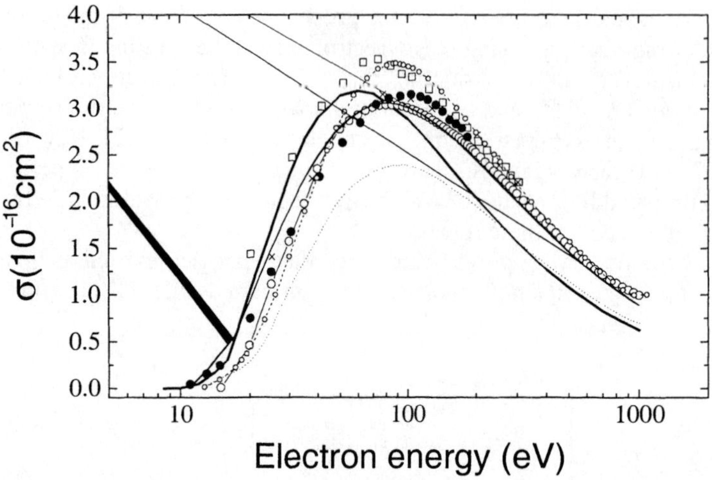

FIGURE 2. Total electron impact ionization cross section for NH_3 as a function of electron energy after [18]. The points denote experimental results from various groups and the lines denote calculations using the DM, BEB and JK approach (see below).

However, in the past ten years major progress has been achieved in both experiment and theory. Reliable cross sections are now available for many systems, and the data sets of several groups agree to within a few percent. Deutsch et al. [18] have recently made extensive comparisons between available reliable experimental cross sections and the predictions of theoretical models for 31 molecules. Fig.2 shows the total ionization cross section for NH_3 as known today and demonstrates the enormous progress made since 1986 (see Figure1).

Progress in the field of partial ionization cross sections has been particularly striking in the past 15 years with experimentalists developing new instruments following two different philosophies. Researchers at Innsbruck (and later on in Greifswald) have modified commercial machines (two sector field spectrometers) using the penetrating–field–extraction and ion beam–deflection method [19–21,22]. Others have constructed dedicated experimental set–ups specifically designed for the use in this field (see references given below). In each case the most important feature is the possibility to detect parent and fragment ions without mass and/or kinetic energy discrimination by using controlled ion source extraction and mass spectrometer transmission techniques. In addition metastable decay must be taken into account for certain type of instruments [21].

Fig.3 shows as an example of the first type of instruments the production of CF^+ by electron impact ionization of CF_4 as discussed and measured in [19]. Previous experiments (also shown in Fig.3) discriminated against highly energetic fragment ions resulting in cross sections that underestimated the ionization yield. An important feature in this context is obviously the measurement of the kinetic energy (distribution) of the fragment ions produced (see Fig.4) as information on this is a prerequisite for accurate cross section determinations [19–21].

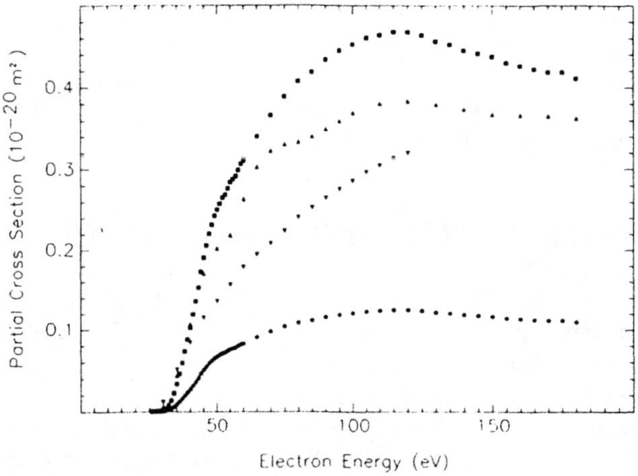

FIGURE 3. Partial electron impact ionization cross section for the production of CF^+ by electron impact ionization of CF_4 after [19]. Full circle: Stephan et al. [23], full inverted triangles: Poll et al. [24], full triangles: Ma et al. [25] and full squares: Poll et al. [19].

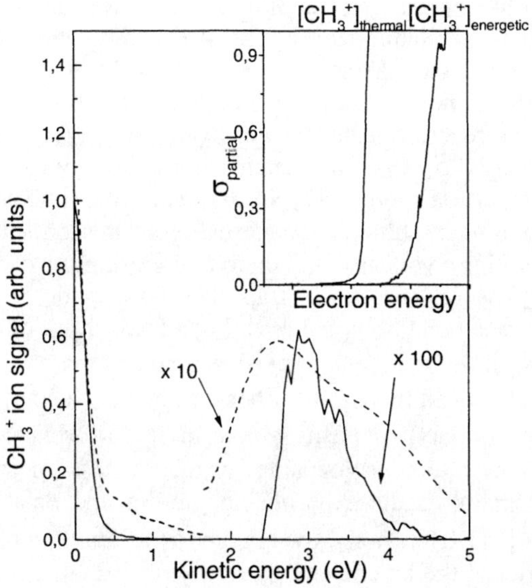

FIGURE 4. Kinetic energy distribution for the fragment ion CH_3^+ produced by electron impact ionization of C_3H_8 as measured with the ion beam deflection method [26] (dashed line) and the mass analyzed retarding potential method [27] (full line). In the inset also shown the corresponding cross section threshold curves for the thermal ions and the energetic ions (peaking at around 3 eV).

Examples of the second type of instruments are (i) the fast–neutral–beam apparatus of Becker and co–workers [28,29] allowing also the study of radical targets, (ii) the time–of–flight position sensitive detection system of Stebbings and co–workers

[30,31] and (iii) the focussing time–of–flight apparatus of Vidal and co–workers [32,33] which also provides information on ion pair information. All these instruments have been tested against some benchmark molecules and there is good agreement (see below). In addition several other groups have during the past 15 years reported valuable cross section data sets [34–42].

ELECTRON IMPACT IONIZATION OF MOLECULAR IONS

To date studies of dissociative collisions involving molecular ions have been dominated by investigations of dissociative recombination. Dissociative ionization, dissociative excitation and direct ionization of molecular ions by electron impact have in contrast received little attention. The data that is available is largely restricted to diatomic and other small ions with the goal of measuring absolute cross sections for fragmentation and ionization [43]. Due to their importance in the edge plasmas of thermonuclear fusion reactors sections for electron induced dissociation of molecular ions are required for a much larger range of (heavier) targets. In addition information on the kinetic energy release (KER) is urgently needed [44].

Electron induced reactions also offer the possibility of analyzing the KER for ions with very short lifetimes. There is considerable interest in the time dependence of the KER [45,46]. Statistical reactions without a reverse activation energy feature a KER which increases with decreasing ion lifetime because, at the same time, the statistically distributed excitation energy in the parent ion increases. However, reactions with a large reverse barrier are characterized by an essentially constant, or even decreasing, KER. Hence, the time dependence of the KER provides information about the nature of the reaction pathway. For example the presence of a strong time dependence in the KER of C_{60}^+ dissociating into C_{58}^+ signals the absence of a significant reverse activation energy [47]. In past work, the time at which dissociation was measured (relative to the time of ion formation in the ion source) was varied by varying the acceleration voltages in double focusing two sector field mass spectrometers [48]. The advent of ion traps has now made it possible to greatly increase the ion lifetimes being sampled [49], but a lower limit of the order of 1 to 10 μs still exists because of the requirement to accelerate and mass select the parent ions.

Recently, we have reduced this limit further (to reactions occurring within less than 0.75 μs after electron impact) by positioning a high–performance electron gun near the intermediate slit of a double focusing magnetic mass spectrometer of reverse geometry [47,50]. Details of the experimental set–up and data analysis have been published elsewhere [51]. Here we show (Figure 5) as an example results of a study where we have measured the kinetic energy release (KER) distributions for electron–induced dissociation of mass–selected molecular parent and fragment ions of propane [50]. They are compared with distributions measured in the same study for spontaneous (metastable) dissociation. The KER distributions are readily derived from the Gaussian fits to the MIKE (mass analyzed ion kinetic energy) spectra after deconvolution with the parent ion peaks. The shift in the distribution for the electron–induced reaction is obvious. The average KER for electron induced dissociation of $C_3H_8^+$ into $C_3H_7^+$ is 13.2 ± 1.2 meV, about 42 % larger than for the spontaneous reaction. This large difference may be attributed to the dramatically reduced time at which the induced reaction can be sampled.

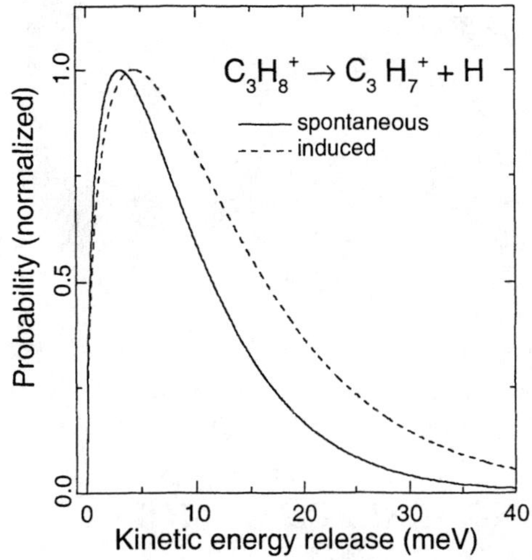

FIGURE 5. Kinetic energy release distribution (KERD) for spontaneous and electron–induced dissociation of $C_3H_8^+$ into $C_3H_7^+$ + H after [50].

In contrast, the KER for dissociation of $C_3H_7^+$ into $C_3H_5^+$, which is characterized by a large reverse activation energy, is hardly affected by the experimental time scale. For more details see [50].

It is interesting to note that we have recently extended our experimental set–up by adding a third field free region with a high performance electron gun and a third analyzer (electric sector field) thus allowing us to study in more detail electron induced dissociation and ionization of mass selected molecular ions (see Fig.6).

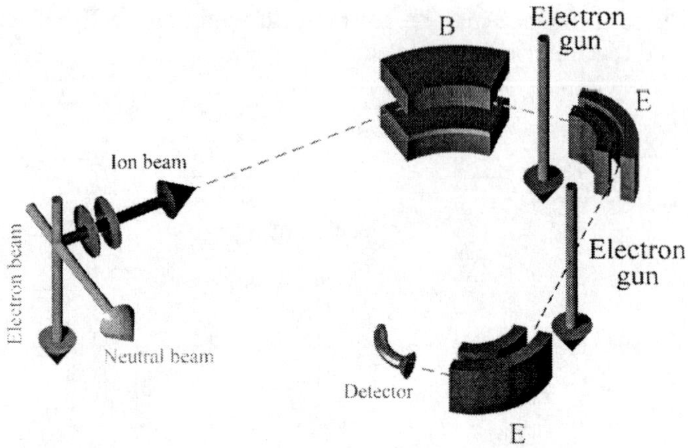

FIGURE 6. Schematic view of recently constructed three sector field mass spectrometer used for the study of electron induced dissociation and ionization reactions in Innsbruck.

APPEARANCE ENERGY OF MOLECULAR IONS

The production of ionic fragments by electron induced dissociative ionization of parent molecules is the dominant physical process underpinning the technique of mass spectrometry. The energy at which each fragment ion is first produced is known as the *appearance energy*, AE. Such energies may be measured using either photons or electrons as the incident projectile.

Due to the higher resolution available photon sources have been used most commonly to measure the appearance energies of dissociative ionic fragments. Modern photoionization experiments use either laser or synchrotron radiation sources [52–55]. The later are preferable since they provide a continuous radiation source in contrast to laser sources which operate within a confined wavelength region, albeit with an order of magnitude higher resolution. Synchrotron light sources are coupled with monochromators to produce well resolved photon beams from the continuous radiation spectrum of the synchrotron. According to Rosenstock [52] the resolution of such monochromators ranges from 0.04 nm to 0.3 nm and the wavelength scale is known typically to an accuracy of 0.02 nm. This represents a "best energy resolution" of about 5 meV at 12 eV [53]. However in practice most experiments have been performed with somewhat lower resolution. This is particularly true for cluster targets where low signal rates require a reduction in incident resolution to 0.2 to 0.4 nm (20 to 40 meV) to obtain an adequate signal to noise ratio [56,57].

The ultimate accuracy of the measured appearance energies depends upon the type of molecular target. Larger molecular systems contain more modes of vibration and rotation both of which are excited at lower energies. Hence for large molecules even at room temperature a considerable proportion of the molecules may be excited in (ro–vibrational states). For example at room temperature the FWHM (full width half maximum) of the rotational envelope of even simple diatomic molecules is approximately 25 meV while that of larger polyatomic systems may be as much as 40 meV [52]. These excitations lead to a shift in the observed appearance energies to lower values since the ionization energy required is less for such rotationally excited states. In order to overcome such effects the molecular target may be cooled using a supersonic beam method, a method also used to produce simple van der Waals clusters [53–55]. However due to the low number density within such sources (and hence poorer signal to noise ratio) the error in the measured appearance energy is not significantly decreased. Finally as the exact shape of the ionization efficiency curve close to the threshold is not known extrapolation to 'zero ion current' in order to determine AE is rather difficult and somewhat ambiguous [58]. All these effects therefore limit the accuracy of AE's of molecular and cluster targets measured by photoionization in general to between 30 to 50 meV.

An alternative methodology would use electrons as the incident (ionizing) projectile. Electron beams are continually tuneable and do not require a monochromator apriori, the energy (wavelength) being set by electric fields. However to date electron sources have had much lower resolutions than photon sources and the use of electrons to measure AE's has been largely limited to those case where photoionization studies appeared not to be feasible, for example large organic molecules and high temperature species effusing from Knudsen cells [52]. However we have recently constructed electron monochromators which produce electron beams with energy resolutions as low as about 5 meV [4–7] while maintaining sufficiently high currents to allow AE's to be measured within good signal to noise ratios.

The apparatus used in these experiments has been described in detail in [59,60] so only the most salient details will be given here. The electron spectrometer consists of an electron gun, a collision chamber and an electron collector. The electrons are monochromatized using a home–built 180° hemispherical analyzer such that the spread of the electron energy distribution is decreased from about 500 meV to a final energy resolution of about 30 meV FWHM with an electron current of approximately 0.5 nA. Ions formed in the collision region were extracted by a weak penetrating electrostatic field and subsequently analyzed by a quadrupole mass spectrometer with a nominal mass range of 2000 amu. The ions were detected by a channel electron multiplier operated in pulse counting mode. Data acquisition during the experiments was controlled by a PC. The apparatus was configured such that it was easy to transfer between collection of anions (produced by dissociative electron impact) and cations. The incident electron beam resolution was therefore calibrated using the well known s–wave attachment cross section for the Cl^- anion formation from CCl_4, while the energy scale was calibrated (i) using the vertical high energy onsets for the production of O^- from CO and NO and (ii) against the appearance energies of cations from the rare gases and simple molecules with an estimated accuracy of <10 meV. This instrument has been used to determine the vertical appearance energies for positively charged molecular and cluster ions [59,60] with accuracy of typically 10 to 20 meV.

As an example of our new technique we will discuss here appearance energy measurements for the ozone molecule and its dimer. Recently, there has been renewed interest in ozone and in particular its molecular aggregates as these are thought to be involved in the ozone destruction in polar stratospheric clouds. No appearance energies for these aggregates have been reported in the literature, in contrast to the monomer case where some studies exist.

FIGURE 7. Ionization efficiency curve measured with an energy resolution of about 100 meV (at about 1 nA electron current) for the production of O_3^+ by electron ionization of neutral O_3 (designated by full squares) after [60]. Also shown for comparison and electron energy calibration purposes corresponding argon data (designated by full triangles). The full lines are fits to the data to extract the corresponding appearance energies as described in the text.

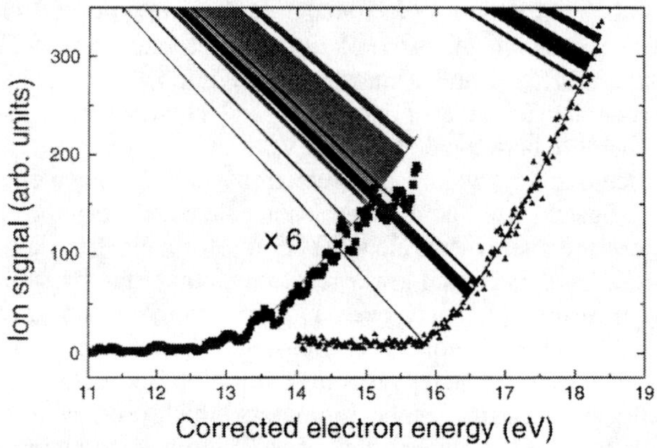

The dependence of the electron impact ionization cross section was measured as a function of electron energy from the appearance energy (AE) to about 3eV above. The appearance energies of the product cations were subsequently derived using a novel data handling procedure described in detail in a recent publication [61] and involving a simultaneous non–linear least squares fit of the raw data by two functions using the Marquart–Levenberg algorithm [61].

Figure 7 shows the measured O_3^+ ionization cross section obtained by electron impact from the ozone monomer. Due to the contact potential drifts that may be induced when adding a reactive gas to the spectrometer, the ozone was mixed with a small amount of argon which was used to provide an in situ electron energy calibration invoking the NIST value of 15.749 eV [62]. Using the fitting procedure discussed above the appearance energy for O_3^+ was determined to be 12.70 ± 0.02eV. The present value is in agreement with the two earlier electron impact determinations for the appearance energy of 12.84 ± 0.10 eV [62] and 12.80 ± 0.05 eV [62]. Nevertheless, it should be noted that adiabatic appearance energy values derived from photon ionization experiments are typically slightly below the present value (outside of the combined errors bars). The most recent photoelectron spectrum (PES) gives an adiabatic ionization energy (referring to the first vibration of the first ionized state consisting altogether of seven peaks) of 12.44 ± 0.01 eV [63]. It is quite interesting to note that the present electron impact value for the appearance energy of 12.70 ± 0.02 eV, however, is in excellent agreement with the values for the vertical ionization energy reported in the photoionization studies, e.g., 12.75 ± 0.01 [63].

The appearance energy for the $(O_3)_2^+$ dimer was measured to be 10.02 ± 0.22 eV, this should be contrasted with the appearance energy for the monomer of 12.70 ± 0.02 eV, the greater error for the dimer arising from the poorer resolution of the incident electron beam. The D (O_3–O_3^+) bond energy may then be derived by using AE (O_3^+) = 12.70 ± 0.02 eV, AE ($(O_3)_2^+$) = 10.02 ± 0.22 eV and a theoretical value for the bond energy of the neutral ozone dimer D (O_3–O_3) = 0.13 eV [64]. The the bond energy of the ozone ionic dimer is then found to be 2.70 ± 0.5eV. This is a very high value compared to the bond energy of rare gas atomic clusters or other molecular clusters, including OClO (see also [53,54,60]). The reason for such a large binding energy is unclear at present but may suggest a cyclic structure for O_6^+. Such a large value for the bond energy of a dimer is, however, not without precedent. Mähnert et al [64] studied the formation of the $(CO)_2^+$ ion by the dissociation of $Ar(CO)_2$ clusters in a threshold photoelectron photoion coincidence experiment and determined a binding energy of the $(CO)_2^+$ dimer ion of 1.80 eV from which they suggested that the CO dimer has a trans–planar structure.

CROSS SECTIONS: THEORETICAL CONSIDERATIONS

Although considerable progress has been made in the experimental determination of cross sections for atomic and molecular targets only some 75 molecules have been quantitatively studied so far, well below the actual need in applied sciences. Therefore theory has been suggested as providing the necessary data for 100's of unmeasured species.

Unfortunately, rigorous quantum mechanical calculations of ionization cross sections for molecular targets are beyond the capability of current quantum–mechanical electron collision theory for essentially all molecules [66–68]. Therefore simplistic *additivity rules* have been adopted to estimate molecular ionization cross sections for many targets. Many variants of the additivity rule, whose concept was first introduced by Ötvos and Stevenson [69] can be found in the literature [70,71]. They all rely on the concept that the molecular ionization cross section is derived by adding in some fashion the ionization cross sections of the individual atomic constituents of the molecule with or without accounting for molecular bonding and/or weighting factors for the atomic cross sections.

The most recent variant of the additivity rule is the so–called Modified Additivity Rule (MAR) of Deutsch et al. [72,73] which includes appropriately chosen weighting factors to account for molecular bonding. Predictions of the MAR have been compared to available experimental cross section data for many molecules with sum formulas of the form AB_X [72] and A_XB_Y, $A_XB_YC_Z$, and $A_pB_sC_tD_u$ [73] and reasonable agreement was found in essentially all cases. In addition to these additivity approaches, there have been other conceptually simplistic (geometric) methods such as

1. The geometric approach of Kistemaker and co–workers [74] which considers different electron–molecule approach geometries and calculates the ionization cross section by averaging over all possible orientations, and
2. A theory based on the calculation of the maximum in the electron–impact ionization cross section as a function of the electron–molecule approach geometry and subsequent averaging over the different orientations [75–77],

In contrast there also exist now more rigorous methods [78–85] including

1. the Deutsch–Märk (DM) formalism which combines a Gryzinski–type energy dependence of the cross section with quantum mechanically calculated molecular structure information based on an additivity concept [78–80],
2. the method of Jain, Khare (JK) and co–workers [81,82] which combines two cross section expressions (the Mott and Bethe cross section) describing ionizing collisions which occur at large impact parameters and at small impact parameters, respectively, and
3. the Binary–Encounter Bethe (BEB) theory of Kim, Rudd and co–workers [83–85] which in addition to combining two cross section expressions (the binary encounter and Bethe cross section) introduces also an additivity concept with quantum mechanically calculated molecular quantities.

Deutsch et al. [18] have recently published an extensive comparison and review of such methods with special emphasis on the DM formalism and the BEB formalism which are the most widely used calculation schemes among those methods that may be labeled "more rigorous" theoretical approaches. Both methods incorporate quantum mechanically calculated molecular structure information, a fact that sets them apart from the more simplistic additivity rules and the purely geometric methods. It is noteworthy to point out in this context, that both the DM formalism and the BEB formalism also make use of an additivity *concept* in the sense that the ionization cross section of a molecule is obtained by summing up the contributions arising from the ejection of an electron from the different molecular orbitals. This has sometimes led people to erroneously refer to the DM and BEB calculations as variants of the additivity *rule*. We would like to make a clear distinction between simplistic additivity rules (like the ones mentioned above [69–75]) and more rigorous methods like the DM formalism and the BEB formalism which include quantum mechanically calculated molecular structure information – even though these approaches also incorporate an additivity concept in some fashion. The method of Jain and Khare and co–workers can also be classified as a more rigorous theoretical approach, but it has only been applied to a limited number of target molecules.

Extensive comparisons has been made by Deutsch et al. [18] between calculated ionization cross sections using the DM formalism and the BEB formalism (including also results from the JK formalism where available) and experimentally determined cross sections for 31 molecules and radicals. In order to demonstrate the present situation and reliability we give below (Figs. 8–11) some typical examples for simple and more complex molecules ranging from H_2, to CO_2, NH_3 (see Fig.2) CF_4, and SF_6.

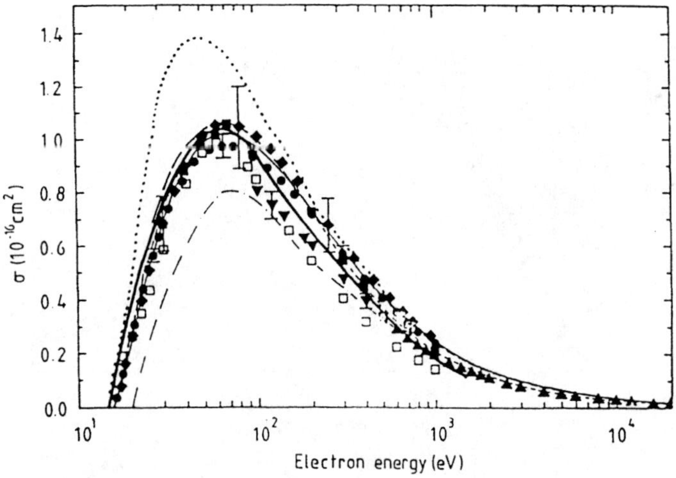

FIGURE 8. Electron impact ionization cross section of H_2 as a function of electron energy after [18]. The experimental data points are from [86] (filled squares), [87] (filled circles), Ref. [39] (filled diamonds), Ref. [88] (filled triangles), and Ref. [89] (filled inverted triangles). The open squares denote CTMC calculations of Ref. [90]. The thick solid line represents the DM calculation [18], the thin solid line and the dashed line denote respectively the BEB and BED calculation [84], the dash–dot line is the calculation of Saksena et al. [91], and the dotted line are results obtained with the method of Khare and co–workers (see [91]).

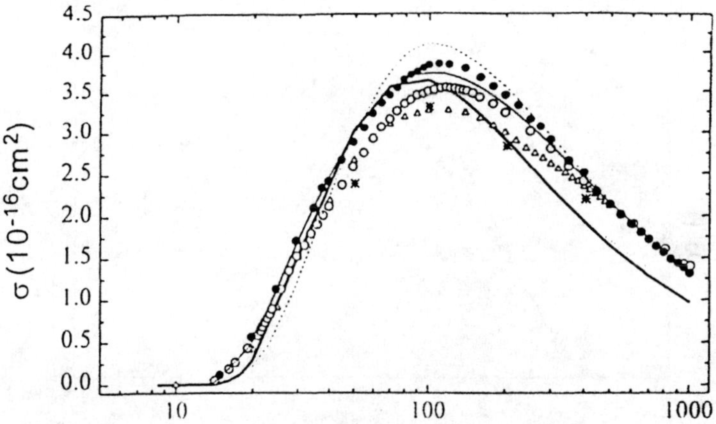

FIGURE 9. Electron impact ionization cross section of CO_2 as a function of electron energy after [18]. The experimental data points are from Ref. [92] (filled circles), Ref. [87] (open circles), Ref. [93] (stars) and Ref. [94] (open triangles). The thick solid line represents the DM calculation [18], the thin solid line denotes the BEB calculation [84], the dotted line represents the calculation of Jain and Khare [81] and the dashed line denotes the calculation of Saksena et al. [91].

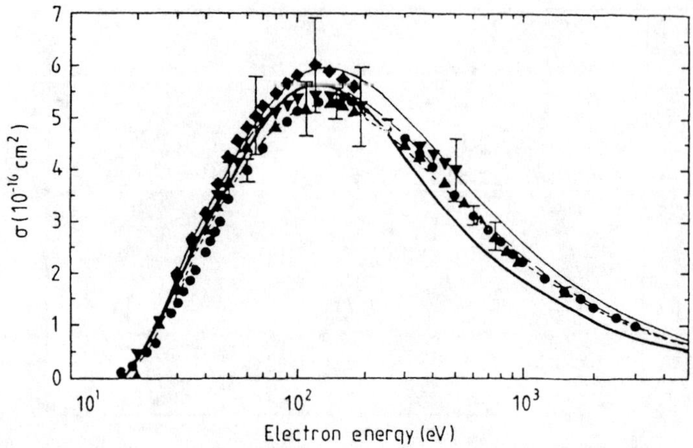

FIGURE 10. Electron impact ionization cross section of CF_4 as a function of electron energy after [18]. The experimental data points are from Ref. [95] (filled circles), Ref. [19] (filled diamonds), Ref. [96] (filled triangles), Ref. [97] (filled inverted triangles), and Ref. [98] (filled squares) . The thick solid line represents the DM calculation [18], the thin solid line and the two dashed lines denote different variants of the BEB calculation [95].

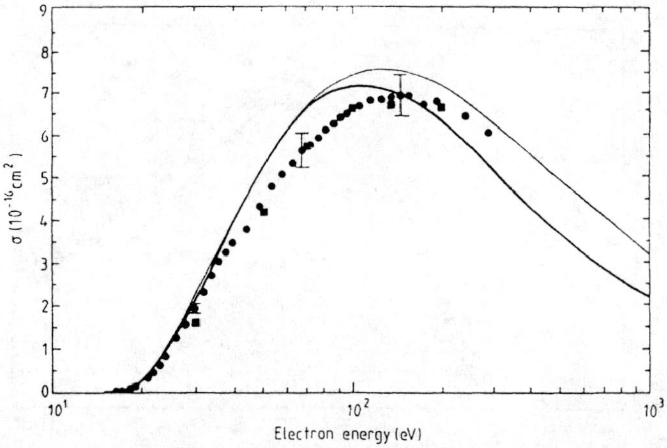

FIGURE 11. Electron impact ionization cross section of SF_6 as a function of electron energy after [18]. The experimental data points are from Ref. [87] (filled squares) and Ref. [99] (filled circles). The thick solid line represents the DM calculation [18], the thin solid line denotes the BEB calculation [84].

CONCLUSIONS

In this paper we have reviewed the present status of our knowledge of electron impact ionization of molecules (and molecular ions). Significant progress has been made in the last decade with the development of new experimental and theoretical techniques. However our knowledge of ionization cross sections and appearance energies of ionic/neutral fragments in particular from clusters remains sparse. Similarly the state of current theory is limited to approximate methods, new techniques to develop a full quantum mechanical treatment of electron impact ionization are needed.

The importance of ionization in many fields of science and technology therefore ensures that the study of electron induced ionization will continue to be a major topic of collision physics in the 21st century.

ACKNOWLEDGMENTS

This work has been carried out within the Association EURATOM–ÖAW and was partially supported by the FWF, Wien, Austria.

REFERENCES

1. Tarnovsky, V., and Becker K., *Plasma Sources Sci. Technol.* **4**, 307 (1995).
2. Janev, R. K., *Atomic And Molecular Processes In Fusion Edge Plasmas*, New York: Plenum Press, 1995.
3. Märk, T. D., "Electron impact ionization of plasma edge atoms," in *Atomic And Molecular Processes In Fusion Edge Plasmas,* edited by R. K. Janev, New York: Plenum Press, 1995, pp. 59–89.
4. Märk, T .D ., *Phys. Rev. Letters* **65**, 2503–2504 (1994).
5. Matejcik, S., Kiendler, A., Stampfli, P., Stamatovic, A., and Märk, T. D., *Phys Rev. Letters* **77**, 3771–3774 (1996).
6. Senn, G., Skalny, J. D., Stamatovic, A., Mason, N. J., Scheier, P., and Märk, T. D., *Phys. Rev. Letters* **82**, 5028–5031 (1999).
7. Denifl, G., Muigg, D., Stamatovic, A., and Märk, T.D., *Chem. Phys. Letters* **288**, 105–110 (1998).
8. Kieffer, L. J., and Dunn, G. H., *Rev. Mod. Physics* **38**, 1–35 (1966).
9. Märk, T. D., "Partial ionization cross sections," in *Electron Impact Ionization,* edited by T.D.Märk and G.H.Dunn, Wien: Springer–Verlag, 1985, pp. 137–197.
10. de Heer, F. J., and Inokuti, M., "Total ionization cross sections," in *Electron Impact Ionization,* edited by T.D.Märk and G.H.Dunn, Wien: Springer–Verlag, 1985, pp. 232–276.
11. Tate, J. T., and Smith, P. T., *Phys. Review* **39**, 270–277 (1932).
12. Märk, T. D., and Dunn, G. H., *Electron Impact Ionization,* Wien: Springer–Verlag, 1985.
13. Märk, T. D., "Ionization of molecules by electron impact", in *Electron Molecule Interactions And Their Applications,* edited by L.G.Christophorou, New York: Academic Press, 1984, pp.251–334.
14. Freund, R. S., "Electron impact ionization cross sections for atoms, radicals and metastables," in *Swarm Studies And Inelastic Electron Molecule Collisions,* edited by L.C.Pitchford, B.V.Mc.Koy, A.Chutjian and S.Trajmar, New York: Springer, 1987, pp. 329–346.
15. Märk, T. D., *Plasma Physics Controlled Fusion,* **34**, 2083–2090 (1992).
16. Märk, T. D., "Mechanisms and kinetics of electron impact ionization of atoms, molecules and clusters", in *Linking The Gaseous And The Condensed Phases Of Matter,* edited by

L.G.Christophorou, E.Illenberger and W.F.Schmidt, New York: Plenum Press, 1994, pp. 155–182.
17. Märk, T.D.,"Partial electron ionization cross sections", Invited review lecture , XIIIth Summer school on the physics of ionized gases (SPIG), Sibenik (1986).
18. Deutsch, H., Becker, K., Matt, S., and Märk, T. D., *Int.J.Mass Spectrometry* **197** 37–69 (2000).
19. Poll, H. U., Winkler, C., Margreiter, D., Grill, V., and Märk, T. D., *Int. J. Mass Spectrometry Ion Processes* **112,** 1–17 (1993).
20. Grill, V., Walder, G., Margreiter, D., Rauth, T., Poll, H. U., Scheier, P., and Märk, T. D., *Z. Physik D* **25,** 217–226 (1993).
21. Foltin, V., Foltin, M., Matt, S., Scheier, P., Becker, K., Deutsch, H., and Märk, T. D., *Chem. Phys. Letters* **289,** 181–188 (1998).
22. Basner, R., Schmidt, M., Becker, K., and Deutsch, H., *Adv. Atomic Molecular Optical Physics* **43,** 147–185 (2000).
23. Stephan, K., Deutsch, H., and Märk, T. D., *J. Chem. Physics* **83,** 5712–5720 (1985).
24. Poll, H. U., and Meichsner, J., *Beiträge Plasmaphysik* **27,** 359–372 (1987).
25. Ma, C., Sporleder, C. H., and Bonham, R. A., *Phys. Review A* **44,** 2921–2934 (1991).
26. Poll, H. U., Grill, V., Matt, S., Abramzon, N., Becker, K., Scheier, P., and Märk, T.D., *Int. J. Mass Spectrometry Ion Processes* **177,** 143–154 (1998).
27. Mair, C., Fiegele, T., Biasioli, F., Wörgötter, R., Grill, V., Lezius, M., and Märk, T.D., *Plasma Sources Sci. Technology* **8,** 191–202 (1999).
28. Freund, R. R., Wetzel, R. C., Shul, R. J., and Hayes, T. R., *Phys.Review A* **41,** 3575–3595 (1990).
29. Tarnovsky, V., and Becker, K., *J. Chem. Phys.* **98,** 7868–7875 (1992).
30. Straub, H. C., Renault, P., Lindsay, B. G., Smith, K. A., and Stebbings, R. F., *Phys. Review A* **52,** 1115–1124 (1995).
31. Straub, H. C., Lindsay, B. G., Smith, K. A., and Stebbings, R. F., *J. Chem. Physics* **108,** 109–116 (1998).
32. Tian, C., and Vidal, C. R., *J. Chem. Physics* **108,** 927–936 (1998).
33. Tian, C., and Vidal, C. R., *Phys. Review A* **59,** 1955–1965 (1999).
34. Almeida, D. P., *Int. J. Mass Spectrometry* **184,** 49–56 (1999).
35. Bruce, M. R., Ma, C., and Bonham, R. A., *Chem .Phys. Letters* **190,** 285–290 (1992).
36. Thompson, W. R., Shah, M. B., and Gilbody, H. B., *J. Phys. B* **28,** 3121–1330 (1995).
37. Jiao, C. Q., Garscadden, A., and Haaland, P. D., *Int. J. Mass Spectrometry* **184,** 83–88 (1999).
38. O'Connor, C. S. S., and Price, S. D., *Int. J. Mass Spectrometry 184,* 11–23 (1999).
39. Krishnakumar, E., and Srivastava, S. K., *Int. J. Mass Spectrometry and Ion Processes* **113,** 1–12 (1992).
40. Syage, J. A., *Phys. Review A* **46,** 5666–5679 (1992).
41. Johnston, M., Fujii, K., Nickel, J., and Trajmar, S., *J. Phys. B* **29** 531–543 (1995).
42. Becker, K., *Novel Aspects Of Electron–Molecule Collisions,* Singapore: World Scientific, 1998.
43. Dunn, G. H., and Djuric, N., "Electron Impact Dissociative Excitation and Ionization of Molecular Ions" in *Novel Aspects Of Electron–Molecule Collisions,* edited by K.Becker, Singapore: World Scientific, 1998, pp.241–281.
44. Matt, S., Scheier, P., Märk, T. D., and Becker, K., " Positive and Negative Ion Formation in Electron Collisions with Fullerenes" in *Novel Aspects Of Electron–Molecule Collsions,* edited by K.Becker, Singapore: World Scientific, 1998, pp.1–56.
45. Cooks, R. G., Beynon, J. H., Caprioli, R. M., and Lester, G. R., *Metastable Ions,* Amsterdam: Elsevier, 1973.
46. Levsen, K., *Fundamental Aspects Of Organic Mass Spectrometry,* Weinheim: Verlag Chemie, 1978.
47. Matt, S., Parajuli, R., Stamatovic, A., Scheier, P., Märk, T. D., Laskin, J., and Lifshitz, C., *Europ. Mass Spectrom.* **5,** 477–484 (1999).
48. Ji, Y., Foltin, M., Liao, C. H., and Märk, T. D., *J. Chem. Physics* **96,** 3624–3627 (1992).
49. Laskin, J., Weickhardt, C., and Lifshitz, C., *Int. J. Mass Spectrometry Ion Processes* **161,** L7–L11 (1997).
50. Matt, S., Echt, O., Stamatovic, A., and Märk, T. D., *J. Chem. Physics* **113,** July 8 issue (2000).

51. Matt, S., Sonderegger, M., David, R., Echt, O., Scheier, P., Laskin, J., Lifshitz, C., and Märk, T. D., *Int. J. Mass Spectrometry Ion Physics* **185/186/187,** 813–823 (1999).
52. Rosenstock, H. M., *Int. J. Mass Spectrometry Ion Physics* **20,** 139–190 (1976).
53. Ng, C. Y., *Adv. Chem. Physics,* **52,** 263–362 (1983).
54. Märk, T. D., and Castleman, Jr., A. W., *Adv. Atomic Molecular Physics* **20,** 66–171 (1985).
55. Haberland, H., *Clusters Of Atoms And Molecules,* Berlin: Springer, 1994
56. Ganteför, P., Bröker, G., Holub–Krappe, E., and Ding, A., *J. Chem. Physics* **91,** 7972–7977 (1989).
57. Steger, H., Holzapfel, J., Hielscher, A., Kamke, W., and Hertel, I.V., *Chem. Phys. Letters* **234,** 455–459 (1995).
58. Kamke, W., deVries J., Krauss, J., Kaiser, E., Kamke, B., and Hertel, I.V., *Z.Physik D* **14,** 339–351 (1989).
59. Muigg, D., Denifl, G., Stamatovic, A., Echt, O., and Märk, T.D., *Chem. Phys. Letter* **239,** 409–416 (1998).
60. Muigg, D., Denifl, G., Fiegele, T., Mason, N.J., Stamatovic, A., and Märk, T.D., *J. Chem. Physics,* in print (2000).
61. Matt, S., Echt, O., Wörgötter, R., Grill, V., Scheier, P., Lifshitz, C., and Märk, T.D., *Chem. Phys. Letters* **264,** 149–156 (1997).
62. NIST Tables, http:// webbook.nist.gov
63. Dyke, J. M., Golob, L., Jonathan, N., Morris, A., and Okuda, M.J., *J. Chem. Soc. Faraday Trans.* **70,** 1828–1836 (1974).
64. Slanina, Z., Adamowicz, L., *J. Atmospheric Chemistry* **16,** 41–46 (1993).
65. Mähnert, J., Baumgärtel, H., and Weitzel, K. M., *J. Chem. Physics* **103,** 7016–7024 (1995).
66. Rudge, M. R. H., *Rev. Mod. Physics* **40,** 564–590 (1968).
67. Younger, S. M., "Quantum theoretical methods for calculating ionization cross sections," in *Electron Impact Ionization,* edited by T. D. Märk and G. H. Dunn, Wien: Springer–Verlag, 1985, pp. 1–23.
68. Younger, S. M., and Märk, T. D., "Semi–empirical and semi–classical approximations for electron ionization," in *Electron Impact Ionization,* edited by T. D. Märk and G. H. Dunn, Wien: Springer–Verlag, 1985, pp. 24–41.
69. Ötvos. J. W., and Stevenson, D. P., *J. Am. Chem. Soc.* **78,** 546–551 (1956).
70. Fitch, W. L., and Sauter, A. D., *Anal. Chem.* **55,** 832–835 (1983).
71. Deutsch, H., and Schmidt, M., *Beitr. Plasmaphysik* **24,** 475–486 (1984).
72. Deutsch, H., Becker, K., and Märk, T. D., *Int. J. Mass Spectrometry Ion Processes* **167/168,** 503–517 (1997).
73. Deutsch, H., Becker, K., Basner, R., Schmidt, M., and Märk, T. D., *J. Phys. Chemistry* **106,** 8819–8826 (1998).
74. Bobeldijk, M., van der Zande, W. J., and Kistemaker, P. G., *Chem. Phys.* **179,** 125–130 (1994).
75. Aitken, C. G., Blunt, D. A., and Harland, P. W., *J. Chem. Physics* **101,** 11074–11076 (1994).
76. Aitken, C. G., Blunt, D. A., and Harland, P. W., *Int. J. Mass Spectrometry Ion Processes* **149/150,** 279–286 (1995).
77. Vallance, A., Harland, P. W., and Maclagan, R. G. A. R., *J. Phys. Chemistry* **100,** 15021–15026 (1996).
78. Margreiter, D., Deutsch, H., Schmidt, M., and Märk, T. D., *Int. J. Mass Spectrometry Ion Processes* **100,** 157–176 (1990).
79. Deutsch, H., Cornelissen, C., Cespiva, L., Bonacic–Koutecky, V., Margreiter, D., and Märk, T. D., *Int. J. Mass Spectrometry Ion Processes* **129,** 43–48 (1993).
80. Deutsch, H., Märk, T. D., Tarnovsky, V., Becker, K., Cornelissen, C., Cespiva, L., and Bonacic–Koutecky, V., *Int. J. Mass Spectrometry Ion Processes* **137,** 77–91 (1994).
81. Jain, D. K., and Khare, S. P., *J. Phys. B* **9,** 1429–1438 (1976).
82. Khare, S. P., and Meath, W. J., *J. Phys. B* **20,** 2101–2116 (1987).
83. Kim, Y. K., and Rudd, M. E., *Phys. Rev. A* **50,** 3954–3967 (1994).
84. Hwang, W., Kim, Y. K., and Rudd, M. E., *J. Chem. Physics* **104,** 2956–2966 (1996).
85. Kim, Y. K., Hwang, W., Weinberger, N. M., Ali, M. A., and Rudd, M. E., *J. Chem. Physics* **106,** 1026–1033 (1997).

86. Straub, H. C., Renault, P., Lindsay, B. G., Smith, K. A., and Stebbings, R. F., *Phys. Rev. A* **54,** 2146–2153 (1996).
87. Rapp, D., and Englander–Golden, P., *J. Chem. Physics* **43,** 1465–1479 (1965).
88. Schram, B. L., de Heer F. J., van der Wiel, M. J., and Kistemaker, J., *Physics* **31,** 94–112 (1965).
89. Schram, B. L., Moustafa, H. R., Schutten, J., and de Heer, F. J., *Physica* **32,** 734–740 (1966).
90. Schultz, D. R., Meng, L., and Olson, R. E., *J. Physics B* **25,** 4601–4618 (1992).
91. Saksena, V., Kushwaha, M. S., and Khare, S. P., *Physica B* **233,** 201–212 (1997).
92. Straub, H. C., Lindsay, B. G., Smith, K. A., and Stebbings, R. F., *J. Chem. Physics* **105,** 4015–4022 (1996).
93. Shyn, T. W., and Sharp, W. E., *Phys Review A* **20,** 2332–2339 (1979).
94. Orient, O. J., and Srivastava, S. K., *J. Physics B* **20,** 3923–3936 (1987).
95. Nishimura, H., Huo, W. M., Ali, M. A., and Kim, Y. K., *J. Chem. Physics* **110,** 3811–3822 (1999).
96. Rao, M. V. V. S., and Srivastava, S. K., in *Book of Abstracts of Contributed Papers*, edited by F. Aumary, G. Betz and HP. Winter, XXth Int. Conference on the Physics of Electronic and Atomic Collisions, Wien: IAP, Universität Wien, 1997.
97. Bruce, M. R., and Bonham, R. A., *Int. J. Mass Spectrometry and Ion Processes* **123,** 97–100 (1993).
98. Beran, J. A., and Kevan, L., *J. Phys. Chemistry* **73,** 3866–3876 (1969).
99. Margreiter, D., Walder, G., Deutsch, H., Poll, H. U., Winkler, C., Stephan, K., and Märk, T. D., *Int. J. Mass Spectrometry and Ion Processes* **100,** 143–156 (1990).

The measurement of absolute Cross Sections for low energy Rearrangement Processes

P. Defrance, F. Brouillard and X. Urbain

Department of Physics, Université catholique de Louvain, Chemin du Cyclotron, 2 B-1348 Louvain-la-Neuve (Belgium)

Abstract. In this paper we show, on a few examples, how the merging beam method is applied to measure absolute cross sections of low energy rearrangement processes such as resonant electron transfer, transfer ionisation and associative ionisation.

INTRODUCTION

The lack of reliable collisional data in the low energy range, already pointed out by Mitchell and McGowan [1] has remained an obstacle for a realistic modelling of low temperature plasmas (astrophysical plasmas, plasma edge or divertor plasmas of thermonuclear reactors). The relevant processes are those with a large cross section at low collisional energies as resonant electron transfer or transfer ionisation and processes that involve a rearrangement of the nuclei as associative ionisation or reactive scattering.

The present paper is dealing with the measurement of the (absolute) cross section of such low energy processes using merging beams, i.e. the only experimental method capable of measuring cross sections at energies below 1eV.

In the following sections, we first give a brief description of the method, showing its possibilities and intrinsic limitations. Then we present some of the results recently obtained for electron transfer, transfer ionisation and associative ionisation.

MERGING BEAMS

In merged beam experiments the absolute cross section (σ) is derived [2] from the number of reactions N(T) observed during a time T through the relation :

$$N(T) = \sigma \frac{v}{q_1 q_2 v_1 v_2} \int_0^T dt \int I_1(t) I_2(t) F(t) dt \qquad (1)$$

CP543, *Atomic and Molecular Data and Their Applications*, edited by K. A. Berrington and K. L. Bell

where v_1, v_2, q_1, q_2, $I_1(t)$, $I_2(t)$ are the velocities, charges and intensities in beams 1 and 2 respectively. v is the relative velocity. F(t) is the so called form factor, a quantity accounting for the beam overlap and defined as :

$$F(t) = \int_L F'(z,t)dz \qquad (2)$$

with

$$F'(z,t) = \frac{\int_{-\infty}^{+\infty}\int_{-\infty}^{+\infty} j_1(x,y,z,t)\, j_2(x,y,z,t)dxdy}{I_1(t)I_2(t)} \qquad (3)$$

j_1 and j_2 are the current densities and L is the effective length of the interaction region. The knowledge of the form factor requires the measurement of the density profiles of the two beams. Presently, this is conveniently done thanks to the computer-assisted measurement of the beam profiles [2]. This measurement is not easy and has to be performed separately, before or after the measurement of the reaction rate. There thus always remains a doubt about the actual value of the form factor during the measurement.

To circumvent this problem, the so-called "Animated-Beam Method" (AMB) was developped [2], first for crossed-beams [3] and more recently for merged beams. For crossed beams, one beam is swept at a constant velocity across the second one, so that the cross section only depends on the sweep velocity and not on a the form-factor. In the case of merging beams, one of the beams is moved against the other in order to scan the entire cross section of this second one (TV-screen).

Measurements with animated beams are time consuming and can therefore not be systematically carried out in the case of small cross sections. In this case, they are used as a calibration tool.

The interaction length L is defined by applying a voltage V on the observation cell. As a result, the kinetic energy of products formed in this cell is raised or reduced according to their initial and final charges. A magnetic analyzer separates the products from the primary beams and from ions formed outside the so-defined collision cell.

The observation voltage also determines the barycentric energy E_{cm}:

$$E_{cm} = \frac{\mu}{m_1}E_1 + \frac{\mu}{m_2}E_2 - 2\sqrt{\frac{\mu}{m_1}E_1\frac{\mu}{m_2}E_2}\cos\theta \qquad (4)$$

with $E_1 = q_1(A_1 - V)/m_1$ and $E_2 = q_2(A_2 - V)/m_2$ where A_1, A_2 and m_1, m_2 are the acceleration voltages and masses of each colliding partner, respectively and θ is the angle between the interacting particle velocities.

In principle, the zero CM-energy can be reached by setting $v_1 = v_2$ and $\theta = 0$. In practice however the energy resolution must be taken into account. And this is limited (i) by the energy spread and (ii) by the angular distribution of the particles in each of the beams.

The angular spread is the main limiting factor as it cannot, practically, be made smaller than 1 milliradian and therefore contributes to about 0.01 eV in the energy resolution, while the contribution of the energy spread can be kept below 0.001 eV. The values of the cross section measured at very low energies are thus average values that do not reflect the true energy dependence but tend to a ceiling value, which is the average value of the cross section over the range 0–0.01 eV.

EXPERIMENTAL SET-UP AND PROCEDURE

The set-up presently used for the study of associative ionisation in the collision of a positive and a negative ion is schematized in Fig. 1. A more detailed description can be found elsewhere [2].

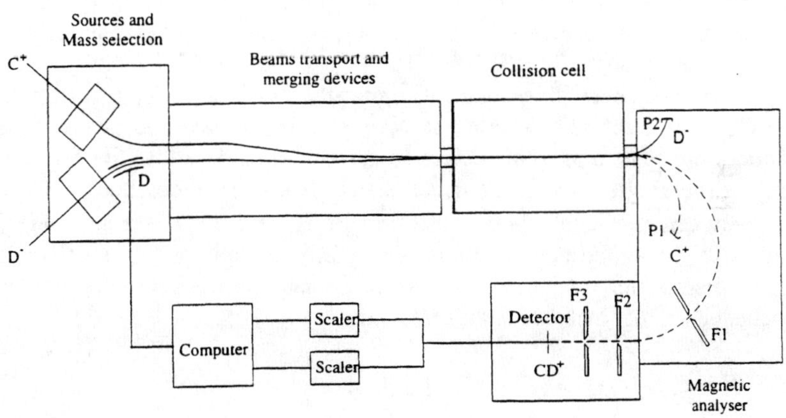

FIGURE 1. Schematic block diagram of the experimental set-up.

Four sections can be distinguished. The first one includes the ion sources and the beam forming optical devices. Positive ions (A^+) and negative ions (B^-) are produced by an Electron Cyclotron Resonance (ECR) ion source and by a duoplasmatron source, respectively. In the second section, the mass selected beams are merged by a set of electrostatic plane deflectors and the interaction takes place in the third section. The interaction region consists of an electrically biased cell called the observation cell and contains some probes and beam profile and position measuring devices. The last section contains the magnetic analyser where the reaction products are separated from the primary beams. Detectors (channel electron multipliers) are placed at various positions for the detection of the reaction products (ions or atoms) The diaphragms are adjusted so that a total transmission of the products to the detectors is achieved.

CHARGE EXCHANGE : $H^- + He^+ \rightarrow H + He$

Among many other reactions, the mutual neutralization between H^- and He^+ is interesting for thermonuclear research as well as for astrophysical investigations because hydrogen and helium (including their ions) are two of the main components of the corresponding plasmas.

The cross section of that reaction was remeasured using both the usual static beam method and the animated beam method [2]. Absolute cross sections obtained in both ways are in good agreement with each other over the whole energy range (5-4000 eV). They are in good agreement with previous measurements by Peart and Hayton [6] but in the low energy range only (E<200eV).

The reaction has been investigated theoretically by Olson et al [19], Ermolaev et al [7] and Chibisov et al, [5]. The last study, which is the most recent and complete one, considers the system as a three electron system which includes the formation of He*(1s,nl) in the singlet and in the triplet states. It also includes the possibility of exchange between the losely bound electron of H^- and the electron of He^+. The agreement with the experimental results is not fully satisfactory: at small collision energies, the discrepancy may be due to transfer ionisation processes, competing with charge exchange and not included in the calculation:

$$\begin{aligned} H^- + He^+(1s) &\rightarrow H^+ + e^- + He(1s^2) \\ &\rightarrow H^+ + e^- + He(1s,nl) \rightarrow H^+ + e^- + He(1s^2) \end{aligned} \qquad (5)$$

At large collision energies, the direct ejection of the loosely bound electron of H- should also be considered. A detailed analysis of this reaction is given by Chibisov et al [5].

TRANSFER IONISATION : $H^- + He^{2+} \rightarrow H^+ + He^+ + e^-$

The cross section for transfer ionisation in the collision of H^- with He^{++} has been measured by Cherkani et al [8] in a merged beam experiment using the set-up described above. In addition to the cross section, the kinetic energy transfered to the ionic products was also measured by a time of flight analysis. The latter measurement confirmed that the transfer ionisation essentially proceeds in two steps : (i) a first electron transfer at large internuclear distance leading to the formation of He^+ (n=4,5) and H(1s); (ii) a second electron transfer, at smaller distance, thus operated in the field of the two nuclei and proceeding as the resonant transfer

$$H(1s) + He^{++} \rightarrow H^+ + He^+(n=2) \tag{6}$$

The result is a doubly excited state of He that quickly autoionises.

ASSOCIATIVE IONISATION : $C^+ + D^- \rightarrow CD^+ + e^-$

The merged beam set-up has been intensively used for the measurement of associative ionisation in the collision of H^- or D^- with positive ions [4,12]. A case of particular interest is that of the reaction :

$$C^+ + H^- \rightarrow CH^+ + e$$

because of the discrepancy existing between the observed and the calculated abundances of this molecular ion in interstellar clouds. The solution of this problem requires the knowledge of the relative importances of all the individual processes involved both in the formation and in the destruction of CH^+. Its formation by associative ionisation between atomic species has however remained unexplored, both theoretically and experimentally. The processes by which CH^+ can be formed are the following

$$C^+ + H^- \rightarrow CH^+ + e$$
$$C^- + H^+ \rightarrow CH^+ + e$$
$$C^+ - + H^+ \rightarrow CH^+ + h\nu$$
$$C - + H^+ \rightarrow CH^+ + h\nu$$
$$C^+ - + H_2 \rightarrow CH^+ + H$$
$$C^+ + H^- \rightarrow CH^+ + e$$

The destruction of CH^+ mainly occurs through dissociative and dielectronic recombination :

$$CH^{+} + e \rightarrow C + H$$
$$CH^{+} + e \rightarrow CH^{*} + h\nu$$

Dissociative excitation or recombination has been the subject of many theoretical studies that were traced by Amitay et al [9]. It has also been investigated experimentally first by Mul et al [10] and more recently by Amitay et al [11].

We report here on the measurement of the cross section for the reaction :

$$C^{+} + D^{-} \rightarrow CD^{+} + e$$

which, although carried out with deuterium, gives a measure of the importance of associative ionisation in the formation of CH^{+}

A nice feature with merged beam measurements on associative ionisation is that the molecular ion can only be produced by the collisions between the beams and not with the residual gas. The measurement is therefore practically free of background. Backgrounds produced by elastic scattering of the primary ions or by the photons emitted at the impact on surfaces are easily dealt with by a chopping of the beams. In the present case the D^{-} beam was chopped by applying a crenel voltage on deflector D (see fig. 1). It had been verified that the negative beam by itself did not produce any background. Thus, the reaction rate was obtained from the difference between the counting rates with the negative beam on and off.

The cross section has been measured at collision energies ranging from 0.01 to 2 eV. Typical experimental conditions and the main factors limiting the accuracy of the present measurements are given in table 1 together with the corresponding errors.

TABLE 1. Typical experimental conditions and errors.

Beam acceleration voltages	18 kV (C^{+})
	4.5 kV (D^{-})
Observation voltage	-1.3 kV
Beam diameters (common)	0.15 cm
C^{+} beam intensity	40 nA
D^{-} beam intensity	40 nA
Background	10 Hz
Signal count rate from 0.01 to 2 eV	5 to 0.4 Hz
Interaction length L	6.8 cm
Beam divergence	2 mrad
Error on the detection efficiency	2%
Error on the interaction length	3%
Error on the beam intensities	2%

In figure 2 we show the cross section as a function of the barycentric energy E. It appears that the energy dependence is compatible, at low energy, with an E^{-1} law. The deviation from this law, observed at very low energy, can be explained by the limited energy resolution of our measurements (see above).

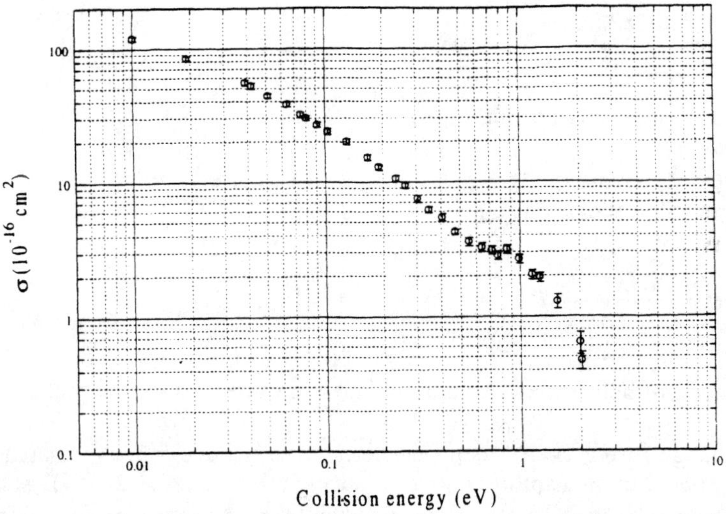

FIGURE 2. Cross section as a function of the barycentric energy.

The energy dependence, as derived from the experimental data between 0.04 and 0.3 eV is :

$$\sigma = \frac{2.1x10^{-16}}{E(eV)} cm^2 \qquad (16)$$

Regarding again the associative ionisation as a two-step process, consisting of an electron transfer :

$$C^- + D^+ \rightarrow C(n) + D(1s) \tag{17}$$

occuring at the internuclear distance R_c with a probability P_c and leading to the formation of a molecular state that autoionises at distances smaller than R_a with an average probability P_a, the cross section can be shown, from a simple classical analysis of the relative motion, to become, at low energy :

$$\sigma = \pi a_o^2 x 27.2 \frac{R_a^2}{R_c} \frac{P_a P_c}{E(eV)} \tag{18}$$

where R_a and R_c are in atomic units (a_o).

If the first step is assumed to take place at $R_c = 16a_o$, corresponding to n=3, and R_a is given a typical value of $3a_o$, one obtains :

$$\sigma = \frac{13.4x10^{-16}}{E(eV)} P_a P_c \tag{19}$$

which compared with the experimental result, gives a value 0.15 for the product P_aP_c.

The change in the energy dependence observed above 0.6 eV (plateau) could possibly be related to the capture in n=2 producing the $(1s)^2(2s)^2(2p)^2\ {}^1D$ state of the carbon and a doubly excited state of the molecule with an energy curve passing above that of the molecular ion at $R < 2a_o$.

All the above considerations do not provide an explanation of the experimental results but are just preliminary qualitative comments. A proper theoretical treatment is desirable as is desirable an intensification of the theoretical activity to clarify the mechanisms of fundamental rearrangement processes in simple molecular systems.

ASSOCIATIVE IONISATION : $H(1s) + H(s) \rightarrow H_2^+ + e^-$

It is obvious that the cross section for the associative ionisation reaction :

$$H(1s) + H(ns) \rightarrow H_2^+ + e$$

is depending on the quantum number n.

Due to the binding energy of the molecular ion and to the atoms internal energy, this reaction is endothermic for n = 1, 2, and exothermic for n > 2. This means that, in the low energy region, the cross section will have a threshold in the first case, while it will follow the E^{-1} power law in the second one. Measurements have been performed for these reactions for n = 2, 3 and 4.

The apparatus used for these experiments differs from the preceding one mainly because the two beams are produced from a single ion (proton) source. The proton beam is partially neutralised in two different gas cells, biased by different electric potentials to produce two atomic beams with different kinetic energies. The atoms formed in the first cell, (Argon) are essentially in the ground state while those produced in the second cell (cesium vapor) are essentially in the n=2 states with 25% of them in the metastable 2s state.

Higher excited states (3s,4s) are obtained by laser excitation of the metastable in presence of an electric field.

The technique of rapid adiabatice passage is used to increase the excitation yield (up to 90%). Measurements have been carried out for the following reactions :

$$H(1s) + H(2s) \rightarrow H_2^+ + e$$

$$D(1s) + D(2s) \rightarrow D_2^+ + e$$

$$H(1s) + H(3s) \rightarrow H_2^+ + e$$

$$D(1s) + D(3s) \rightarrow D_2^+ + e$$

$$H(1s) + H(4s) \rightarrow H_2^+ + e$$

over the barycentric energy range 0.01-10eV.

The results are shown in Fig. 3. Part of them have been previously published [4, 12, 15].

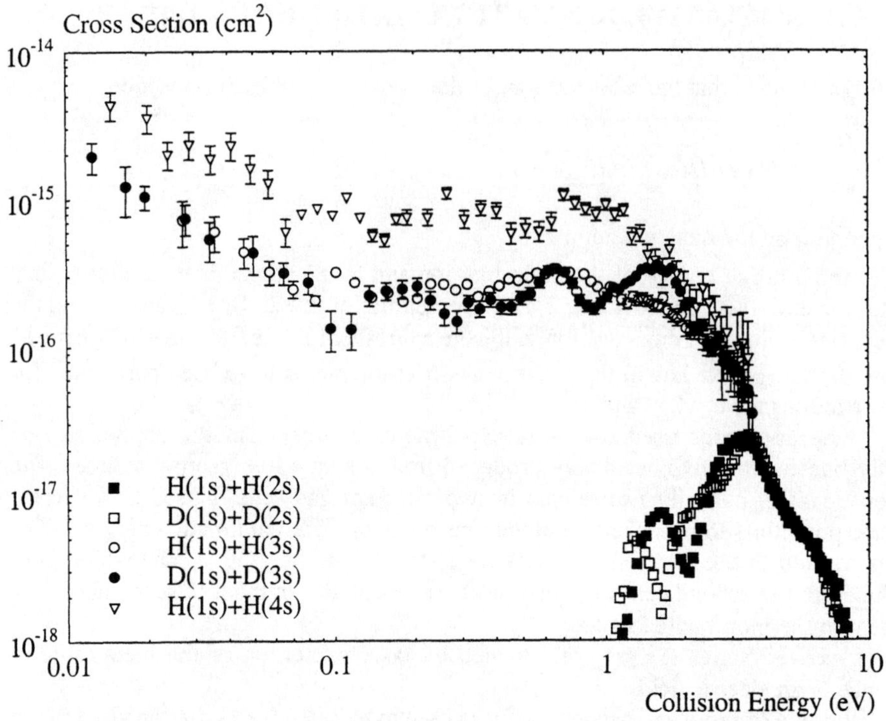

FIGURE 3. Cross section as a function of the barycentric energy for associative ionization between hydrogen or deuterium atoms in various excited states.

Three regions are clearly observed. At low energy (below 0.1eV), the cross sections follow the E^{-1} law. A high energy (above 3eV), they fall much more rapidly, reflecting the increasing importance of dissociation channels.

At intermediate energies (from 0.1 to 3eV), the cross section shows a oscillatory dependence on the collision energy which is most probably due to the interference of two (or more) reaction channels.

The cross section for associative ionisation in the collision of two metastable atoms has been measured previously by X. Urbain et al [14]

$$H(2s) + H(2s) \rightarrow H_2^+ + e$$

$$D(2s) + D(2s) \rightarrow D_2^+ + e$$

Those reactions have been found to have unexpectedly large cross sections.

REFERENCES

1. Mitchell, J. B. A., and McGowan, J. Wm., Ap. J. **222** 1978 L77-7.
2. Olamba, K., Szücs, S., Chenu, J. P., Naji, A., and Brouillard, F., *J. Phys. B: At. Mol. Opt. Phys.* **29,** 2837-2846 (1996)
3. Defrance, P; Brouillard, F., Claeys, W. and Van Wassenhove, G., *J. Phys. B: At. Mol* **4**, 103-110 (1981)
4. Naji, A., Olamba,K., Chenu, J. P., Szücs, S., Chibisov, M., and Brouillard, F., *J. Phys. B: At. Mol. Opt. Phys.* **31**, pp. 2961–2970,(1998).
5. Chibisov, M., Brouillard, F., Chenu, J. P., Cherkani, M.H., Fussen, D., Olamba, K., and Szücs, S., *J. Phys. B: At. Mol. Opt. Phys.* **31**, pp. 2961–2970 (1997).
6. Peart, B. and Hayton, D.A., *J. Phys. B: At. Mol. Opt. Phys.* **27**, pp. 2551 (1994)
7. Ermolaev, A.M., *J. Phys. B: At. Mol. Opt. Phys.* **25**, 3133 (1992)
8. Cherkani, M.H., Szücs, S., Hus, H., and Brouillard, F., *J. Phys. B: At. Mol. Opt. Phys.* **24**, pp. 2367–2377 (1991).
9. Amitay, Z., Zajfman, D., Forck, P., Hechtfischer, U., Seidel, B., Grieser, M., Habs, D., Repnow, R., Schwalm, D., and Wolf, A., Phys. Rev. A **54**, 4032 (1996).
10. Mul, P. M., Mitchell, J. B. A., Defrance, P., D'Angelo, V. S., McGowan, J. Wm., Froelich, H.R., *J. Phys. B **14**, 1353 (1981).*
11. Amitay, Z., Zajfman, D., Forck, P., Hechtfischer, U., Grieser, M., Habs, D., Schwalm, D., and Wolf, A., (1998) private communication.
12. Naji, A., Olamba,K., Chenu, J. P., Szücs, S., and Brouillard, F., *J. Phys. B: At. Mol. Opt. Phys.* **31**, pp. 4887–4894 (1998).
13. Rearrangement Processes Involving Hydrogen and Helium Atoms and Ions, 1991, p. 1685 Brouillard, F. and Urbain, X., Atomic and Molecular Processes in Fusion Edge Plasmas, R. Janev, ed., Plenum Press, New York and London (1995), pp 309-339
14. Urbain, X., Cornet, A., Brouillard, F., and Giusti-Suzor, A., Phys. Rev. Lett. **66**, p. 1685 (1991).
15. Urbain, X., Nehari, D. Masukidi, L.S. and Jureta J., «□Absolute total cross section for associative ionization in D(1s)+D(3s) collisions□» in *6th EPS Conference on Atomic and Molecular Physics, Contributed Papers,* edited by C. Biancalana et al., Europhysics Conference Abstracts 22D, Siena, Italy, 1998, p. 4-148
16. Nehari, D., Andrianarijaona, V., and Urbain, X., private communication, to be presented at the «□□Colloque de la division de Physique Atomique, Moléculaire et Optique de la Société française de Physique, PAMO-2000□» Lyon, France, 2000
17. Urbain, X., Cornet, A., and Jureta, J., *J. Phys. B: At. Mol. Opt. Phys.* **25**, pp. L189-192 (1992).
18. Karangwa, P.-C., Essarroukh, N., Jureta, J., and Urbain, «□Associative ionization between metastable deuterium atoms in the energy range 0.02-2 eV□» in *XXth International Conference on the Physics of Electronic and Atomic Collisions, Scientific Program and Abstracts of Contributed Papers, volume 2,* edited by Aumayr et al., Vienna, Austria, 1997 p. TU-178
19. Olson, R.E., Peterson, J., and Moseley, J., J. Chem. Phys. **53**, p. 3391 (1970).

Electron-Impact Ionization Cross Sections for Polyatomic Molecules, Radicals, and Ions

Yong-Ki Kim and Karl K. Irikura

National Institute of Standards and Technology, Gaithersburg, Maryland 20899 U.S.A.

Abstract. The binary-encounter-Bethe (BEB) model for calculating electron-impact ionization cross sections for molecules is described with emphasis on polyatomic molecules and their radicals and ions. The BEB model provides an analytic formula for the cross section. The formula is for each molecular orbital, and requires only three numbers from the initial state of the target molecule: the electron occupation number, the orbital binding energy, and the orbital kinetic energy. The sum of the orbital cross sections provides the molecular cross section. The BEB model produces reliable ionization cross sections from the threshold to several keV in incident electron energy for neutral molecules and ions. The BEB cross sections for some fluorine-containing radicals are almost a factor of two larger than available experimental values. Molecular constants are included to generate BEB cross sections for H_2, C_2F_6, H_2^+, CH_3, CF_x, BF_x, BCl_x for x=1-3, and several unsaturted molecules relevant to hydrocarbon combustion.

1. INTRODUCTION

Electron-impact ionization cross sections of atoms and molecules are important not only for the development of basic collision theory, but also in a wide range of practical applications. For instance, the modeling of semiconductor manufacturing by plasma processing requires ionization cross sections of fluorocarbons used as the feed gas, their ions and fragments. Hydrocarbon molecules, such as methane, are found in the vicinity of divertors in a fusion device, making it necessary to include ionization of such molecules to model low-temperature plasmas in tokamaks. Ionization cross sections near the ionization threshold are also used in modeling combustion of flammable gases. In this article, we describe a simple, effective theory—the binary-encounter-Bethe (BEB) model—to calculate electron-impact ionization cross sections for large and small molecules, including their fragments and ions.

The BEB model combines a modified form of the Mott cross section and the dipole-interaction term of the plane-wave Born approximation. The original Mott cross section [1] is for the collision of two free electrons. This formula must be modified to include the binding energy B, so that the cross section does not diverge, as the original Mott cross section does, for the ejection of an electron with a very low kinetic energy W. Using the plane-wave Born approximation, Bethe has shown [2] that the leading term of the Born cross section at high incident electron energy T represents the

CP543, *Atomic and Molecular Data and Their Applications,* edited by K. A. Berrington and K. L. Bell
2000 American Institute of Physics 1-56396-971-8

dipole interaction between the incident and the target electrons. The dipole interaction is characteristic of a bound electron, and the Mott cross section lacks this feature.

Since the Mott cross section is derived using a Coulomb function for the incident electron while the Born approximation uses a plane wave, these two theories cannot be simply added. In the past, many attempts were made to combine the Mott and Born cross sections mostly by introducing empirical parameters. Kim and Rudd [3] have succeeded in combining the two types of cross sections without introducing any adjustable or empirical parameters. They required that both the combined ionization cross section and the matching stopping cross section (integral of energy transfer times ionization cross section) satisfy the correct asymptotic (high *T*) forms derived by Bethe [2,4].

Kim and Rudd [3] combined differential ionization cross sections, which used continuum dipole oscillator strength *f(E)* with energy transfer *E*. In principle *f(E)* can be deduced from photoionization cross sections. What is required in their theory, however, is *f(E)* for each molecular orbital, not just the total photoionization cross section. Often details of *f(E)* for each orbital are available only for simple molecules such as H_2. When the weighted integral of *f(E)* shown in Eq. (2) is known, we get a simple expression for the integrated cross section—i.e., integrated over the secondary electron energies W—which we refer to as the BEQ cross section:

$$\sigma_{\mathrm{BEQ}}=\frac{S}{t+u+1}\left[\frac{Q}{2}\ln t\left(1-\frac{1}{t^2}\right)+(2-Q)\left(1-\frac{1}{t}-\frac{\ln t}{1+t}\right)\right], \qquad (1)$$

$$Q=\frac{2BM_i^2}{NR}, \quad \text{with} \quad M_i^2=\frac{R}{B}\int_B^{\infty}\frac{B}{E}f(E)\mathrm{d}E, \qquad (2)$$

where, $S=4\pi a_0^2N(R/B)^2$, a_0 is the Bohr radius, R is the Rydberg energy, N is the electron occupation number, $t = T/B$, $u = U/B$, and U is the orbital kinetic energy.

When *f(E)* is unknown, we can approximate $Q = 1$, and obtain the BEB equation:

$$\sigma_{\mathrm{BEB}}=\frac{S}{t+u+1}\left[\frac{\ln t}{2}\left(1-\frac{1}{t^2}\right)+1-\frac{1}{t}-\frac{\ln t}{1+t}\right]. \qquad (3)$$

The first logarithmic term in Eq. (3) represents the dipole interaction from the Born approximation, $1-1/t$ originates from the direct and exchange collision terms in the Mott cross section, and the last logarithmic term stands for the interference between the direct and exchange collision terms. Unlike conventional collision theories, the BEB model does not explicitly use the final state of the target. The only quantity linked to the final state of the ion is the binding energy *B*. As in the case of photoionization, vertical ionization energy should be used for *B*.

The BEB cross section is sensitive to the ionization energies for outer valence orbitals, while inner shell binding energies need not be very accurate. For the examples shown in this article, the binding energies for inner-shell orbitals were approximated by the orbital energies from restricted Hartree-Fock or equivalent molecular wave functions, while experimental values or results from correlated wave functions were used for outer valence orbitals.

In Sec. 2 we compare BEB cross sections to experimental data for neutral molecules, molecular ions, and radicals to demonstrate the utility of the BEB model. Computational details for the molecular wave functions are described in Sec. 3, our conclusion is presented in Sec. 4, and molecular constants to generate BEQ/BEB cross sections for a number of molecules are listed in Table 1.

2. COMPARISONS TO EXPERIMENTS

In this section, we present comparisons of the BEB cross sections to available experimental data. In general the total ionization cross section is measured by two different methods: (a) measure the total ion current, or (b) measure partial cross sections for fragment ions and then the partial cross sections are summed to deduce the total ionization cross section. Method (a) produces the gross total ionization cross section, in which multiply charged ions are weighted by their net charge *n*:

$$\sigma_{\text{gross}} = \Sigma\, n\sigma_{n+}\,,$$

where σ_{n+} is the cross section for producing an *n*-times charged ion. On the other hand, method (b) leads to the counting total ionization cross section:

$$\sigma_{\text{count}} = \Sigma\, \sigma_{n+}.$$

A straightforward sum of BEB orbital cross sections corresponds to the counting total ionization cross section. To emulate the gross ionization cross section, we must multiply BEB orbital cross sections for some inner shells by appropriate net charge *n*. For instance, a hole created by ionizing an inner orbital whose binding energy is larger than the energy needed to produce a doubly charged ion will decay by the Auger process. A hole in a very deep inner shell may even lead to the production of triply and quadruply charged ions. Such a highly charged molecular ion, however, is unstable and eventually breaks up into singly charged fragments. Besides, the cross sections for such deep inner-shell orbitals are insignificant compared to the large ionization cross sections of valence orbitals. Hence, we found that it is sufficient in most cases to account only for the production of doubly charged ions for comparison to experimental gross ionization cross sections. Whether a doubly charged molecular ion breaks up into two singly charged fragments or not, we found that doubling the cross sections for orbitals, whose binding energies are in the range of about 30 to 200 eV and exceed the double ionization energy, led to reliable gross ionization cross sections.

The gross ionization cross section deduced from the method (b) by multiplying partial cross sections by *n* before summing them does not necessarily agree with the gross ionization cross section from the method (a) because some σ_{n+} are too small to be measured.

2.1. Comparisons for Neutral Molecules

In Fig. 1, we compare the BEQ cross section for H_2 to other theories and available experimental data. The triangles marked DM in Fig. 1 represent the semiempirical theory by Deutsch et al. [5], which is based on a modified form of Gryzinski's classical theory [6] and commonly referred to as the DM formalism.

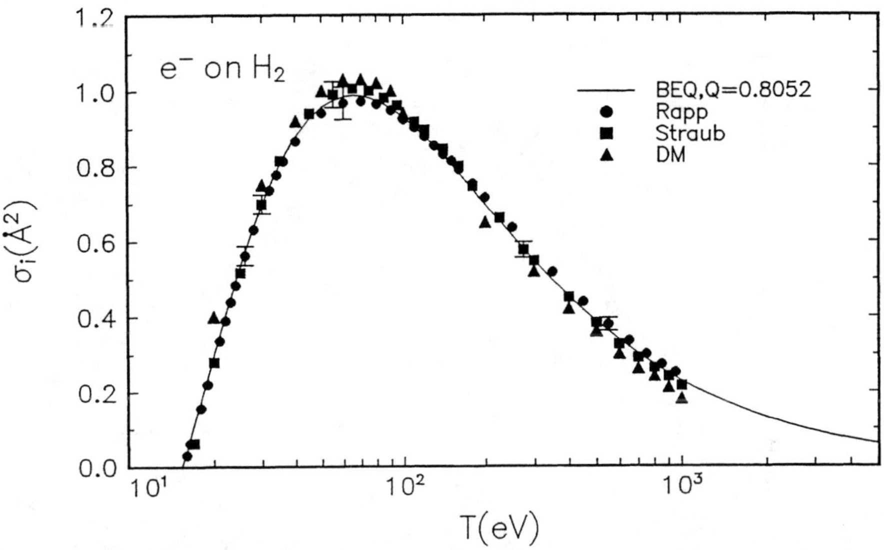

FIGURE 1. Electron-impact total ionization cross section of H_2. Solid curve, the BEQ cross section, Eq. (1); circles, experimental data by Rapp and Englander-Golden [7]; squares, data by Straub et al. [8]; triangles, semiempirical theory by Deutsch et al. [5].

The BEQ cross section in Fig. 1 is in excellent agreement with the benchmark experiments by Rapp and Englander-Golden [7] (with total uncertainties of ±4%) and by Straub et al. [8] (with total uncertainties of ±3.5%). *We recommend use of the BEQ cross section as the normalization standard for the total ionization cross section of* H_2. The recommended cross section should be reliable to ±5%. The molecular constants for H_2 along with those for other molecules are listed in Table 1.

Large molecules with many bound electrons tend to have many valence orbitals with low ionization energies. For such molecules, the orbital binding energies B are sensitive to the quality of the molecular wave function, though the orbital kinetic energy $U=\langle \mathbf{p}^2/2m \rangle$ with the usual notation are not because it is an expectation value of a one-electron operator. As an example, we compare in Fig. 2 available experimental and semiempirical cross sections for C_2F_6 to the BEB cross sections calculated from a restricted Hartree-Fock (RHF) wave function and from correlated, complete-active-space (CAS) wave functions.

The solid curve in Fig. 2 includes theoretical estimates for double ionization. The correlated wave functions lead to lower ionization energies for valence orbitals, which in turn produce higher orbital cross sections. The effect of electron correlation can be as much as 20% [9] for large molecules. The semiempirical DM formalism tends to produce low cross sections at high T because the classical theory, upon which the DM cross sections are based, lacks the dipole interaction which is dominant at high T. The molecular constants for the BEB cross section of C_2F_6 are listed in Table 1.

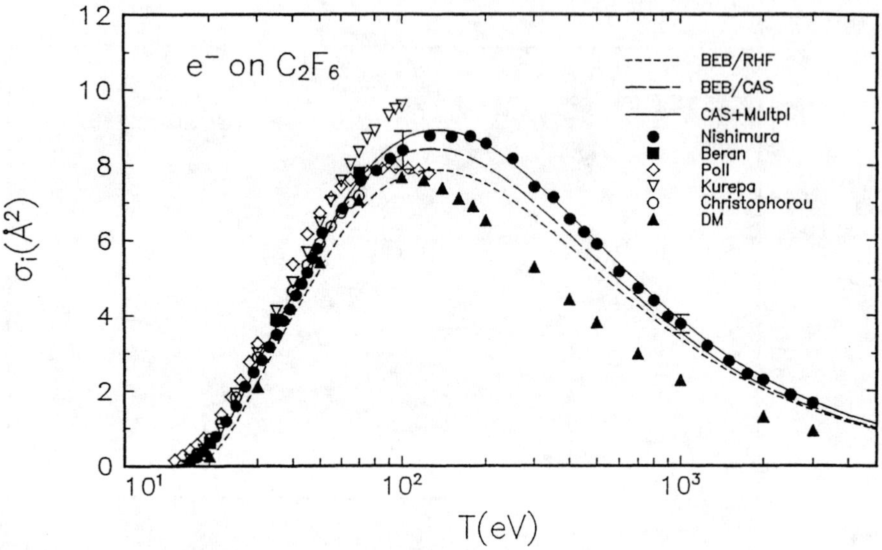

FIGURE 2. Electron-impact total ionization cross section of C_2F_6. Solid curve, BEB σ_{gross} from correlated wave functions; long-dashed curve, BEB σ_{count} from correlated wave functions; short-dashed curve, BEB σ_{count} from an uncorrelated wave function; filled circles, experimental σ_{gross} by Nishimura et al. [9]; filled squares, experimental σ_{gross} by Beran and Kevan [10]; open diamonds, experimental σ_{count} by Poll and Meichsner [11]; open triangles, σ_{gross} by Kurepa [12]; open circles, σ_{gross} adopted by Christophorou and Olthoff [13]; filled triangle, semiempirical DM formalism [5].

Hwang et al. [14] reported that known experimental cross sections for valence orbitals with $n \geq 3$ and primarily atomic character--as indicated by a Mulliken population of 85% or higher--are reproduced well when u in the denominators of Eqs. (1) and (3) is reduced. Currently, *we recommend replacing* $u+1$ *by* $(u+1)/n$ *for* $n \geq 3$.

For molecules containing very heavy elements, such as WF_6, orbital data from relativistic wave functions are desirable. Although computer codes to calculate relativistic wave functions for atoms are readily available, relativistic wave functions for polyatomic molecules are scarce and time consuming to calculate. Instead, one can use molecular orbital data obtained from an effective core potential as is commonly done. The use of an effective core potential has another advantage: for atomic orbitals with high principal quantum numbers n, the orbital kinetic energies remain low because inner radial function nodes have been eliminated in such a calculation [15].

In Fig. 3, we compare the BEB σ_{count} calculated using a relativistic effective core potential for WF_6 with recent experimental and semiempirical cross sections. This is one area where we need more experimental data to assess the validity of the BEB model for molecules containing heavy atoms.

Heavy atoms have many inner-shell orbitals with $n \geq 3$. Except for the outermost valence orbitals, Kim et al. [16] found an alternative to dividing u+1 by n, i.e., take a simple average of Eq. (3) and the same equation with the denominator t+u+1 replaced by t,

$$\sigma_{\mathrm{BEBav}} = \tfrac{1}{2}[1 + (t+u+1)/t] \times \text{Eq. (3)}, \tag{4}$$

for inner orbitals with $n \geq 3$.

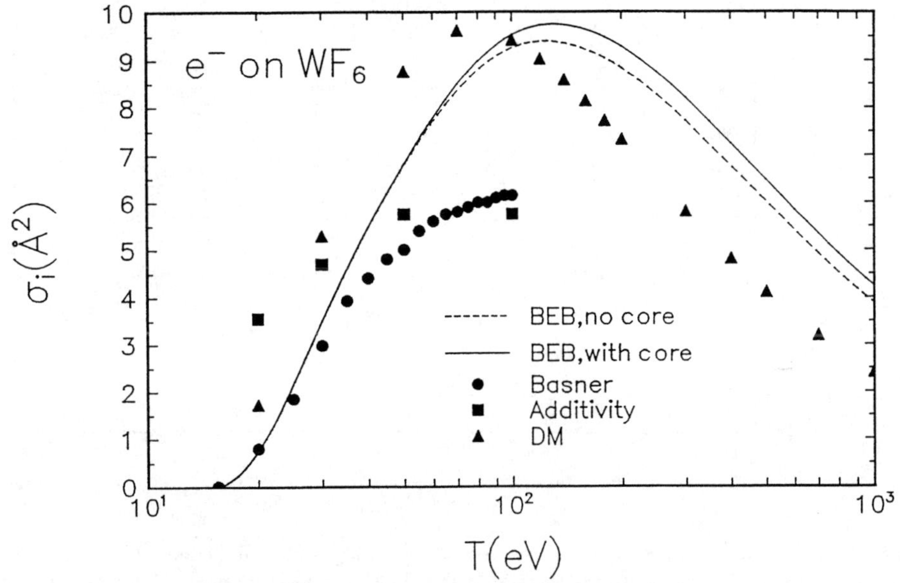

FIGURE 3. Electron-impact ionization cross section of WF_6. Solid curve, BEB σ_{count} calculated from an effective core potential plus core cross sections calculated from atomic core wave functions for W; dashed curve, same as the solid curve minus the contribution from the W core; circles, experimental data by Basner et al. [17]; squares, semiempirical additivity-rule cross section by Deutsch et al. [18]; triangles, semiempirical DM formalism by Deutsch et al. [5].

2.2 Comparison for Molecular Ions

Unlike atomic ions, molecular ions with a net charge of 2 or higher are unstable, and hence we need to adapt the BEB model only to singly-charged molecular ions. Kim and Rudd [3] noted that u+1 had to be reduced for atomic ions because the correlation between the incident electron and a target ion is different than a neutral target. For singly-charged ions, *we recommend replacing* u+1 *in Eqs.* (1) *and* (3) *by* (u+1)/2. For molecular ions containing heavy atoms, we do not have any experimental data to compare. Judging from our limited experience with heavy atomic ions, we suggest using σ_{BEBav} defined by Eq. (4) for the core orbitals ($n \geq 3$) of heavy atoms in molecules.

We compare in Fig. 4 the experimental data by Peart and Dolder [19] for H_2^+ to the BEQ and BEB cross sections using (u+1)/2. Comparisons of the BEB cross sections to experimental data for N_2^+ and CD^+ also show good agreement [20]. Since a simple dissociation of a molecular ion without ionizing any electron still produces an ion, the determination of experimental data for the ionization of a molecular ion is much more difficult than that for the ionization of a neutral molecule.

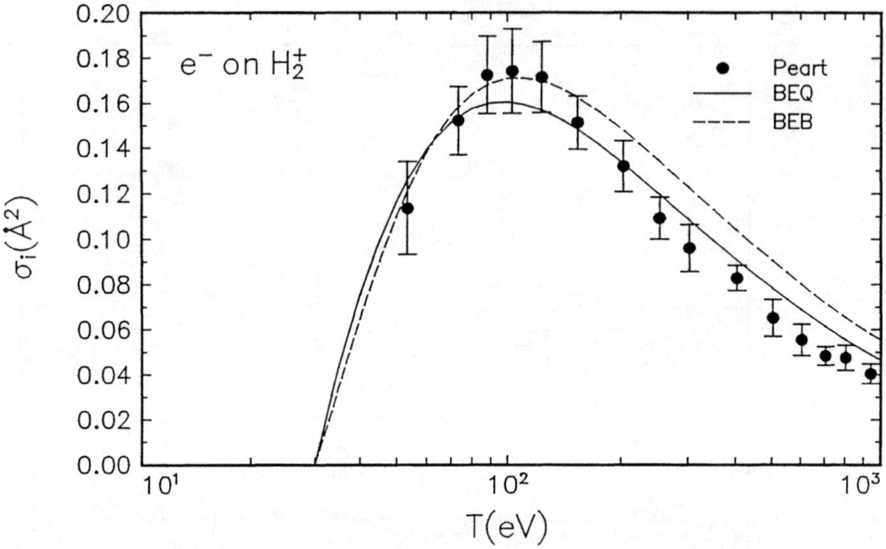

FIGURE 4. Electron-impact total ionization cross section of H_2^+ [20]. Solid curve, BEQ cross section with $(u+1)/2$; dashed curve, BEB cross section with $(u+1)/2$; circles, experimental data by Peart and Dolder [19].

2.3 Comparisons for Radicals

The BEB model treats a neutral radical or fragment the same as a neutral, stable molecule, i.e., Eqs. (1) or (3) are used. Radicals often have open-shell configurations, and one must be careful in generating molecular data for such a system. Also, experimental data on radicals are rare and often unreliable because of the difficulty in making enough target molecules.

In Fig. 5, we compare the BEB cross section for CD_3 to available experimental and semiempirical data. Although experiments used CD_3 [21,22], the BEB cross section is for CH_3 because the theory does not distinguish isotopes. We found that the BEB cross sections on other hydrocarbon radicals are in excellent agreement with available experiments [14,23].

In contrast, we find nearly a factor of two disagreement between the BEB cross sections and experimental data on fluorine containing molecules, such as CF_x, and NF_x. We compare in Fig. 6 the BEB cross section and experimental data by Becker [24] and by Tarnovsky et al. [25] for CF_3. The experimental data in Figs. 5 and 6 were taken by the same group using the same apparatus. At present, we do not understand the large discrepancy in Fig. 6, except for the possibility that the resulting ions of CF_3

dissociate in such a way as to escape detection by the apparatus used. The dashed curve in Fig. 6 is the BEB cross section which includes only those orbitals with $B <$ 19 eV. The threshold for dissociative ionization of CF_3 is approximately 19 eV.

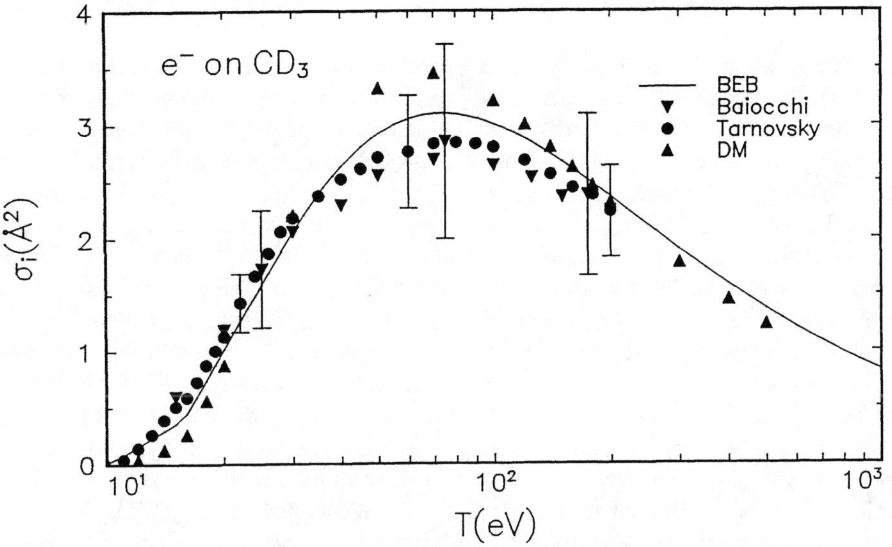

FIGURE 5. Electron-impact total ionization cross section of CD_3. Solid curve, BEB cross section for CH_3; inverted triangles, experimental data by Biocchi et al. [21]; circles, experimental data by Tarnovsky et al. [22]; upright triangles, semiempirical DM formalism by Deutsch et al. [5].

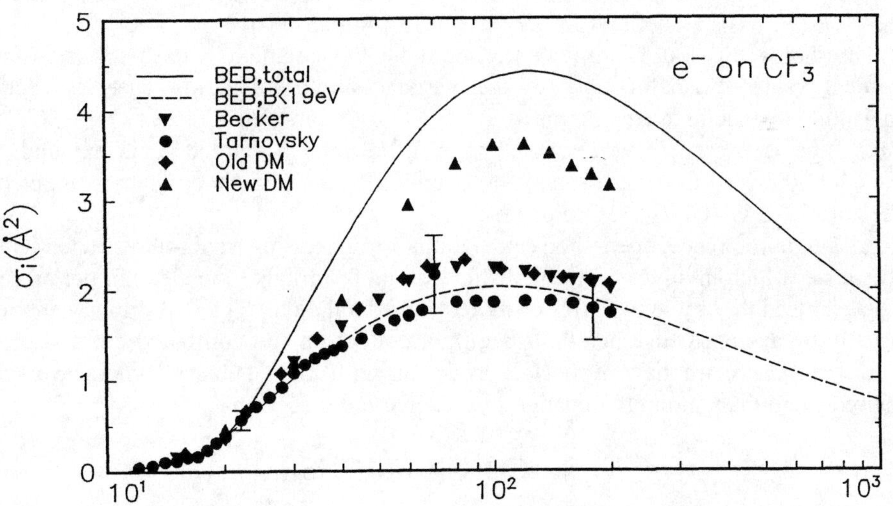

FIGURE 6. Electron-impact total ionization of CF_3. Solid curve, the BEB cross section; dashed curve,

the BEB cross section for orbitals with $B < 19$ eV; inverted triangles, experimental data by Becker [24]; circles, experimental data by Tarnovsky et al. [25], diamonds, semiempirical additivity rule by Deutsch et al. [18]; upright triangles, semiempirical DM formalism [5].

3. DETAILS OF MOLECULAR CALCULATIONS

Different ab initio methods yield different values for molecular properties. In particular, orbital binding energies (B) depend upon the theoretical method employed. Thus, the details of the molecular orbital calculations affect the final predictions for ionization cross sections. In general, more thorough calculations include more electron correlation, which reduces values of B and increases the predicted cross sections.

Binding and kinetic energies are insensitive to molecular geometry. We generally use geometries that are calculated using the nonlocal B3LYP density functional (a parameterized exchange functional [26] and a fitted correlation functional [27]) in combination with the 6-31G(d) basis of atom-centered Gaussian functions [28].

Since the final cross sections are especially sensitive to the value of the lowest binding energy, we use either experimental vertical ionization energies, typically from photoelectron spectra, or theoretical values computed using the CCSD(T) method (coupled-cluster theory including the single- and double-excitation operators [29] and a perturbative estimate of connected triples [30]) combined with the polarized, valence-triple-zeta basis designated cc-pVTZ [31,32]. We expect these CCSD(T) binding energies to be reliable to about 0.1 eV or better.

Binding energies for the other valence orbitals are computed either using the outer-valence Green's function (OVGF) method [33,34] or the uncorrelated Hartree-Fock method (HF), both with the diffuse 6-311+G(d,p) basis as defined in the Gaussian 98 computer code [35]. Orbital kinetic energies are computed using HF theory and the 6-311G(d,p) basis as defined in the GAMESS program [36].

Thresholds for double ionization should be the adiabatic values instead of the vertical values. Analogously to the vertical single ionization energies, double ionization thresholds were computed at the CCSD(T)/cc-pVTZ level using B3LYP/6-31G(d) geometries. In several cases, however, the doubly-charged ion is unbound, and we adopt the vertical double ionization energy instead. This is typically computed at the correlated OVGF/6-311G(d,p) level.

All single-reference, open-shell calculations are done using a spin-unrestricted (UHF) reference, which involves spatially distinct α and β orbitals (one electron per orbital). This can lead to very long tables of molecular orbital data. The α/β orbital pairs may usually be averaged together with negligible effect on the computed cross sections. For conciseness, we have done this in preparing Table 1, except where averaging caused more than about 1% change in the final cross sections.

4. CONCLUSIONS

We have presented comparisons of the BEB cross sections to available experimental and other theoretical data on small and large neutral molecules, molecular ions, and

Table 1. Molecular orbital constants to generate the BEQ/BEB cross sections [see Eqs. (1) and (3)]. Q=1 unless noted in the table. Only orbital cross sections marked Dbl=Y are doubled to account for double ionization. Orbitals that are essentially atomic with $n \geq 3$ are indicated in the "AO" column.

Molecule	Orbital	B(eV)	U(eV)	N	Q	Dbl	AO	Remarks
H_2^+	1σ	30.00	16.40	1	0.5275			
H_2	1σ	15.43	15.98	2	0.8052			Adiabatic B
CD_3/CH_3	$1a_1'$	305.45	436.19	2		Y		UHF (averaged)
	$2a_1'$	22.76	34.12	2				B from OVGF
	1e'	15.09	26.31	4				B from OVGF
	$1a_2''$	9.84	29.81	1				Vertical B from exper.
C_2F_6	$1e_g$	718.05	1013.49	4				CASSCF
	$1e_u$	718.05	1013.50	4				
	$1a_{2u}$	718.05	1013.51	2				
	$1a_{1g}$	718.05	1013.53	2				
	$2a_{1g}$	314.56	436.28	2				
	$2a_{2u}$	314.55	436.53	2				
	$3a_{1g}$	48.90	88.82	2		Y		
	$3a_{2u}$	48.16	94.16	2		Y		
	$2e_u$	46.05	103.39	4		Y		
	$2e_g$	45.60	107.04	4		Y		
	$4a_{1g}$	29.74	69.84	2				
	$4a_{2u}$	26.56	83.75	2				
	$5a_{1g}$	24.98	65.27	2				
	$3e_u$	24.12	74.26	4				
	$3e_g$	23.17	78.30	4				
	$5a_{2u}$	20.69	84.03	2				
	$4e_g$	20.86	82.12	4				
	$4e_u$	19.82	87.85	4				
	$5e_u$	18.42	88.57	4				
	$5e_g$	16.38	95.04	4				
	$6a_{1u}$	16.58	95.50	2				
	$6a_{2g}$	17.41	93.54	2				
	$7a_{1g}$	14.5	79.89	2				Vertical B from exper.
BF	1σ	717.85	1013.53	2		Y		
	2σ	210.08	297.18	2		Y		
	3σ	46.25	104.14	2		Y		
	4σ	21.71	84.12	2				B from OVGF
	1π	18.78	80.94	4				B from OVGF
	5σ	11.13	27.52	2				B from CCSD(T)
BF_2	$1b_2$	716.80	1013.50	2		Y		UHF (averaged)
	$1a_1$	716.80	1013.51	2		Y		
	$2a_1$	211.09	296.86	2		Y		

	$3a_1$	45.41	101.31	2	Y		
	$2b_2$	44.65	105.56	2	Y		
	$4a_1$	21.49	80.69	2			B from OVGF
	$3b_2$	20.00	81.97	2			B from OVGF
	$1b_1$	18.62	75.74	2			B from OVGF
	$5a_1$	18.59	77.06	2			B from OVGF
	$1a_2$	17.11	85.72	2			B from OVGF
	$4b_2$	16.84	89.92	2			B from OVGF
	$6a_1$	10.28	40.51	1			B from CCSD(T)
BF_3	1e'	716.93	1013.47	4	Y		
	$1a_1'$	716.93	1013.48	2	Y		
	$2a_1'$	212.33	296.73	2	Y		
	$3a_1'$	46.01	98.98	2	Y		
	2e'	44.86	105.18	4	Y		
	$4a_1'$	21.77	82.01	2			B from OVGF
	3e'	20.48	78.08	4			B from OVGF
	$1a_2''$	19.20	72.16	2			B from OVGF
	4e'	17.05	87.51	4			B from CCSD(T)
	1e"	16.77	85.72	4			B from CCSD(T)
	$1a_2'$	15.99	92.90	2			B from CCSD(T)
BCl	1σ	2853.82	3731.10	2	Y		
	2σ	288.57	593.27	2	Y		
	3σ	219.65	561.36	2	Y		
	1π	219.60	561.81	4	Y		
	4σ	210.31	297.39	2	Y		
	5σ	31.42	77.17	2	Y	3s	
	6σ	17.09	56.50	2			B from OVGF
	2π	13.49	57.24	4		3p	B from OVGF
	7σ	10.20	31.95	2			B from CCSD(T)
BCl_2	$1b_2$	2853.45	3731.10	2	Y		UHF (averaged)
	$1a_1$	2853.45	3731.10	2	Y		
	$2b_2$	288.19	593.26	2	Y		
	$2a_1$	288.19	593.27	2	Y		
	$3b_2$	219.30	561.36	2	Y		
	$3a_1$	219.30	561.36	2	Y		
	$1b_1$	219.22	561.81	2	Y		
	$1a_2$	219.22	561.82	2	Y		
	$4b_2$	219.20	561.86	2	Y		
	$4a_1$	219.20	561.86	2	Y		
	$5a_1$	210.64	297.01	2	Y		
	$6a_1$	31.52	73.63	2	Y		
	$5b_2$	30.52	80.44	2	Y	3s	
	$7a_1$	17.29	57.81	2			B from OVGF

	$6b_2$	15.02	59.26	2			B from OVGF
	$8a_1$	13.59	51.77	2			B from OVGF
	$2b_1$	13.59	53.19	2		3p	B from OVGF
	$2a_2$	12.31	61.36	2		3p	B from OVGF
	$7b_2$	12.06	64.28	2		3p	B from OVGF
	$9a_1$	8.75	45.19	1			B from CCSD(T)
BCl_3	1e'	2853.46	3731.1	4	Y		
	$1a_1'$	2853.46	3731.1	2	Y		
	2e'	288.2	593.24	4	Y		
	$2a_1'$	288.2	593.26	2	Y		
	3e'	219.3	561.2	4	Y		
	$3a_1'$	219.3	561.2	2	Y		
	1e"	219.24	561.82	4	Y		
	$1a_2''$	219.24	561.81	2	Y		
	4e'	219.23	561.98	4	Y		
	$1a_2'$	219.23	561.98	2	Y		
	$4a_1'$	211.86	297	2	Y		
	$5a_1'$	32.19	70.01	2	Y		
	5e'	30.56	80.15	4	Y		
	$6a_1'$	17.78	60.57	2			B from OVGF
	6e'	15.42	56.01	4			B from OVGF
	$2a_2''$	14.09	49.92	2			B from OVGF
	7e'	12.54	62.79	4		3p	B from CCSD(T)
	2e"	12.34	61.43	4		3p	B from CCSD(T)
	$2a_2'$	11.68	67.06	2		3p	B from CCSD(T)
C_2H_3	1a'	306.18	435.63	2	Y		UHF (averaged)
(vinyl)	2a'	305.51	436.63	2	Y		
	3a'	28.13	42.5	2			
	4a'	19.23	33.39	2			B from OVGF
	5a'	15.84	30.43	2			B from OVGF
	6a'	14.9	29.73	2			B from OVGF
	1a"	10.61	27.82	2			B from OVGF
	7a'	9.54	33.52	1			B from CCSD(T)
C_3H_3	$1a_1$	306.36	435.55	2	Y		UHF (averaged)
(propargyl)	$2a_1$	305.99	436.01	2	Y		
	$3a_1$	305.57	436.54	2	Y		
	$4a_1$	28.98	45.42	2	Y		
	$5a_1$	22.68	42.88	2			B from OVGF
	$6a_1$	18.11	33.63	2			B from OVGF
	$7a_1$	16.12	36.64	2			B from OVGF
	$1b_2$	15.58	26.27	2			B from OVGF
	$1b_1$	11.52	28.38	2			B from OVGF
	$2b_2$	10.38	29.8	2			B from OVGF
	$2b_1$	8.55	32.44	1			B from CCSD(T)

C_3H_4	$1a_1$	306.53	436.08	2	Y	
(allene)	$1b_2$	305.88	436.06	2	Y	
	$2a_1$	305.88	436.13	2	Y	
	$3a_1$	29.27	43.81	2	Y	
	$2b_2$	26.11	41.25	2	Y	
	$4a_1$	17.64	35.54	2		B from OVGF
	$3b_2$	15.47	37.09	2		B from OVGF
	1e	15.1	26.1	4		B from OVGF
	2e	10.22	28.21	4		B from CCSD(T)
C_3H_5	$1a_1$	305.82	435.71	2	Y	UHF (averaged)
(allyl)	$1b_2$	305.50	436.13	2	Y	
	$2a_1$	305.37	436.52	2	Y	
	$3a_1$	29.06	40.11	2	Y	
	$2b_2$	22.22	38.53	2		B from OVGF
	$4a_1$	18.90	32.55	2		B from OVGF
	$5a_1$	16.38	26.98	2		B from OVGF
	$3b_2$	15.04	32.24	2		B from OVGF
	$4b_2$	13.65	32.38	2		B from OVGF
	$6a_1$	12.96	30.58	2		B from OVGF
	$1b_1$	10.94	26.63	2		B from OVGF
	$1a_2$	8.02	31.57	1		B from CCSD(T)
C_3H_4	$1a_1$	306.24	436.09	2	Y	
(propyne)	$2a_1$	305.82	435.95	2	Y	
	$3a_1$	305.38	436.14	2	Y	
	$4a_1$	28.80	42.70	2	Y	
	$5a_1$	26.10	42.28	2		
	$6a_1$	17.80	33.68	2		B from OVGF
	$7a_1$	15.39	37.09	2		B from OVGF
	1e	15.02	25.75	4		B from OVGF
	2e	10.48	29.49	4		B from CCSD(T)
C_3H_6	1a'	305.79	436.13	2	Y	
(propene)	2a'	305.45	436.14	2	Y	
	3a'	305.34	436.15	2	Y	
	4a'	28.97	38.93	2	Y	
	5a'	22.58	37.25	2		B from OVGF
	6a'	18.68	32.93	2		B from OVGF
	7a'	15.90	27.01	2		B from OVGF
	8a'	14.57	32.04	2		B from OVGF
	1a"	14.50	25.17	2		B from OVGF
	9a'	13.29	31.14	2		B from OVGF
	10a'	12.36	31.12	2		B from OVGF
	2a"	9.95	27.25	2		B from CCSD(T)

C_4H_2	$1\sigma_g$	306.70	435.59	2	Y	
(diacet-	$1\sigma_u$	306.67	435.92	2	Y	
ylene)	$2\sigma_g$	306.48	436.23	2	Y	
	$2\sigma_u$	306.47	436.28	2	Y	
	$3\sigma_g$	29.78	47.66	2	Y	
	$3\sigma_u$	27.84	49.54	2	Y	
	$4\sigma_g$	20.55	43.58	2		B from OVGF
	$4\sigma_u$	17.97	33.91	2		B from OVGF
	$5\sigma_g$	17.24	36.45	2		B from OVGF
	$1\pi_u$	12.51	27.78	4		B from OVGF
	$1\pi_g$	10.19	30.47	4		B from CCSD(T)
C_4H_4	1a'	306.59	436.06	2	Y	
(vinyl-	2a'	306.20	436.16	2	Y	
acetylene)	3a'	305.98	435.86	2	Y	
	4a'	305.73	436.23	2	Y	
	5a'	29.76	42.66	2	Y	
	6a'	27.41	46.97	2	Y	
	7a'	23.01	38.87	2		
	8a'	17.79	32.83	2		B from OVGF
	9a'	16.54	30.84	2		B from OVGF
	10a'	15.32	35.72	2		B from OVGF
	11a'	13.53	31.07	2		B from OVGF
	1a"	11.99	26.34	2		B from OVGF
	12a'	10.47	29.97	2		B from OVGF
	2a"	9.63	29.11	2		B from CCSD(T)
C_4H_6	$1a_g$	305.97	435.95	2	Y	
(trans-	$1b_u$	305.95	436.22	2	Y	
butadiene)	$2a_g$	305.69	436.17	2	Y	
	$2b_u$	305.69	436.18	2	Y	
	$3a_g$	29.76	40.25	2	Y	
	$3b_u$	27.27	41.50	2	Y	
	$4a_g$	22.56	37.60	2		
	$4b_u$	18.51	30.46	2		B from OVGF
	$5b_u$	15.85	31.33	2		B from OVGF
	$5a_g$	15.58	28.72	2		B from OVGF
	$6a_g$	13.72	37.67	2		B from OVGF
	$6b_u$	13.62	29.62	2		B from OVGF
	$7a_g$	12.31	31.34	2		B from OVGF
	$1a_u$	11.54	24.83	2		B from OVGF
	$1b_g$	9.14	27.78	2		B from CCSD(T)
C_4H_8	1a	305.71	436.11	2	Y	
(1-butene)	2a	305.57	436.18	2	Y	

	3a	305.35	436.16	2	Y	
	4a	305.28	436.16	2	Y	
	5a	29.36	37.44	2	Y	
	6a	26.89	38.18	2	Y	
	7a	23.29	35.89	2		
	8a	18.58	31.86	2		B from OVGF
	9a	15.97	26.32	2		B from OVGF
	10a	15.08	29.01	2		B from OVGF
	11a	14.76	27.37	2		B from OVGF
	12a	13.52	30.33	2		B from OVGF
	13a	13.04	31.07	2		B from OVGF
	14a	12.38	29.95	2		B from OVGF
	15a	11.92	32.02	2		B from OVGF
	16a	9.86	27.84	2		B from est. CCSD(T)
C_4H_8	$1a_g$	305.50	435.89	2	Y	
(trans-	$1b_u$	305.45	436.37	2	Y	
2-butene)	$2a_g$	305.33	436.15	2	Y	
	$2b_u$	305.33	436.15	2	Y	
	$3a_g$	29.29	38.61	2	Y	
	$3b_u$	26.73	35.53	2	Y	
	$4a_g$	23.90	38.10	2		
	$4b_u$	18.27	30.91	2		B from OVGF
	$5b_u$	15.06	30.49	2		B from OVGF
	$5a_g$	15.09	29.42	2		B from OVGF
	$1a_u$	14.53	24.71	2		B from OVGF
	$1b_g$	14.14	25.68	2		B from OVGF
	$6b_u$	13.27	28.92	2		B from OVGF
	$6a_g$	12.77	33.73	2		B from OVGF
	$7a_g$	11.98	32.70	2		B from OVGF
	$2a_u$	9.38	28.64	2		B from est. CCSD(T)
C_4H_8	$1a_1$	305.91	436.16	2	Y	
(isobutene)	$1b_2$	305.39	436.12	2	Y	
	$2a_1$	305.39	436.14	2	Y	
	$3a_1$	305.09	436.13	2	Y	
	$4a_1$	29.62	38.10	2	Y	
	$2b_2$	22.76	35.04	2		B from OVGF
	$5a_1$	25.21	37.71	2	Y	
	$6a_1$	17.68	34.93	2		B from OVGF
	$3b_2$	15.36	26.62	2		B from OVGF
	$7a_1$	15.22	28.96	2		B from OVGF
	$1b_1$	15.00	23.88	2		B from OVGF
	$1a_2$	13.71	26.96	2		B from OVGF

	$4b_2$	13.27	28.77	2		B from OVGF
	$8a_1$	12.90	33.96	2		B from OVGF
	$5b_2$	11.89	34.60	2		B from OVGF
	$2b_1$	9.50	27.88	2		B from est. CCSD(T)
C_6H_2	$1\sigma_g$	307.14	435.46	2	Y	
(triacet-	$1\sigma_u$	307.05	436.32	2	Y	
ylene)	$2\sigma_g$	306.85	435.66	2	Y	
	$2\sigma_u$	306.84	435.67	2	Y	
	$3\sigma_u$	306.69	436.25	2	Y	
	$3\sigma_g$	306.69	436.25	2	Y	
	$4\sigma_g$	30.43	47.40	2	Y	
	$4\sigma_u$	29.23	48.09	2	Y	
	$5\sigma_g$	27.74	49.92	2	Y	
	$5\sigma_u$	21.30	45.58	2		B from OVGF
	$6\sigma_g$	19.36	44.32	2		B from OVGF
	$6\sigma_u$	17.65	35.03	2		B from OVGF
	$7\sigma_g$	17.35	37.02	2		B from OVGF
	$1\pi_u$	12.94	27.66	4		B from OVGF
	$1\pi_g$	11.42	28.91	4		B from OVGF
	$2\pi_u$	9.55	31.73	4		B from est. CCSD(T)
C_6H_6	$1a_1$	306.22	435.99	2	Y	
(fulvene)	$2a_1$	306.07	436.11	2	Y	
	$3a_1$	305.76	435.88	2	Y	
	$1b_2$	305.74	436.11	2	Y	
	$2b_2$	305.59	436.08	2	Y	
	$4a_1$	305.59	436.09	2	Y	
	$5a_1$	31.67	38.87	2	Y	
	$6a_1$	27.97	41.44	2	Y	
	$3b_2$	26.42	42.70	2	Y	
	$7a_1$	18.21	39.00	2		B from OVGF
	$4b_2$	17.95	34.27	2		B from OVGF
	$8a_1$	17.51	33.99	2		B from OVGF
	$9a_1$	16.76	30.25	2		B from OVGF
	$5b_2$	14.99	28.67	2		B from OVGF
	$6b_2$	13.82	34.76	2		B from OVGF
	$10a_1$	13.44	41.07	2		B from OVGF
	$11a_1$	12.66	35.97	2		B from OVGF
	$1b_1$	12.57	23.77	2		B from OVGF
	$7b_2$	12.00	36.33	2		B from OVGF
	$2b_1$	9.40	27.74	2		B from OVGF
	$1a_2$	8.59	29.67	2		B from est. CCSD(T)

C_6H_6 (benzene)	$1a_{1g}$	305.86	435.70	2	Y	
	$1e_{1u}$	305.84	435.86	4	Y	
	$1e_{2g}$	305.81	436.23	4	Y	
	$1b_{1u}$	305.80	436.41	2	Y	
	$2a_{1g}$	31.32	39.01	2	Y	
	$2e_{1u}$	27.64	42.09	4	Y	
	$2e_{2g}$	19.70	38.78	4		B from OVGF
	$3a_{1g}$	17.31	25.22	2		B from OVGF
	$2b_{1u}$	15.79	34.02	2		B from OVGF
	$1b_{2u}$	14.79	39.65	2		B from OVGF
	$3e_{1u}$	14.41	32.22	4		B from OVGF
	$1a_{2u}$	12.31	23.81	2		B from OVGF
	$3e_{2g}$	12.09	37.55	4		B from OVGF
	$1e_{1g}$	9.37	28.12	4		B from est. CCSD(T)
CF	1σ	718.06	1013.66	2	Y	UHF (averaged)
	2σ	309.67	436.92	2	Y	
	3σ	46.66	101.39	2	Y	
	4σ	23.40	79.50	2		B from OVGF
	1π	19.13	81.28	4		B from OVGF
	5σ	14.51	49.63	2		B from OVGF
	2π	9.45	43.27	1		B from CCSD(T)
CF_2	$1b_2$	717.93	1013.31	2	Y	
	$1a_1$	717.93	1013.32	2	Y	
	$2a_1$	312.22	436.69	2	Y	
	$3a_1$	47.66	95.30	2	Y	
	$2b_2$	45.44	106.89	2	Y	
	$4a_1$	24.30	81.58	2		B from OVGF
	$3b_2$	20.82	81.24	2		B from OVGF
	$5a_1$	19.60	75.64	2		B from OVGF
	$1b_1$	19.44	74.54	2		B from OVGF
	$1a_2$	17.31	88.09	2		B from OVGF
	$4b_2$	16.80	94.96	2		B from OVGF
	$6a_1$	12.18	56.75	2		B from CCSD(T)
CF_3	1e	717.68	1013.31	4	Y	UHF (averaged)
	$1a_1$	717.67	1013.33	2	Y	
	$2a_1$	313.60	436.43	2	Y	
	$3a_1$	47.82	92.38	2	Y	
	2e	45.39	105.15	4	Y	
	$4a_1$	24.70	82.37	2		B from OVGF
	3e	21.44	77.03	4		B from OVGF
	$5a_1$	20.07	69.93	2		B from OVGF
	4e	17.99	86.59	4		B from OVGF

	5e	17.07	90.60	4			B from OVGF
	$1a_2$	16.50	94.64	2			B from OVGF
	$6a_1$	11.02	73.10	1			B from CCSD(T)
CF_4	$1t_2$	717.96	1013.34	6	Y		
	$1a_1$	717.96	1013.36	2	Y		
	$2a_1$	316.38	436.64	2	Y		
	$3a_1$	49.21	88.29	2	Y		
	$2t_2$	45.85	104.55	6	Y		
	$4a_1$	25.49	86.13	2			B from OVGF
	$3t_2$	22.26	75.09	6			B from OVGF
	1e	18.59	83.64	4			B from OVGF
	$4t_2$	17.27	89.32	6			B from CCSD(T)
	$1t_1$	16.19	94.97	6			B from CCSD(T)
SiF	1σ	1873.43	2510.14	2	Y		UHF (averaged)
	2σ	715.46	1013.44	2	Y		
	3σ	168.38	360.56	2	Y		
	4σ	116.74	332.22	2	Y		
	1π	116.69	331.90	4	Y		
	5σ	43.23	104.29	2	Y		
	6σ	19.74	78.49	2			B partly from OVGF
	2π	16.50	80.80	4			B from OVGF
	7σ	12.52	45.02	2		3sp	B from OVGF
	3π	7.39	27.54	1		3p	B from CCSD(T)
SiF_2	$1a_1$	1874.53	2509.98	2	Y		
	$1b_2$	715.87	1013.41	2	Y		
	$2a_1$	715.87	1013.41	2	Y		
	$3a_1$	169.40	360.22	2	Y		
	$2b_2$	117.79	331.82	2	Y		
	$1b_1$	117.74	332.55	2	Y		
	$4a_1$	117.74	331.41	2	Y		
	$5a_1$	44.03	102.29	2	Y		
	$3b_2$	43.44	106.34	2	Y		
	$6a_1$	19.99	79.39	2			B from OVGF
	$4b_2$	17.98	84.31	2			B from OVGF
	$7a_1$	17.23	77.86	2			B from OVGF
	$2b_1$	16.99	77.39	2			B from OVGF
	$1a_2$	16.18	83.52	2			B from OVGF
	$5b_2$	15.80	87.63	2			B from OVGF
	$8a_1$	11.09	46.61	2		3sp	B from CCSD(T)
SiF_3	$1a_1$	1875.92	2509.91	2	Y		
	$2a_1$	716.48	1013.42	2	Y		
	1e	716.48	1013.42	4	Y		

	$3a_1$	170.67	359.78	2	Y	
	2e	119.19	331.92	4	Y	
	$4a_1$	119.15	332.32	2	Y	
	$5a_1$	44.83	101.12	2	Y	
	3e	44.11	106.11	4	Y	
	$6a_1$	21.20	80.10	2		B from OVGF
	4e	19.05	82.02	4		B from OVGF
	$7a_1$	18.19	76.02	2		B from OVGF
	5e	17.54	82.63	4		B from OVGF
	6e	16.84	85.63	4		B from OVGF
	$1a_2$	16.41	88.47	2		B from OVGF
	$8a_1$	10.61	53.01	1		B from CCSD(T)
SiF_4	$1a_1$	1876.92	2510.04	2	Y	
	$2a_1$	717.04	1013.45	2	Y	
	$1t_2$	717.04	1013.44	6	Y	
	$3a_1$	171.63	359.77	2	Y	
	$2t_2$	120.20	332.06	6	Y	
	$4a_1$	45.75	99.63	2	Y	
	$3t_2$	44.77	106.18	6	Y	
	$5a_1$	21.93	82.47	2		B from OVGF
	$4t_2$	19.73	81.00	6		B from OVGF
	1e	18.15	80.59	4		B from OVGF
	$5t_2$	17.29	84.73	6		B from CCSD(T)
	$1t_1$	16.31	88.76	6		B from CCSD(T)
OH	1σ (α)	561.69	792.91	1	Y	UHF
	2σ (α)	37.76	80.49	1		
	3σ (α)	17.79	58.59	1		B from OVGF
	1π (α)	15.97	68.94	1		B from OVGF
	1π (α)	13.94	68.94	1		B from OVGF
	1σ (β)	560.59	795.52	1	Y	
	2σ (β)	33.42	71.79	1		
	3σ (β)	16.68	56.16	1		B from OVGF
	1π (β)	12.79	52.55	1		B from CCSD(T)
HO_2	1a' (α)	563.10	793.52	1	Y	
(hydro-	2a' (α)	562.63	794.31	1	Y	
peroxyl)	3a' (α)	43.02	74.60	1	Y	
	4a' (α)	28.85	85.88	1		B from OVGF
	5a' (α)	19.46	57.55	1		B from OVGF
	1a" (α)	17.30	64.42	1		B from OVGF
	6a' (α)	17.27	66.74	1		B from OVGF
	2a" (α)	12.29	77.50	1		B from OVGF
	7a' (α)	12.50	73.07	1		B from OVGF

	1a' (β)	562.43	794.62	1	Y	
	2a' (β)	561.97	795.64	1	Y	
	3a' (β)	41.29	71.78	1	Y	
	4a' (β)	27.45	80.99	1		B from OVGF
	5a' (β)	19.01	57.35	1		B from OVGF
	6a' (β)	16.70	66.31	1		B from OVGF
	1a" (β)	14.77	58.38	1		B from OVGF
	7a' (β)	11.35	71.24	1		B from CCSD(T)
CH_2O	$1a_1$	560.04	794.47	2	Y	
(formal-	$2a_1$	308.79	436.04	2	Y	
dehyde)	$3a_1$	38.38	73.38	2	Y	
	$4a_1$	21.71	44.02	2		B from OVGF
	$1b_2$	17.19	36.50	2		B from OVGF
	$5a_1$	16.24	60.83	2		B from OVGF
	$1b_1$	14.47	47.36	2		B from OVGF
	$2b_2$	10.81	54.70	2		B from CCSD(T)
CHO	1a'	561.08	794.48	2	Y	
(formyl)	2a'	308.96	436.11	2	Y	
	3a'	39.42	75.22	2	Y	
	4a'	20.61	50.28	2		B from OVGF
	5a'	17.30	53.81	2		B from OVGF
	1a"	15.39	50.69	2		B from OVGF
	6a'	14.84	53.32	2		B from OVGF
	7a'	9.20	50.07	1		B from CCSD(T)
C_2H_3O	1a'	560.12	794.56	2	Y	
(vinoxyl)	2a'	308.01	436.01	2	Y	
	3a'	306.49	435.96	2	Y	
	4a'	37.99	72.42	2	Y	
	5a'	26.12	43.16	2		B partly from OVGF
	6a'	19.80	42.09	2		B from OVGF
	7a'	16.98	40.86	2		B from OVGF
	8a'	15.78	50.53	2		B from OVGF
	9a'	14.81	34.80	2		B from OVGF
	1a"	13.74	45.58	2		B from OVGF
	10a'	10.72	57.58	2		B from OVGF
	2a"	10.20	35.04	1		B from CCSD(T)

radicals. Except for fluorine-containing radicals, we find good to excellent agreement. For H_2, the BEQ cross section in Fig. 1 is accurate enough to serve as the normalization standard for experiments. Comparisons of the BEB model to a large collection of molecules and some atoms are available from a public NIST web site: *http://physics.nist.gov/ionxsec*. The orbital constants and tables of BEB cross sections can be downloaded from the web site in the ascii format.

Encouraged by the good agreement between the BEB model and the experimental data for hydrocarbon radicals, we have included in Table 1 molecular constants for many hydrocarbons of interest to combustion modeling. Moreover, to stimulate more experimental and theoretical work and to provide data for modeling plasma processing of semiconductors with fluorine-containing molecules, we also have included in Table 1 molecular constants to generate BEB cross sections for such molecules.

ACKNOWLEDGMENTS

The present work was supported in part by the Advanced Technology Program of NIST and by the Office of Fusion Energy Sciences, the U.S. Department of Energy. We are grateful for the help we have received from Dr. W. M. Huo and Prof. M. A. Ali in generating the input data for some of the molecules presented in this article.

REFERENCES

[1] Mott, N.F., *Proc. R. Soc. London Ser. A* **126**, 259 (1930).
[2] Bethe, H., *Ann. Phys. (Leipzig)* **5**, 325 (1930).
[3] Kim, Y.-K., and Rudd, M.E., *Phys. Rev. A* **50**, 3954 (1994).
[4] Inokuti, M., *Rev. Mod. Phys.* **43**, 297 (1972).
[5] Deutsch, H., Becker, K., Matt, S., and Märk, T.D., Int. J. Mass Spectrom. **197**, 37 (2000).
[6] Gryzinski, M., *Phys. Rev.* **138**, A305 (1965); **138**, A322 (1965); **138**, A336 (1965).
[7] Rapp, D. and Englander-Golden, P., *J. Chem. Phys.* **43**, 1464 (1965).
[8] Straub, H.C., Renault, P., Lindsay, B.G., Smith, K.A., and Stebbings, R.F., *Phys. Rev. A* **54**, 2146 (1996).
[9] Nishimura, H., Huo, W.M., Ali, M.A., and Kim, Y.-K., *J. Chem. Phys.* **110**, 3811 (1999).
[10] Beran, A. and Kevan, L., *J. Phys. Chem.* **73**, 3866 (1969).
[11] Poll, H.U. and Meichsner, J., *Contrib. Plasma Phys.* **27**, 5359 (1987).
[12] Kurepa, M.V., *3rd Cz. Conf. Electronics and Vacuum Transacts.* (1965).
[13] Christophorou, L.G., and Olthoff, J.K., *J. Phys. Chem. Ref. Data* **27**, 1 (1998).
[14] Hwang, W., Kim, Y.-K., and Rudd, M.E., *J. Chem. Phys.* **104**, 2956 (1996).
[15] Huo, W.M., and Kim, Y.-K., *Chem. Phys. Lett.* **319**, 576 (2000).
[16] Kim, Y.-K., Santos, J.P., and Parente, F., submitted to *Phys. Rev. A* (2000).
[17] Basner, R., Schmidt, M., and Deutsch, H., private communication (1999).
[18] Deutsch, H., Märk, T.D., Tarnovsky, V., Becker, K., Cornelissen, C., Cespiva, L., and Bonacic-Kutecky, V., *Int. J. Mass Spectrom. Ion Processes*, **137**, 77 (1994).
[19] Peart, B., and Dolder, K.T., *J. Phys. B* **6**, 2409 (1973).
[20] Kim, Y.-K., Irikura, K.K, and Ali, M.A., *J. Res. NIST* **105**, 285 (2000).
[21] Biocchi, F.A., Wetzel, R.C., and Freund, R.S., *Phys. Rev. Lett.* **53**, 771 (1984).
[22] Tarnovsky, V., Levin, A., Deutsch, H., and Becker, K., *J. Phys. B* **29**, 139 (1996).
[23] Kim, Y.-K., Ali, M.A., and Rudd, M.E., *J. Res. NIST* **102**, 693 (1997).
[24] Becker, K., private communication (1999).

[25] Tarnovsky, V., Kurunczi, P., Rogozhinikov, D., and Becker, K., *Int. J. Mass Spectrom. Ion Processes*, **128**, 181 (1993).
[26] Becke, A.D., *J. Chem. Phys.* **98**, 5648 (1993).
[27] Lee, C., Yang, W., Parr, R.G., *Phys. Rev. B* **37**, 785 (1988).
[28] Francl, M.M., Pietro, W.J., Hehre, W.J., Binkley, J.S., Gordon, M.S., DeFrees, D.J., Pople, J.A., *J. Chem. Phys.* 77, 3654 (1982).
[29] Purvis, G.D.I., Bartlett, R.J., *J. Chem. Phys.* **76**, 1910 (1982).
[30] Raghavachari, K., Trucks, G.W., Pople, J.A., Head-Gordon, M., *Chem. Phys. Lett.* **157**, 479 (1989).
[31] Dunning, T.H., Jr., *J. Chem. Phys.* **90**, 1007 (1989).
[32] Woon, D.E., Dunning, T.H., Jr., *J. Chem. Phys.* **98**, 1358 (1993).
[33] von Niessen, W., Schirmer, J., Cederbaum, L.S., *Comput. Phys. Rep.* **1**, 57 (1984).
[34] Zakrzewski, V.G., Ortiz, J.V., *J. Phys. Chem.* **100**, 13979 (1996).
[35] Frisch, M.J., Trucks, G.W., Schlegel, H.B., Scuseria, G.E., Robb, M.A., Cheeseman, J.R., Zakrzewski, V.G., Montgomery, J.A., Jr., Stratmann, R.E., Burant, J.C., Dapprich, S., Millam, J.M., Daniels, A.D., Kudin, K.N., Strain, M.C., Farkas, O., Tomasi, J., Barone, V., Cossi, M., Cammi, R., Mennucci, B., Pomelli, C., Adamo, C., Clifford, S., Ochterski, J., Petersson, G.A., Ayala, P.Y., Cui, Q., Morokuma, K., Malick, D.K., Rabuck, A.D., Raghavachari, K., Foresman, J.B., Cioslowski, J., Ortiz, J.V., Stefanov, B.B., Liu, G., Liashenko, A., Piskorz, P., Komaromi, I., Gomperts, R., Martin, R.L., Fox, D.J., Keith, T., Al-Laham, M.A., Peng, C.Y., Nanayakkara, A., Gonzalez, C., Challacombe, M., Gill, P.M.W., Johnson, B., Chen, W., Wong, M.W., Andres, J.L., Gonzalez, C., Head-Gordon, M., Replogle, E.S., Pople, J.A., Gaussian 98, Gaussian, Inc., Pittsburgh, PA, 1998.
[36] Schmidt, M.W., Baldridge, K.K., Boatz, J.A., Elbert, S.T., Gordon, M.S., Jensen, J.H., Koseki, S., Matsunaga, N., Nguyen, K.A., Su, S.J., Windus, T.L., Dupuis, M., Montgomery, J.A., *J. Comput. Chem.* **14**, 1347 (1993).

Towards accurate wave functions and oscillator strengths

Alan Hibbert

Department of Applied Mathematics and Theoretical Physics, The Queen's University of Belfast, Belfast BT7 1NN, Northern Ireland

Abstract. Recent advances in the resolution of observational data have required a much greater degree of accuracy in the determination of oscillator strengths than has been necessary in the past. We discuss how corresponding theoretical advances are leading to improved accuracy in the calculation of wave functions of atomic states, resulting in oscillator strengths whose uncertainties are within acceptable limits. We show the importance of using experimental energies in refining *ab initio* calculations, and we illustrate these developments by considering difficult transitions in Mg II, Na III and Fe II.

INTRODUCTION

The calculation of oscillator strengths can appear to be easy. All one needs is access to one of a number of large-scale computer codes, some knowledge of how to set up the input data, and results flow. But the crucial question is: how accurate are these results? Indeed, can one set bounds on the extent of their inaccuracies? For certainly, no calculation or experiment gives exact oscillator strengths.

One of the difficulties is that, unlike the calculation of energies, there is no monotonic improvement in the calculated oscillator strength values as the wave functions improve (or at least become more complex). A relatively simple calculation may turn out to give a significantly more accurate result than a more elaborate calculation. Again, two seemingly extensive calculations may result in quite different oscillator strength values.

Of course, it is more common for the larger-scale calculations to give the more accurate results, but exceptions are so numerous that a careful analysis of all results is essential if their accuracy is to be assessed. That assessment may be undertaken by the original researchers working on a particular problem, or by scientists such as those at NIST who are dedicated to a study of the accuracy of various results and the determination of the degree of reliability.

Two examples, to which we shall return below, suffice as illustrations.

CP543, *Atomic and Molecular Data and Their Applications,* edited by K. A. Berrington and K. L. Bell

TABLE 1. Oscillator strengths of the Mg II 3s ^{2}S – 4p ^{2}P^o transition

Authors	Method	f-value
Black *et al* [1]	Central field + core polarisation	14.6×10^{-4}
Saraph [2]	Close coupling, no core pol	2.2×10^{-4}
Froese Fischer [3]	Simple MCHF	9.7×10^{-4}
Biémont [4]	Hartree-Fock	4.0×10^{-4}
Hibbert *et al* [5]	Simple CI	4.0×10^{-4}
Butler *et al* [6]	Close coupling + polarised states	6.5×10^{-4}
Laughlin [7]	Model Potential	8.0×10^{-4}
Opacity Project [8]	R-matrix	5.9×10^{-4}
Sigut and Pradhan [9]	R-matrix	7.2×10^{-4}
Fleming *et al* [10]	CI	8.3×10^{-4}
Godefroid and Froese Fischer [11]	MCHF	8.3×10^{-4}
Godefroid and Froese Fischer [11]	MCHF+BP	7.4×10^{-4}
Sofia *et al* [12]	Derived from observations	18.8×10^{-4}
Fitzpatrick [13]	Derived from observations	9.6×10^{-4}

1. The oscillator strength of the 3s ^{2}S – 4p ^{2}P^o transition in Mg II, which is unusually weak, has been the focus of many studies. A selection of results is presented in Table 1. If we accept the most recent calculations of Fleming *et al* [10] and Godefroid and Froese Fischer [11] as rather accurate, then we can see that the conceptually simple of model potential method used by Laughlin [7] fares rather better than the much more extensive R-matrix calculations. We note too the wide variation in the recent derivations from observations.

2. Scientists at NIST present their comparisons between different sets of results in the form of scatter diagrams, usually as log-log graphs. A frequently appearing comparison is made between the present author's calculations and results derived from the data in the Opacity Project database. Some examples of these comparisons appear in the monograph by Wiese *et al* [14] for ions of C, N, O. Others can be found for example for Na III in Kelleher *et al* [15]. In these sources, it is argued that even for some quite strong transitions, there are major discrepancies between the two sets of data.

Both these examples suggest that there are problems in obtaining reliable atomic data, accurate enough to allow the interpretation of recent observations from space-based telescopes. But is the situation as bad as would appear at first sight? The aim of this paper is to show that it is not, and that we are indeed moving towards accurate wave functions and oscillator strengths.

There are several indicators by which we can assess the accuracy of a particular calculation. By far the easiest is the comparison between the calculated and experimental energy spectrum. In a general sense, the closer the agreement the better the calculation. The comparison between length and velocity forms of the oscillator strength is a useful but not a reliable guide to accuracy. For exact wave functions, the two forms agree, but they can also agree when very approximate wave functions (such as those which ignore exchange effects) are used. Some measure of confidence can be obtained by considering the convergence of the results as the complexity of the calculation is increased, especially when this is combined with one of a range of processes which make semi-empirical adjustments to the wave functions in order to improve the calculated energy spectrum (see below). This does not rule out the systematic omission of some important effect. Such a problem might be overcome by different (groups of) researchers working independently on a problem and comparing their results. Finally, a comparison between theory and experiment or, in the context of astrophysical or other applications, the consistency of for example element abundances derived using different transitions provides another indicator of the accuracy or otherwise of the results.

CONFIGURATION INTERACTION

Many of the methods of calculation of wave functions incorporate the concept of configuration interaction (CI). The basic idea is that a wave function may be expanded in terms of other known (or determinable) functions which have the same symmetry (of angular momentum and parity):

$$\Psi = \sum_{i=1}^{M} a_i \Phi_i \tag{1}$$

Finite summations over complete sets of functions are a classic way of approximating a function. Its usefulness in the context of atomic structure lies in the fact that many atomic states are quite well represented by the Hartree-Fock (HF) approximation, and the HF wave function can be chosen as Φ_1.

Common features of CI methods

In the HF approximation, the N electrons are associated with a set of N one-electron orbital functions which are customarily expressed as products of a radial function, a spherical harmonic and a spin function :

$$u(\mathbf{r}, \sigma) = \frac{1}{r} P_{nl}(r) Y_l^{m_l}(\theta, \phi) \chi_{m_s}(\sigma) \tag{2}$$

While this may be a good first approximation to the wave function, the parameters n, l, m_l, m_s do not constitute "good" quantum numbers (unlike their hydrogenic counterparts); only the total angular momenta and parity are constants of the motion. Hence other configuration state functions (CSFs) Φ_i may be constructed with these or other orbitals of a similar form and used in the expansion (1) in order to improve the representation of the atomic state functions (ASFs).

For a specific choice of the M CSFs in (1), the variationally optimal coefficients a_i form one of the eigenvectors of the Hamiltonian matrix whose typical element is $H_{ij} =< \Phi_i|H|\Phi_j >$, where H is the Hamiltonian of the atomic system. Moreover the ordered eigenvalues, which form approximations to the energies of the corresponding states, satisfy the inequalities

$$E_i > E_i^{\text{exact}} \qquad (i = 1, 2, \cdots) \tag{3}$$

which follows from the Hylleraas-Undheim theorem. These inequalities provide a set of variational principles from which the other unknown quantities – the radial functions $P_{nl}(r)$ – can be determined optimally.

Differences between CI methods

It is in the determination of the radial functions that the various CI-based methods differ. The author's code CIV3 [16] expresses the radial functions in analytic form

$$P_{nl}(r) = \sum_{j=1}^{k} c_{jnl} r^{I_{jnl}} \exp(-\zeta_{jnl} r) \tag{4}$$

and treats the c_{jnl} and ζ_{jnl} as variational parameters. So a particular eigenvalue E_i in (3) is chosen, and that eigenvalue is minimised with respect to the parameters of a particular radial function. It is customary, and in most cases essential, that different radial functions are optimised on different energies, so that all the important atomic states can be represented at an equivalent level of accuracy. It is also customary to construct the orbitals sequentially, in the sense that once P_{nl} has been determined in a particular calculation, it is not usually changed when further functions $P_{n'l'}$ are added. This is however not a requirement of the process and in sensitive calculations may have to be lifted (see e.g. [10]).

The MCHF method of Froese Fischer [17] uses the inequalities (3) in an alternative variational process, in which coupled integro-differential equations for the P_{nl} are constructed (generalising the usual HF process). These equations are solved numerically in an iteration procedure, at each iteration stage the coefficients a_i being the eigenvector components of the current Hamiltonian matrix. It is customary for all the radial functions to be optimised simultaneously, though the facility for fixing some of them at pre-determined functions is available, and this is useful for very large-scale calculation when some of the orbitals are rather insensitive to further extension of the set of orbitals. Additionally, configurations not included in the MCHF variational process can be added and the a_i determined in a no-variation CI calculation.

A number of other CI methods determine radial functions by processes which are not directly variational but seek to incorporate the essential physics of the atomic system. In the SUPERSTRUCTURE code [18], the radial functions are eigenfunctions of a Schrödinger-type equation involving a scaled Thomas-Fermi potential. The Cowan code [19] adopts a similar procedure with a range of different one-particle potentials involving a statistical approximation to the exchange potential.

Again, other methods provide alternatives to the numerical tabulation or Slater-type orbital expansion (4) of the radial functions. Hansen [20] and Froese Fischer [21] and their co-workers have successfully used a B-spline representation of the radial functions. The R-matrix approach [22] augments a set of bound orbitals (typically produced with CIV3 or SUPERSTRUCTURE) with "continuum" orbitals, suitable for representing continuum states or interactions between low and high lying bound states.

Some methods use model or parametric potentials to incorporate specific physical effects. Of particular usefulness is the replacement of the effect of a large atomic core by a model potential representing both the electron correlation within the core and the core-valence effects, especially core polarisation. Recent applications of these quite long established procedures are provided by Laughlin [7] and Migdalek [23]. Perhaps the most extensive of methods of this genre is the orthogonal operators process. Initiated by Judd [24], it has been developed by Hansen and co-workers who have applied it with considerable success to very complex atoms and ions [25]. A large number of physical effects can be included by this method, with the result that the energy spectrum can be modelled very accurately.

Many of these processes also allow (in various ways) for the inclusion of relativistic effects, in order to treat fine and hyperfine structure and the calculation of transitions which would otherwise be forbidden. The non-relativistic codes are most easily modified to incorporate these possibilities by adding to the Hamiltonian some or all of the operators of the Breit-Pauli approximation. Normally these are included *after* the calculation of the radial functions, but such a procedure fails to allow for the relativistic contraction (or expansion) of the radial functions. This is important for heavy elements and can be incorporated within the Cowan code, or in an unpublished form of CIV3 [26]. However, the correct way to introduce relativistic effects is through the use of the Dirac rather than the Schrödinger Hamiltonian. The MCDF or GRASP [27] method parallels that of MCHF but is based on the Dirac-Fock process rather than HF. For light elements, the Breit-Pauli approximation generally works well and avoids the need to consider large and small components of the wave function separately. In these cases, the implied identity of the radial functions describing the two j-values of any orbital is generally less serious an approximation than the inevitably limited inclusion of electron correlation (the finiteness of M in (1)).

Semi-empirical corrections to ab initio calculations

Any wave function of the form (1) with M finite will be approximate. Clearly the approximation can be improved by increasing M, but usually this is a very slowly converging process: a large increase in M may produce only a slightly better approximation. The various methods described in the previous section can be augmented by different techniques for speeding up the convergence, so as to construct wave functions which more accurately reproduce the energy spectrum and thence provide more accurate oscillator strengths. One simple improvement to an *ab initio* calculation is to use the experimental rather than the calculated transition energy. For transitions between isolated (non-interacting) states, this often represents the major correction to a purely *ab initio* calculation. However, when the states involved in the transition interact with other states, the coefficients a_i in (1) could also change with a larger calculation. More elaborate procedures are needed to extrapolate the a_i to be close to their correct values. In the Cowan code, the *ab initio* radial integrals can be multiplied by factors to improve the agreement between the calculated and experimental energies. The SUPERSTRUCTURE code and the method of orthogonal operators also include parameters which may be adjusted to obtain a close fit to the experimental energy spectrum. CIV3 incorporates an option of "fine-tuning", whereby the diagonal elements of the Hamiltonian matrix are adjusted to ensure that, as closely as possible, the eigenvalue differences agree with the experimental energy differences. It must be stressed that none of these processes result in exact wave functions, but generally they do produce oscillator strengths which are more accurate than those from an *ab initio* calculation undertaken with the same set of CSFs.

THE 3s – 4p TRANSITION IN MG II

This slowness of convergence is one of the principal difficulties in the calculation of this electric dipole (E1) allowed transition, which is anomalously weak. The radial integral in the dipole matrix element suffers substantial cancellation, with positive and negative parts being of almost equal magnitude. It is therefore difficult to determine the oscillator strength accurately, as Table 1 demonstrates. For strong transitions, after an extensive calculation of the oscillator strengths has been undertaken, further configurations have only a small proportionate effect. However, for this weak transition, the same size of correction is proportionately much greater. Convergence can be achieved only after a much more extensive calculation than is needed for strong transitions.

Recently two groups of researchers [10,11], though working collaboratively, undertook two independent calculations of this transition using the CIV3 and MCHF codes. In both cases, a sequence of calculations was carried out to permit a study of the convergence of the oscillator strength as the calculations became more extensive, though the approach is different in the two cases.

TABLE 2. CIV3 results for the Mg II 3s ^{2}S – 4p ^{2}P^o transition

	Configurations included	ΔE (cm^{-1})	f_l	f_v
1	$1s^2 2s^2 2p^6 nl$, nl=3s,4s,3p,4p	78549.96	0.00047	0.00049
2	1 plus main core-valence and core-core effects	80169.68	0.00066	0.00067
3	More orbitals, all SD[a] replacements from 1	80046.78	0.00078	0.00077
4	3 plus main triple excitations	80395.74	0.00083	0.00080
5	4 plus opening of $1s^2$	80391.35	0.00083	0.00080
	Experiment	80639.85		

[a] SD denotes single and double

TABLE 3. MCHF results for the Mg II 3s ^{2}S – 4p ^{2}P^o

	Max nl	ΔE (cm^{-1})	f_l	f_v
MCHF-SD	$4f$	81979.47	0.00049	0.00043
	5g	80395.54	0.00050	0.00046
	$6g$	80261.95	0.00067	0.00064
	$7g$	80261.73	0.00087	0.00075
	10g	80275.09	0.00085	0.00070
CI-SD[10g]∪SDT[7g]∪SDTQ[4f]	10g	80542.35	0.00084	0.00083

The CIV3 results [10] are shown in Table 2. Each line in the table corresponds to the largest of a sequence of calculations of increasing complexity within a specific limitation. The smoothness of the convergence, and the agreement between length and velocity forms, was achieved only when the valence electron radial functions were reoptimised during Calculation 2, i.e. with the main correlation effects included during the optimisation.

In the CIV3 calculation, each of the different radial functions is optimised so as to focus on a particular effect, e.g. core polarisation. In the MCHF calculation, this selection is undertaken automatically, to concentrate on the most significant correlation effects. Godefroid and Froese Fischer [11] undertook an extensive range of trial calculations with the MCHF code and, like Fleming *et al* [10], found quite a wide range of oscillator strength values during those trials. As a result of their investigations, they were able to establish a converging pattern of results which is shown in Table 3. The length form for the MCHF calculation converges to a value similar to the CIV3 result, and when further CI is added without additional optimisation, the length and velocity values in both calculations agree very well. For a transition which is so sensitive to small correlation contributions, this agreement of two independent *ab initio* calculations is a remarkable result, and shows the benefit of two groups working independently on a difficult problem.

All these calculations were non-relativistic. Godefroid and Froese Fischer went on to add the spin-independent Breit-Pauli operators, though without reoptimising the radial functions. They found a somewhat reduced f-value (see Table 1). With such a weak transition, the normal perturbative treatment of these operators may not be valid. For example, the optimal radial functions (especially those for the valence electrons) may be a little different if these relativistic operators are included, so that the dipole radial integral could then change again, such is the sensitivity to the $3s$ and $4p$ functions. Hence we would treat with some caution their final calculated value in Table 1.

TRANSITIONS IN FLUORINE-LIKE IONS

Some years ago, Blackford and Hibbert [28] published extensive tables of transitions in many fluorine-like ions, calculated using the CIV3 code. Recently, Kelleher *et al* [15] have made comparisons between these data and those provided in the

Opacity Project (OP) database [29]. Those comparisons showed that while reasonable agreement (up to 20%) was achieved between length and velocity forms of the oscillator strengths for many of the transitions, much greater disagreement occurred in others, surprisingly up to a factor of five even for some strong transitions.

McPeake and Hibbert [30] used CIV3 to reexamine all the transitions previously considered, for Na III. They extended the configuration set beyond that used by Blackford and Hibbert, and also used the fine-tuning process to improve the CI mixing. (Blackford and Hibbert had not fine-tuned their *ab initio* calculation, but had used experimental transition energies.) Following McPeake and Hibbert, we make the following points:

1. Kelleher *et al* did not actually compare the earlier CIV3 and OP calculations, since the former are given in LSJ coupling while the latter are in LS coupling. Instead they first transformed the OP oscillator strengths to LSJ results by using the formula

$$f_l(L_iS_iJ_i \to L_fS_fJ_f) = \frac{\Delta E_{LSJ}}{\Delta E_{LS}}(2J_f+1)(2L_i+1)\ \delta_{S_iS_f} \times [W(J_i,1,S_f,L_f;J_f,L_i)]^2\ f_l(L_iS_i \to L_fS_f) \quad (5)$$

where ΔE_{LSJ}, ΔE_{LS} denote the transition energies in LSJ and LS coupling respectively, and W is a Racah coefficient. Then they compared their values with the CIV3 results. The above formula ignores any mixing between states with symmetry LSJ and $L'S'J$, and especially for the $2p^43d$ levels, this mixing turns out to be very strong.

2. The only correct comparison which can be made between the CIV3 and OP results is demonstrated by the examples shown in Table 4, where we have used firstly the same configurations as Blackford and Hibbert and then the extended configurations of McPeake and Hibbert, but in purely LS coupling. The transitions selected are those which lead to the greatest deviations in the comparisons given by Kelleher *et al.* The new results give much better length/velocity agreement, but the length values (which are the ones used in the Kelleher *et al* comparisons) have changed only a little, and both sets of CIV3 results agree with the OP results to within a few per cent. The OP results therefore do not disagree substantially with the CIV3 results, when comparisons are made on a "like-with-like" basis.

3. That situation is completely changed if the actual CIV3 results are compared with the results obtained by transforming the OP data using (5), as demonstrated in Table 5. The LSJ coupling is introduced into the CIV3 calculations through the explicit inclusion of the operators of the Breit-Pauli

Hamiltonian, and consequently the CIV3 calculations include the mixing between LSJ and $L'S'J$ symmetries, which the use of (5) omits. It turns out that there is strong mixing between $2p^43p\ ^2P^o_{1/2}$ and $^2S^o_{1/2}$ and also between $2p^43d\ ^2P_{3/2}$ and $^2D_{3/2}$. These mixings induce a substantial redistribution of the oscillator strength. This can be seen in Table 5, for while the f-values obtained in different calculations for a single transition differ considerably, the totals and sub-totals are quite close.

4. When strong mixings occur, even a large *ab initio* calculation may not give energies in sufficiently good agreement with experiment to provide accurate mixing coefficients. McPeake and Hibbert [30] fine-tuned both their new calculations and the earlier CIV3 calculations in order to improve the mixings. The results of this process are also shown in Table 5. It may be seen that the two fine-tuned sets are in fairly good agreement with each other and that for the $^2P^o_{1/2} - ^2P_{3/2}$ transition, the f-value now disagrees with the "Opacity" result by a factor of eight, yet the totals still remain essentially in agreement.

We conclude that the new CIV3 calculations do make substantial improvements on the earlier CIV3 calculations, partly because of their more extended nature

TABLE 4. Multiplet oscillator strengths: $2p^43p - 2p^43d$ transitions in Na III

Transition				Opacity	CIV3-A	CIV3-C
$^2P^o$	–	^2P	f_l	0.973	1.022	1.049
			f_v		0.877	0.999
	–	^2D	f_l	3.020	2.844	3.026
			f_l		2.918	2.853
$^2S^o$	–	^2P	f_l	1.810	1.885	1.755
			f_v		1.379	1.682
	–	^2D		0.000	0.000	0.000

Opacity: The Opacity Project [29]; CIV3-A: Blackford and Hibbert [28]
CIV3-C: McPeake and Hibbert [30] – CIV3-A extended and fine-tuned

TABLE 5. Line oscillator strengths: $2p^43p - 2p^43d$ transitions in Na III

Transition			"Opacity"	CIV3-A	CIV3-B	CIV3-C
$^2P^o_{1/2}$	–	$^2P_{3/2}$	0.0556	0.2628	0.4778	0.4625
	–	$^2D_{3/2}$	0.5130	0.3401	0.2058	0.1875
Subtotal			*0.5686*	*0.6029*	*0.6836*	*0.6500*
$^2S^o_{1/2}$	–	$^2P_{3/2}$	0.6170	0.4820	0.2678	0.2447
	–	$^2D_{3/2}$	0.0000	0.0781	0.1430	0.1412
Subtotal			*0.6170*	*0.5601*	*0.4108*	*0.3859*
Total			**1.1856**	**1.1630**	**1.0944**	**1.0359**

"Opacity": [15,29]; CIV3-A: [28]; CIV3-B: CIV3-A fine-tuned
CIV3-C: [30] – CIV3-A extended and fine-tuned

(for example, giving good length/velocity agreement) but mostly because of the fine-tuning which we see as an essential process in improving *ab initio* oscillator strengths. We also conclude that where substantial fine-structure mixing between states is likely to occur, transformations using (5) should not be carried out – they are like to lead to erroneous results.

BRANCHING FRACTIONS IN FE II

Transitions in Fe II are of considerable astrophysical importance and hence have received widespread attention, though most of the work has concentrated on transitions in which at least one of the states is the lowest of its symmetry. In this section we will consider two types of transition amongst sextet levels.

The a ^{6}D – y ^{6}P^o UV 8 multiplet

Many of the Fe II excited levels are quite heavily mixed. However, the $y\ ^6P^o$ levels appear to be rather isolated, so that an extensive calculation can focus on just two states. Indeed equation (5) could be applied to an LS calculation, though in fact the calculations we show in Table 6 have involved fine-structure operators explicitly. The calculation of Donnelly and Hibbert [32] was undertaken with CIV3, that of Raassen and Uylings [33] with the orthogonal operators method, while the results of Kurucz [34] largely involved the use of the Cowan code. The f-values are in all cases somewhat smaller than the earlier, pioneering results of Nussbaumer *et al* [35] who used SUPERSTRUCTURE, but with a small number of configurations and fine-tuning. The results appear to be in fairly good agreement with the experimental work of Mullman *et al* [31], though a somewhat different picture of all these results emerges when we consider the branching fractions R, given relative to the largest oscillator strength (for the first transition in Table 6).

TABLE 6. Oscillator strengths for the UV8 multiplet in Fe II

		Experiment		Calculations							
		[31]		[32]		[33]		[34]		[35]	
g_l	g_u	f	R	f	R	f	R	f	R	f	R
10	8	0.058(5)^a	1.00	0.062	1.00	0.055	1.00	0.072	1.00	0.096	1.00
8	8	0.022(5)	0.38	0.022	0.35	0.020	0.36	0.026	0.36	0.035	0.38
8	6	0.041(4)	0.71	0.038	0.61	0.035	0.64	0.041	0.57	0.061	0.64
6	6	0.028(3)	0.48	0.037	0.60	0.034	0.62	0.038	0.54	0.058	0.60
6	4	0.018(3)	0.31	0.018	0.29	0.016	0.29	0.021	0.29	0.029	0.30
4	4	0.032(5)	0.55	0.041	0.66	0.037	0.67	0.047	0.66	0.066	0.69
2	4	0.049(5)	0.84	0.058	0.94	0.053	0.96	0.067	0.93	0.095	0.99

^{a}: Experimental error in brackets
f: Oscillator strengths; R: Branching fractions

From Table 6 it can be seen that the calculations are in very substantial agreement, particularly the most recent CIV3 and orthogonal operators results. Of course, this is merely indicative of an absence of mixing with other levels, particularly in the upper state. In the CIV3 work, other symmetries were included in the calculation, but no significant mixing with other levels was found. By contrast, some of the experimental values for these fractions differ substantially from the theoretical results, and this may indicate that mixing with other levels is indeed important. A new experimental determination of these fractions is in progress (J.C. Pickering, private communication) and if the experimental values [31] are confirmed, a new theoretical search for the causes of these discrepancies would be needed.

TABLE 7. Oscillator strength ratios for sextet transitions in Fe II

			f-values	relative to f-value of [36]			
	g_l	g_u	[36]	[37]	[33]	[34]	[38]
$a\,^6D - z\,^6D^o$	10	8	0.070	0.73	0.99	0.91	1.27
	8	8	0.127	1.01	1.02	0.94	0.97
	6	8	0.085	0.99	1.01	0.96	0.68
	8	6	0.108	1.00	0.99	0.92	1.14
	6	6	0.049	0.98	1.04	0.96	1.10
	4	6	0.127	0.97	1.02	0.95	0.73
	6	4	0.124	1.05	1.00	0.93	1.05
	4	4	0.004	1.00	1.00	1.00	1.33
	2	4	0.177	1.02	1.02	0.94	0.78
$a\,^6D - z\,^6F^o$	10	8	–	–	–	–	–
	8	8	0.088	1.10	1.02	0.98	0.93
	6	8	0.270	1.14	1.07	1.02	1.11
	8	6	0.006	–	–	–	–
	6	6	0.128	1.08	1.03	0.98	0.98
	4	6	0.233	1.11	1.05	1.00	1.15
	6	4	0.017	1.13	1.06	1.00	0.74
	4	4	0.166	1.12	1.03	0.99	1.03
	2	4	0.201	1.15	1.08	1.00	1.15
$a\,^6D - z\,^6P^o$	10	8	0.131	1.15	1.04	1.02	1.04
	8	8	0.059	1.20	1.11	1.04	1.31
	6	8	0.029	0.85	0.81	0.74	2.90
	8	6	0.080	–	1.04	1.00	0.99
	6	6	0.095	–	1.06	1.01	1.23
	4	6	0.063	–	–	0.87	1.66
	6	4	0.039	1.11	1.05	1.00	0.97
	4	4	0.102	1.11	1.02	0.99	1.15
	2	4	0.168	1.07	0.99	0.95	1.33

The a ^{6}D – z ^{6}D^o, ^{6}F^o, ^{6}P^o multiplets

The odd parity levels of these transitions, though lower in energy than y ^{6}P^o, do in fact mix, sometimes quite strongly. We show in Table 7 the actual oscillator strengths obtained in the fine-tuned CIV3 calculations of Donnelly and Hibbert [36] together with the ratios of these results to those by Raassen and Uylings [33] and by Kurucz [34]. Also included in Table 7 are the ratios with respect to the experimental results of Bergeson *et al* [37], and to the calculation of Nahar and Pradhan [38] who essentially scaled the OP results using equation (5).

The consistency of the ratios in Table 7 within a multiplet shows that there is good agreement for the branching fractions between all the determinations with the exception of Nahar and Pradhan [38]. This provides a further example of the inadequacy of (5) when the lower or upper levels of a transition involve fine-structure mixing. Except for a small number of weak transitions, there is a better than 10% agreement between the extensive calculations of Donnelly and Hibbert [36] and Raassen and Uylings [33]. For the ^{6}D – ^{6}D^o transitions, both these sets of theoretical results agree very well with the experimental result of Bergeson *et al* [37] while for the other two multiplets, the theoretical results are mostly around 10% higher than experiment.

Nonetheless, these results show that present-day calculations are of very satisfactory accuracy, even for rather complex ions.

CONCLUSIONS

In this article, we have tried to show that calculations are now at a level of sophistication that it is possible to obtain oscillator strengths for a wide range of ions that are sufficiently accurate to be of use in the analysis of astrophysical observations. If one is to be assured of that level of accuracy, care must be taken, in arriving at the final results, to employ all available refinements to *ab initio* calculations (for example, the use of known energy levels to determine variable parameters or in the process of fine-tuning). Where possible (usually for the calculation of a small number of transitions), the convergence of the results with respect to an increase in the complexity of the calculation can be analysed. Finally, the comparison between determinations using different methods (theoretical or experimental) is an important means of assessing accuracy, and the interaction between researchers facilitates the deepening of our understanding of how those methods might be best applied. In this context, the work of the NIST scientists is invaluable. It is important though that comparisons are made between original data and not between results obtained by transforming that data. The fact that one *can* transform LS data into LSJ data does not mean that one *should*.

REFERENCES

1. Black, J.H., Weisheit, J.C. and Laviana, E., *Astrophys. J.* **177**, 567-572 (1972).
2. Saraph, H.E., *J. Phys. B: Atom. Mol. Phys.* **9**, 2379-2383 (1976).
3. Froese Fischer, C., in *Beam Foil Spectroscopy*, edited by I.A. Sellin and D.J. Pegg, Plenum, New York, 1976, **1**, pp. 69-76 .
4. Biémont, E., *Astron. Astrophys. Supp.* **31**, 285-290 (1978).
5. Hibbert, A, Dufton, P.L., Murray M.J. and York, D.G., *Mon. Not. R. Astr. Soc.* **205**, 535-541 (1983).
6. Butler, K., Mendoza, C. and Zeippen, C.J., *J. Phys. B: Atom. Mol. Phys.* **17**, 2039-2048 (1984).
7. Laughlin, C., *Physica Scripta* **45**, 238-245 (1992).
8. *The Opacity Project*, edited by M.J. Seaton, IOP Publishing, Bristol, Vol. 1.
9. Sigut, T.A.A. and Pradhan, A.K., *J. Phys. B: Atom. Mol. Opt. Phys.* **28**, 4879-4893 (1995).
10. Fleming, J., Hibbert, A., Bell, K.L. and Vaeck, N., *Mon. Not. R. Astr. Soc.* **300**, 767-772 (1998).
11. Godefroid, M. and Froese Fischer, C., *J. Phys. B: Atom. Mol. Opt. Phys.* **32**, 4467-4483 (1999).
12. Sofia, U.J., Cardelli, J.A. and Savage, B.D., *Astrophys. J.* **430**, 650-666 (1994).
13. Fitzpatrick, E.L., *Astrophys. J.* **482**, L199-L202 (1997).
14. Wiese, W.L., Fuhr, J.R. and Deters, T.M., *J. Phys. Chem. Ref. Data*, Monograph 7 (1996).
15. Kelleher, D.E., Wiese, W.L., Fuhr, J.R. and Podobedova, L.I., "Critical Evaluation of Transition Probabilities, H–Ca", in *ASOS6 Abstracts*, edited by J. Tatum, 1998, pp 83-86.
16. Hibbert, A., *Comput. Phys. Commun.* **9**, 141-172 (1975).
17. Froese Fischer, C., *Comput. Phys. Commun.* **64**, 431-454 (1991).
18. Eissner, W., Jones, M. and Nussbaumer, H., *Comput. Phys. Commun.* **8**, 270-306 (1974).
19. Cowan, R.D., *The Theory of Atomic Structure and Spectra*, University of California Press, Berkeley, 1981.
20. van der Hart, H.W. and Hansen, J.E., *J. Phys. B: Atom. Mol. Opt. Phys.* **25**, 41-59 (1992).
21. Xi, J. and Froese Fischer, C., *Phys. Rev. A.* **53**, 3169-3177 (1996).
22. Berrington, K.A., Burke, P.G., Butler, K., Seaton, M.J., Storey, P.J., Taylor, K.T. and Yu Yan, *J. Phys. B: Atom. Mol. Phys.* **20**, 6379-6397 (1987).
23. Migdalek, J. and Garmulewicz, M., *J. Phys. B: Atom. Mol. Opt. Phys.* **33**, 1735-1743 (2000).
24. Judd, B.R., Hansen, J.E. and Raassen, A.J.J., *J. Phys. B: Atom. Mol. Phys.* **15**, 1457-1472 (1982).
25. Hansen, J.E., Judd, B.R., Raassen, A.J.J. and Uylings, P.H.M., *Phys. Rev. Lett.* **78**, 3078-3081 (1997).
26. Mohan, M. and Hibbert, A., *J. Phys. B: Atom. Mol. Phys.* **20**, 907-918 (1987).
27. Parpia, F.A., Froese Fischer, C. and Grant, I.P., *Comput. Phys. Commun.* **94**, 249-271 (1996).
28. Blackford, H.M.S. and Hibbert, A., *Atom. Data Nucl. Data Tables* **58**, 101-164 (1994).
29. Cunto, W., Mendoza, C., Ochsenbein, F. and Zeippen, C.J., *Astron. Astrophys.* **275**, L5-L8. (1993).
30. McPeake, D. and Hibbert, A., *J. Phys. B: Atom. Mol. Opt. Phys.*, submitted (2000).
31. Mullman K.L., Sakai M. and Lawler J.E., *Astron. Astrophys. Supp.* **122**, 157-161 (1997).
32. Donnelly, M.P. and Hibbert, A., *Mon. Not. R. Astr. Soc.* **302**, 413-415 (1999).
33. Raassen, A.J.J. and Uylings, P.H.M., *J. Phys. B: Atom. Mol. Opt. Phys.* **31**, 3137-3146 (1998).
34. Kurucz, R.L., *Trans IAU* **20B**, 168 (1988).
35. Nussbaumer H., Pettini M. and Storey P.J., *Astron. Astrophys.* **102**, 351-358 (1981).
36. Donnelly, M.P. and Hibbert, A., *Mon. Not. R. Astr. Soc.*, submitted (2000).
37. Bergeson S.D., Mullman K.L., Wickliffe M.E., Lawler J.E., Litzen, U. and Johansson, S., *Astrophys. J.* **464**, 1044-1049 (1996).
38. Nahar, S.N. and Pradhan, A.K., *J. Phys. B: Atom. Mol. Opt. Phys.* **27**, 429-446 (1994).

The Quantum Chemistry Alternative

Yves Ellinger

Laboratoire d'Etude Théorique des Milieux Extrêmes, Ecole Normale Supérieure
24, rue Lhomond, 75005 Paris, France

Abstract. This review paper presents illustrative cases showing how Quantum Chemistry calculations can be an alternative to laboratory experiments. After a short description of the basic concepts and methods used, the presentation is centered on the ability of computational chemistry to produce data pertinent to solving problems of astrophysical interest ranging from identification of new molecules to IR emission analysis and rate constants determinations.

INTRODUCTION

Many of the roles played by laboratory work in the sequence going from the characterisation of an unstable species, reaction intermediate or excited state by its spectral signatures to the understanding of the possible mechanisms of formation and destruction, can be played by theoretical methods. Astrophysics is an ideal field for applications of Quantum Chemistry as it is the natural complement of observations, laboratory experiments and chemical modeling (see Fig. 1).

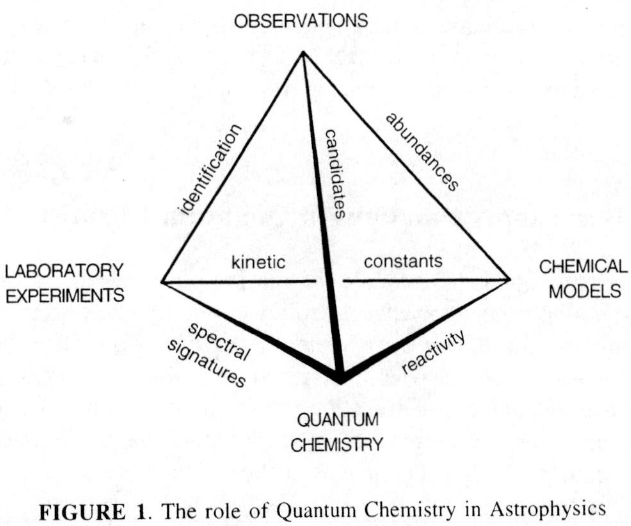

FIGURE 1. The role of Quantum Chemistry in Astrophysics

CP543, *Atomic and Molecular Data and Their Applications,* edited by K. A. Berrington and K. L. Bell

Quantum Chemistry is well adapted to Astrophysics as it refers for a large part to isolated systems and can be used to examine unstable species and chemical processes that are difficult, if not impossible to study in the laboratory.

It can support or even replace spectroscopy experiments by *a-priori* calculations of the spectral signatures without any restriction in the electronic spectrum (Radio, IR, Visible-UV). This ability can profitably be exploited for identification of a specific molecule which has been generated in a mixture of transient species. The prediction of the spectral signatures of unknown molecules can also be an incentive for new observations.

Predictions of the reaction paths and branching ratios, without limitations concerning the nature of the processes involved, are also possible. There is no basic difference between the methods used to study ion-neutral or neutral-neutral reactions. In the same way the study of dissociative electronic recombinations requires strategies very similar to those used for photochemical processes or radiative associations. More recently, the methods of Quantum Chemistry have reached a point where large systems can be apprehended, making possible studies of the depletion of atoms and molecules on dust particles and new investigations of reactions at the solid-gas interface. These studies can provide the rate constants which cannot be obtained in the laboratory, but are necessary to the modeling of the reaction networks governing the molecular abundances observed in space.

What is true for Astrophysics is also true for a number of fields such as combustion chemistry, lighting, plasmas, surface collisions or atmospheric processes. Applications of Quantum Chemistry to these topics are worth their own reviews and will not be presented here.

THE METHODS OF QUANTUM CHEMISTRY

Although reporting the details of the calculations performed is far beyond the scope of this review, it seems important to make clear to the reader what is the philosophy behind computational chemistry, including its strengths, weaknesses and limits of applications.

Basic approximations in Quantum Chemistry

The first approximation is to decouple the motion of the electrons from that of the nuclei in the Schrödinger equation that describes the complete system of particles in interaction. Known as the *Born-Oppenheimer approximation*, it relies on the intuitive reason that nuclei are much heavier than electrons, which means an instantaneous adjustment of the electronic wavefunction Ψ to the position of the nuclei. The Schrödinger equation can then be reduced to an electronic equation at fixed positions of the nuclei. The motion of the nuclei that is at the origin of infrared and rovibrational observables is treated *a posteriori* by solving a Schrödinger equation in which the total electronic energy serves as a potential.

The second approximation considers that the electrons are non interacting particles. Known as the *independent particle approximation*, it means that an electron is moving in the average field of the others and can be treated with a one-electron Schrödinger equation. The one-electron function φ describing the motion of the electron is named electronic spin-orbital and depends on the spatial and spin coordinates of one electron only. A description of n-electron systems can be obtained in the form of an antisymmetrized product of spin-orbitals better known as a Slater determinant Φ or electronic configuration.

The third approximation is a direct consequence of the second one that considers the electrons as independent particles, which obviously is not valid for a number of problems, such as bond breaking and bond formation or electronic spectra and photochemistry. The motion of an electron not only depends upon the average electrostatic field in the molecule but also on the instantaneous position of all other electrons. This effect, known as *interaction of configuration*, can be taken into account by mixing the ground state configuration with appropriate excited configurations. An important notion in this respect, and in chemistry in general, is the concept of valence space. It means the space spanned by the union of the atomic orbitals that are occupied in the atomic ground states of the constitutive atoms. From a molecular point of view, it is the space describing the bonding, antibonding and electron pairs levels; finally, it is the space where chemistry takes place.

The fourth approximation is directly related to the practical aspect of Quantum Chemistry. For pragmatic reasons, spin-orbitals are given analytical forms that are easily handled in numerical calculations. They are expanded as linear combinations of atomic orbitals. Known as the *LCAO approximation*, it means that the analytical function describing the isolated atom is also a good representation of the atom in the molecule. Atomic orbitals are then defined and optimized in specific calculations. These optimized functions will be used as basis sets in polyatomic calculations. For Photophysics purposes, such as those related to Rydberg states, basis functions of diffuse character must be added to properly describe the regions far from the nuclei.

Levels of theory

How close the calculated wavefunction will be to the exact wavefunction depends on a quadruple choice than can be expressed in the form of four equations interlocked as a nest of Russian dolls:

- Choice of the atomic basis set:

  $$X_p = \Sigma_{(g)}\ G_g$$

  where the G_i are generally Gaussian type functions centered on the atoms and previously determined in atomic calculations.
- Choice of the spin-orbitals:

  $$\varphi_i = \Sigma_{(p)}\ c_{pi}\, X_p$$

where the c_{pi} are the coefficients defining the contributions of the different atoms to the molecular levels.

- Choice of the electronic configuration:

$$\Phi_K = 1/(n!)^2 \| \varphi_{1(1)} \varphi_{2(2)} \ldots\ldots \varphi_{n(n)} \|$$

where the φ_i are the one-electron levels defining the electronic state in terms of spatial symmetry and multiplicity.

- Choice of the basis of electronic configurations:

$$\Psi = \Sigma_{(K)} d_K \Phi_K$$

where the total electronic wavefunction is expressed in the form of a linear combination of electronic configurations built on the φ_i previously calculated.

The energy ***E*** and associated wavefunction Ψ are calculated by

$$\mathbf{E} = \langle \Psi | \mathbf{H} | \Psi \rangle / \langle \Psi | \Psi \rangle$$

using two types of approaches, i.e., variational methods in which **E** is minimized or perturbational methods designed to improve as best as possible a zeroth order solution. The whole machinery of computational chemistry can now easily be grasped using the variational approach as an example. Two minimum conditions must be satisfied:

$$\partial \mathbf{E} / \partial c_p = 0$$

$$\partial \mathbf{E} / \partial d_K = 0$$

which define the level of calculation done according to the choices that are made:

- By fixing the coefficients d_K of the basis of electronic configurations

$\{ d_K \} = \{ 1, 0, 0, \ldots 0\}$ $\quad\quad$ $\partial \mathbf{E} / \partial c_p = 0$

one obtains the category of Hartree-Fock methods (RHF, UHF, ...).

- By fixing the coefficients of the atomic orbitals in the molecular spin-orbitals φ_i

$\{ c_{pi} \} = \{ \text{fixed } c_p \text{ in all } \varphi\}$ $\quad\quad$ $\partial \mathbf{E} / \partial d_K = 0$

one obtains the category of configuration interaction methods (CI).

- By relaxing both the coefficients of the atomic orbitals in the molecular spin-orbitals and the coefficients of the configuration expansion

$\{ c_{pi} \text{ relaxed}\}$ and $\{ d_K \text{ relaxed} \}$ $\quad\quad$ $\partial \mathbf{E} / \partial c_p = 0$ and $\partial \mathbf{E} / \partial d_K = 0$

one obtains the class of multiconfiguration self consistent field methods (MCSCF).

Wavefunction based methods

All methods start with an initial guess of the wavefunction. The most common starting point in a large majority of computational strategies (see Fig. 2) is the Hartree-Fock wavefunction. Correlation effects due to configuration interaction are taken into account by making the SCF ground state couple with configurations in which at least two electrons have been excited into unoccupied electronic orbitals. When limited to single and double excitations from the ground state one has a typical SDCI calculation. Interaction with the same doubly excited configurations, as represented on Fig. 2 for the SDCI diagram, is the most popular method in the perturbative approach and known as MP2 (Moller-Plesset perturbation to 2nd order). In the CAS methods, all possible configurations that can be generated by distributing the valence electrons in the valence orbitals are considered. It is an MCSCF approach which can be further refined by considering single (FOCI) or double

(SDCI) excitations from the valence space. Here also, perturbation theory can be applied on the MCSCF function, the CASPT2 or MC/P acronyms being self explicative. Higher order perturbative methods, MP3 or MP4 are also available.

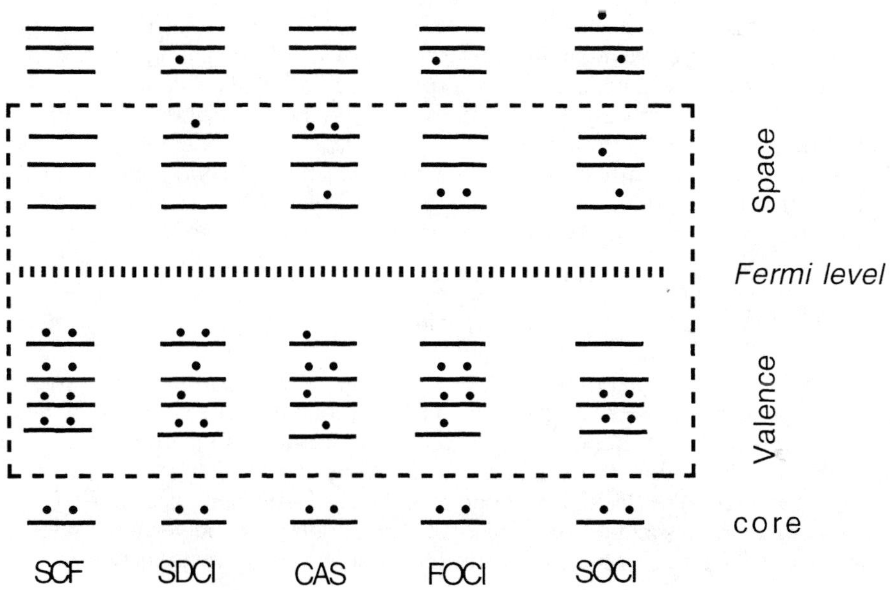

FIGURE 2. Electronic distributions generated in widespread correlated methods

Another powerful technique to perform large scale configuration interaction in a systematic way relies on the Coupled Cluster methodology. It permits to define a hierarchy of improvement on energy and observables by progressively incorporating more and more excited configurations, namely CCSD (single and double), CCSDT (single, double and triple), CCSDTQ (single, double, triple and quadruple).

DFT based methods

A completely different approach is that of Density Functional Theory (DFT). It relies on the fact that it is not necessary to know the wavefunction Ψ ($\mathbf{r}_1$, $\mathbf{r}_2$ ... $\mathbf{r}_n$) of the n electrons to determine the energy and ground state properties but only the electronic density $\rho(\mathbf{r})$. Hohenberg et Kohn have shown that the kinetic energy $\mathbf{T}[\rho]$, the potential energy $\mathbf{V}[\rho]$ and the total energy $\mathbf{E}[\rho]$ are functionals of the electronic density. In this formalism, the electronic correlation is no longer a theoretical problem since it is automatically included. However, the lack of an analytic expression for the energy functional makes it impossible to determine the exact density. The problem has then to be approximated in another way.

By analogy with the Hartree-Fock scheme a guess of $\rho(\mathbf{r})$ can be derived from a set of orbitals to which ocupation numbers (2, 1 or 0) are associated. Considering *non interacting particles*, it is possible to derive a new expression of the energy functional in the form of a sum of functionals of kinetic energy $\mathbf{T}_s[\rho]$, Coulomb energy $\mathbf{J}[\rho]$, plus another functional representing the difference between the system of non interacting electrons and the exact functional. This new functional $\mathbf{E}_{XC}[\rho]$, named the exchange-correlation functional, contains a part of kinetic energy (the difference between the kinetic energies $\mathbf{T}[\rho]$, and $\mathbf{T}_s[\rho]$ of the interacting of non interacting electrons) contrary to the purely bielectronic correlation term of the wavefunction based methods.

The correlation problem which is treated by interaction of configurations in the wavefunction based methods[1] has been replaced by the problem of finding the best approximation of the exchange-correlation functional in DFT methods[2].

Limitations

The choice of the method to be employed (see Fig. 3) results from a compromise critically dependent of the two dimensions of the problem, chemical and practical. The first is given by the number of electrons; for chemical properties, it can be reduced by using effective core potentials for electrons that are only spectators in the process considered. The second is the number of basis functions necessary to the obtention of the orbitals and/or the number of configurations, defined by the level of excitation, that have to be generated to accurately describe the wavefunction.

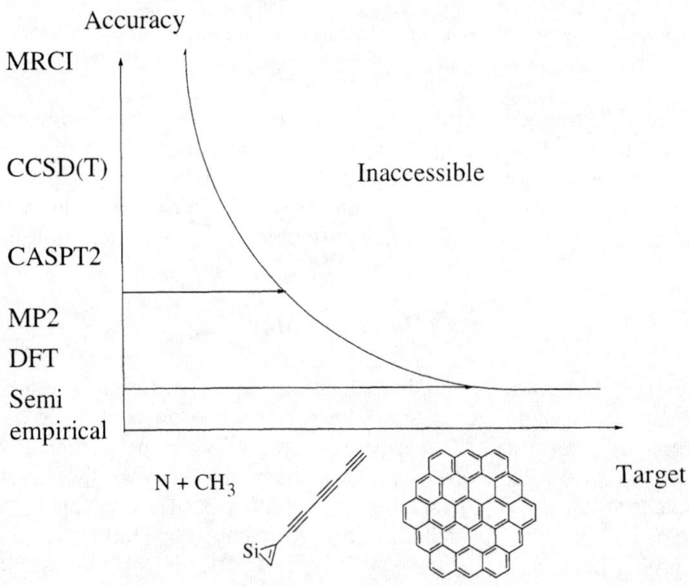

FIGURE 3. Methods used as a function of the size of the target molecule

For small molecules or for the study of reaction processes where a high accuracy is required, state of the art methods like MP4, MRCI, CCSDT or CASPT2 are employed with large basis sets. The quality of the calculations (error bar) can be checked by making sure that the results are stable with respect to any extension of both the atomic basis and the configuration space.

When the chemical dimension increases beyond ≈10 atoms, perturbative methods are more advantageous, the order of perturbation decreasing from MP4 to MP3 and MP2 with the size of the molecule. Large systems, up to ≈50 atoms are the area of predilection of DFT methods. The point to stress here is that, once the level of theory has been chosen, there are no more approximations in the calculations: the only input data are the fundamental constants of Physics, the number of electrons and the atoms constituting the molecule.

For very large systems over ≈100 atoms, only semi-empirical methods in which parameters are adjusted on experimental data can be used.

THE DATA PRODUCED BY QUANTUM CHEMISTRY

The data provided by Quantum Chemistry are related to three types of properties useful for Astrophysics: structural, spectroscopic and kinetic.

- Structural:
  Geometries and conformations can be obtained with high precision and it is possible to predict the most stable arrangements of any polyatomic system up to medium size molecules.
- Spectroscopic:
  Spectral signatures can be calculated with enough accuracy to be used as guidelines for laboratory experiments aimed at the identification of new molecules or classes of molecules in space.
- Kinetic:
  Rate constants can be evaluated from potential energy surfaces for reaction as different as ion-molecule, dissociative recombination or photodissociation.

The important point is that numerical simulations are able to do much more than what can be reasonably done in the laboratory, even test the craziest ideas.

Data pertinent to microwave and Radio

The radio signature is a fingerprint, specific of a molecule, and over 110 species have been identified in this way. It is directly related to the moment of inertia, which means the geometry (bond lengths and bond angles), to the electronic multiplicity and spin properties and to the electronic distribution since the intensity of the lines observed scales with the square of the dipole moment. The situation is different according the molecule is the next member of a series whose leading terms have already been identified or is a leading term in itself for which nothing is experimentally known. In this latter case it is important to have structural information on parent systems.

Best calculation on individual molecules

When the attention is focused on a precise candidate, the best level of theory has to be used. The search for negative ions in the C_n^- series is illustrated in Table 1 on the example of C_6^-. The first step in this approach is to determine, as best as possible, the structure of the parent molecule. Using a very large basis set and the Coupled Cluster method, it was possible to reach the convergence of B_e for the $^3\Sigma_g^-$ ground state at the CCSD(T) level of wave function. Calibration on C_5 showed that even at this level, bond lengths are still too long by 0.007 Å. After correction, a best estimate of the rotational constant of B_e can be derived in remarkable agreement with experiment[3].

TABLE 1. Observed and computed spectroscopic constants (Mhz) for C_6 and C_6^-

Method	C_6	C_6^-
RHF	1488.3	1472.0
RCCSD	1454.1	1440.0
RCCSD[T]	1435.2	1425.6
RCCSD-T	1437.2	1427.2
RCCSD(T)	1437.7	1426.9
Best estimate	1452	1442
Experiment	1453.43	

Taking the same level of wavefunction and transferring the correction on the bond lengths, it is now possible to estimate the rotational constant for the $^2\Pi_u$ ground state of the unknown molecule. The prediction is expected to be accurate within 0.2%.

Best estimate in a series of analogous molecules

Since the precision needed for identification is at the very limit or beyond the capability of the actual methods of Quantum Chemistry, a current strategy for obtaining rotational constants is to combine theoretical results with observed data on the leading terms of the series under consideration. The principle of these estimations is to derive an analytical, or numerical, error function that links B_e (theory) to B_0 (observed). This strategy is illustrated in Table 2.

TABLE 2. Observed and computed spectroscopic constants (MHz) for the C_nH and C_nO series.

	Polyynyl radicals				Polycarbon monoxides	
	B_0 Observed	B_e Calculated	B_e Best estimate		B_0 Observed	B_e Calculated
C_2H	43675	43322		C_2O	11545.5970	11578.94
C_3H	11186	11067		C_3O	4810.88638	4801.812
C_4H	4759	4350		C_4O	2351.26250	2349.321
C_5H	2395	2373.2		C_5O	1366.84709	1364.760
C_6H	1386.2	1375.7		C_6O	849.75709	849.2431
C_7H	874.4869	867.34	873 ± 0.002	C_7O	572.94105	572.6526
C_8H	586.6707	582.23	586 ± 0.001	C_8O	400.64183	400.3246
C_9H	413.2579	409.62	412 ± 0.001	C_9O	293.73611	293.7576

Calculations made for the polyynyl radicals, C_nH, beyond n=6 have predicted the rotational constants of the higher up to n=9. In this type of study[4], the basic requirement is to use strictly the same level of theory (here Hartree-Fock methods) for all the members of the series even if it is trivial to make much better calculations for the first terms. These values, obtained in 1991, have been used very recently to identify C_7H, C_8H and C_9H in laboratory experiments[5]. The measured rotational constants match the theoretical values within the error bar. This theory-experiment agreement immediately led to the assignment[6] of some interstellar lines to C_7H and C_8H. No evidence was found for C_9H in the available astronomical surveys, but observation conditions will certainly progress in a close future.

The progress of Quantum Chemistry methods is illustrated with the polycarbon monoxide series. These results are obtained using the DFT/B3LYP method with an extended basis set and need no longer be corrected[7]. The predictions are expected to be accurate within 0.2% for this type of problem in homogeneous series.

Data pertinent to Infra Red

The IR signature is also a finger print of small systems but mostly a transferable characteristics, specific of functional groups, that is found whatever the architecture of the molecule. Once more the situation is different according the molecule is part of a series of known species or is an exotic system for which no experiment has been successful in the past. In this latter case it is also important to have spectroscopic information on parent systems.

Best calculation on the individual molecules

For a small system, the IR signature is complementary to the Radio spectrum. Once the geometry has been determined, the vibrational frequencies can be evaluated beyond the harmonic approximation by taking into account the anharmonicity of the potential surface and the coupling between vibrations. Using a very large basis set and the Coupled Cluster method, it is also possible to reach the convergence for the IR frequencies of the $^2\Pi_u$ ground state of the yet unknown C_6^-. The results[3] presented in Table 3 should be accurate within a few cm^{-1}. Such a precision cannot actually be obtained for IR intensities which depend on the derivatives of the dipole moment along the normal coordinates. In favorable circumstances they are evaluated within 10% .

TABLE 3. Computed IR stretching frequencies (cm^{-1}) for C_6^-

Method	ν_1	ν_2	ν_3
RHF	2376.5	2014.2	686.7
RCCSD	2274.0	1873.4	650.3
RCCSD[T]	2140.0	1782.8	635.6
RCCSD-T	2146.6	1806.9	638.0
RCCSD(T)	2145.0	1804.2	637.3

Best estimate in a series of analogous molecules

A formidable challenge in Astrophysics is the interpretation of the IR emission of the galaxy between 2 and 15 μm. The same type of emission has been observed in objects submitted to a variety of environments such as reflection nebulae, planetary and protoplanetary nebulae, HII regions and galactic nuclei. These bands, known as the unidentified infrared bands (UIR bands) are ubiquitous in the interstellar medium. They appear at 3.3, 6.2, 7.7, 8.6, 11.3 and 12.7 μm (respectively, 3030, 1610, 1300, 1160, 885 and 790 cm^{-1}). The strong similarity of these features to the IR signatures of aromatic compounds has led to consider that the Polycyclic Aromatic Hydrocarbon (PAH) were the carriers of the UIR bands[8], although the detail of the spectral match is not completely satisfactory neither for the band positions nor for their relative intensities.

Because of the strong radiation field in the PAH environment, several important questions remain unsettled. For example, the structure, linear, compact, irregular, ... is not kown. PAHs may be neutral, positive or negative ions; they may have substituents attached at the periphery, be partially dehydrogenated, or perhydrogenated, or any combination of these situations. Quantum Chemistry provides a unique and reliable possibility for probing these hypotheses at moderate cost when compared to the investment required to build the appropriate experiments.

The case of a dehydrogenated PAH generated in a Neon matrix is an illustrative example (see Table 4). The calculation of the frequencies is performed, in the harmonic approximation, at the Hartree-Fock level, due to the dimension of the problem. In order to take into account the correlation effects that are missing at this level of theory, all frequencies are multiplied by a scaling factor previously determined so as to minimize the root mean square deviation between the calculated and experimental spectra of more than a hundred of test molecules. The agreement between an experiment in a solid matrix[9] and theory[10] is remarkable for an exotic species. Systematic investigations of the structure and ionicity of PAHs suggest that, in space, these molecules should be ionized[11] and partly dehydrogenated[10,12] to accomodate the IR interstellar emission.

TABLE 4. IR spectrum of 2,3-didehydrogenated naphthalene (ν cm-1; Intensity km/mol)

Observed in Neon matrix		Calculated at the RHF level		
ν	Relative Intensity	ν	Intensity	Assignment[a]
3098	8	3090	13	$r_{(CH)}$
3084	10	3078	19	$r_{(CH)}$
3069	16	3065	21	$r_{(CH)}$
1213.1	4	1232	23	$\beta_{(CH)}/R_{(CC)}$
1018.3	2	1010	17	$R_{(CC)}/\beta_{(CH)}$
848.6	9	869	24	$\alpha_{(CCC)}$
834.1	37	843	74	$\varepsilon_{(CH)}$
738.6	47	739	51	$\tau_{(CCC)}$
618.1	79	600	54	$\alpha_{(CCC)}$
451.8	9	446	22	$\tau_{(CCC)}$
446.5	92	438	92	$\alpha_{(CCC)}$

[a]Only theoretical calculations can provide the assignment of the nuclear motions

Data pertinent to chemical processes

Molecular abundances in interstellar clouds depend on the balance between a series of formation and destruction mechanisms, the relative importance of which is difficult to establish.The extreme physical conditions prevailing in the interstellar mediium - low densities, low temperatures, radiation field - lead to a peculiar chemistry far from that operating on Earth.

In the gas phase, only 2-body reactions can be efficient. The absence of a 3rd body implies that most processes are dissociative as there is no other possibility to evacuate the excess energy released in the reaction. Due to the low temperature, there is no energy available to pass over a transition barrier (1kcal/mol ≈ 500K) which means that only exothermic reactions are allowed with no activation barrier on the reaction path. The other possibility is to consider photochemical reactions where the energy is provided by the radiation field.

The great unknown of space chemistry is the role played by dust particles. They may act as a cold wall, catalysts or reagent reservoirs. The experimental studies of gas-solid interface processes are still a paramount task and Quantum Chemistry is once more an alternative.

Ion - neutral chemistry

One of the most intriguing problem of interstellar chemistry to day is the apparent inconsistency of the HNC/HCN abundance ratio which was observed to vary from 2 to 1/80 according the type of object observed. One of the key process in the chemical network governing carbon - nitrogen chemistry is the reaction between atomic carbon and ammonia. The potential energy surface is reported in Fig. 4.

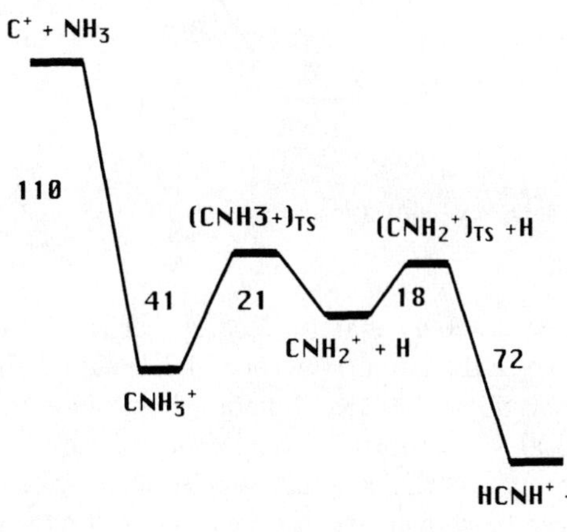

FIGURE 4. Energy profile (kcal/mol) for C^+ (X^2P) + NH_3 → H_2NC^+ ($^1A_1$) +H → $HCNH^+$ ($^1\Sigma^+$) + H

Using a basis set close to the Hartree-Fock limit and 4th order perturbation theory (MP4) to account for correlation effects, it has been shown that this process is exothermic and that the transition states on the reaction path lie much lower in energy than the initial reagents. The vibrationally excited complex H_3NC^+ formed in the first step will easily overcome the barrier opposing the formation of CNH_2^+. Isomerization into linear $HCNH^+$ is a more complex problem for several reasons. First, if the energy released is largely dissipated as translational energy, there will not be enough remaining energy to pass over the barrier on the route to $HCNH^+$. Second, the relaxation of CNH_2^+ *via* internal conversion may be more rapid than isomerization itself. Using an RRKM statistical method, it has been possible to estimate the isomerization rate and determine than 98% of CNH_2^+ is converted into the linear ion[13].

The dissociative recombination of CNH_2^+ which was originally thought to provide the excess of HNC can then be discarded from the chemical models and one is left with the dissociative recombination of $HCNH^+$ that produces equivalent amounts of HCN and HNC[14].

Neutral - neutral chemistry

Direct isomerization being not possible due to a high activation barrier, an external process has to be considered. The hydrogen exchange reaction may be one of those. The reaction path reported in Fig. 5 has been calculated at the MP4 level of theory using the same basis set close to the Hartree-Fock limit.

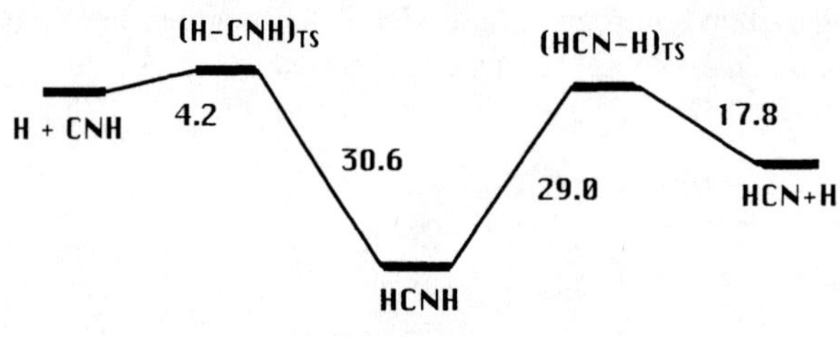

FIGURE 5. Energy profile (kcal/mol) for H + CNH $\leftrightarrow$ HCN + H

The potential energy surface shows that the energy barrier is much lower for HCN formation than in the reverse direction. However, 4.2 kcal/mol for the barrier on the entrance channel represents about 2000 K. Dynamical calculations, including tunnelling effects show that a typical value of 10^{-10} - 10^{-11} cm^3 s^{-1} mol^{-1} of the collisional rate constant is obtained at about 700 K. Assuming a fixed abundance of hydrogen, typical of molecular clouds, it can be shown that this reaction will not affect the HNC/HCN ratio unless temperatures in excess of 300 K are available[15]; the problem is still open.

Solid - gas interface chemistry

The chemical function of dust and grains has only recently been considered in astrochemical models. Solid particles, present in the interstellar medium in the form of silicates, cristalline or amorphous carbon, ices, ... provide various types of surfaces acting as support and/or catalysts for the synthesis of molecules which, otherwise, would not be formed in 2-body reactions. Gas-surface interactions can also affect the composition of the interstellar gas due to sticking of atoms or molecules at the surface of dust particles. Unfortunately, experimental data dealing with the adsorption of small molecules are sparse and do not allow to derive conclusions pertinent to astrophysical conditions.

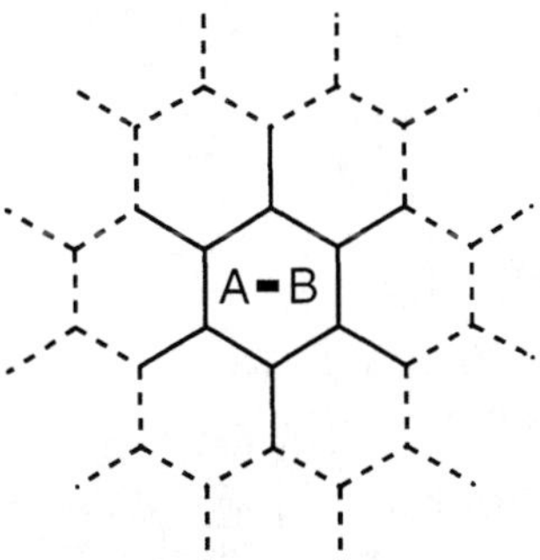

A-B = N_2, CO, O_2

FIGURE 6. Computational model for the study of adsorption on the carbonaceous matter (PAHs)

The size of the surface that should be considered for the carbonaceous matter, at least a coronene molecule (Fig. 6), precludes the systematic use of common correlated ab-initio methods. Alternatively, semi-empirical or molecular mechanics calculations which would be attractive have the built-in drawback that the parametrization is derived from well behaving systems far from the exotic species present in space.

DFT offers an other alternative. Although calculations of that type are widely used for structural studies of large molecules, the usual implementations of DFT lie at the borderline of applicability for weak complexes. A new formulation of the Kohn-Sham equations based on interacting electronic densities constrained to the systems in interaction has been proposed[16]; illustrative applications are presented in Table 5.

TABLE 5. Equilibrium distances (Å) and binding energies (kcal/mol) for C_6H_6 ...A-B complexes

Method	C_6H_6 ... O_2		C_6H_6 ... N_2		C_6H_6 ... CO	
	R_0	E_{int}	R_0	E_{int}	R_0	E_{int}
KSCED/PW91	3.33	-1.10	3.25	-1.37	3.17	1.79
KSCED/PW91[a]	3.30	-1.13	3.15	-1.47	3.12	1.80
Exp[b]		-1.2±0.3	3.53.5		3.24	1.73
Exp.[b]		-1.65±0.32	3.3	-1.43	3.44	

[a]Extended basis set

The basic philosophy of the KSCED approach is to describe adsorption as an interaction between correlated fragments. To obtain reliable results, it has been shown that it is necessary to use properly balanced exchange-correlation functionals, such as PW91, that are well-behaved in the region where Van der Waals complexes are formed.

The KSCED/PW91 equilibrium geometries and binding energies are in a good agreement with available experimental data as well as with available results obtained from high-level post Hartree-Fock ab initio calculations. It is worth mentioning that, when convergence in the observable has been reached, the uncertainty on the theoretical parameters compare favorably to those reported for experimental measurements.

Depletion of molecules and atoms on solid particles is now within reach of Quantum Chemistry.

REFERENCES

1. Szabo, A., and Ostlund, N.S., *Modern Quantum Chemistry,* McGraw-Hill, New-York, 1992.
2. Parr, R.G., and Yang, W., *Density Functional Theory of Atoms and Molecules,* Oxford University Press, Oxford, 1989.
3. Schmatz, S., and Botschwina, P., *Chem. Phys. Letters,* **235**, 5 (1995).
4. Pauzat, F., Ellinger, Y., and McLean, A.D., *Astrophys. J. Letters,* **369**, L13 (1991).
5. Travers, M.J., McCarthy, M.C., Gottlied, C.A., and Thaddeus, P., *Astrophys. J. Letters,* **465**, L77 (1996); McCarthy, M.C., Travers, M.J., Kalmus, P., Gottlieb, C.A., and Thaddeus, P., *Astrophys. J. Letters,* **467**, L125 (1996); McCarthy, M.C., Travers, M.J., Kovacs, A., Gottlieb, C.A., and Thaddeus, P., *Astron. Astrophys. Letters,* **309**, L31 (1996).
6. Cernicharo, J., and Guélin, M., *Astron. Astrophys. Letters,* **309**, L27 (1996); Guélin, M., Cernicharo, J., Travers, M.J., McCarthy, M.C., Gottlied, C.A., Thaddeus, P., Ohishi, M., Saito, S., and Yamamoto, S., *Astron. Astrophys. Letters,* **317**, L1 (1997).
7. Cheikh, F., and Pauzat, F., *Astron. Astrophys.,* **348**, 17 (1999).
8. Léger, A., and Puget, JL., *Astron. Astrophys. Letters,* **137**, L5 (1984); Allamandola, L.J., Tielens, A.G., and Barker, J.R., ., *Astrophys. J. Letters,* **290**, L25 (1985).
9. Weimer, H.A., McFarland, B.J., Li, S., and Weltner, W.Jr., *J. Phys. Chem.,* **99**, 1824 (1995).
10. Pauzat, F., Talbi, D., and Ellinger, Y., *Astron. Astrophys.,* **293**, 263 (1995).
11. DeFrees, D.J., Miller, M., Talbi, D., Pauzat, F., and Ellinger, Y., *Astrophys. J.,* **408**, 530 (1993); Langhoff, S.R., *J. Phys. Chem.,* **100**, 2819 (1996).
12. Pauzat, F., Talbi, D., and Ellinger, Y., *Astron. Astrophys.,* **319**, 318 (1997).
13. Talbi, D., and Herbst, E., *Astron. Astrophys.,* **319**, 318 (1997).
14. Talbi, D., and Ellinger, Y., *Chem. Phys. Letters,* **288**, 155 (1998).
15. Talbi, D., Ellinger, Y., and Herbst, E., *Astron. Astrophys.,* **314**, 688 (1996).
16. Wesolowski, T.A., *J. Chem. Phys.,* **106**, 8516 (1997).
17. Wesolowski, T.A., Ellinger, Y., and Weber, J., *J. Chem. Phys.,* **108**, 6078 (1997).

Oscillator strengths and radiative transition probabilities from the IRON Project.

C.J. Zeippen

UMR 8631 (associée au CNRS et à l'Université Paris 7) et DAEC, Observatoire de Paris, F-92195 Meudon, France.

e-mail address: zeippen@obspm.fr

Abstract. A short summary is given of the methods used by the IRON Project team in order to supply users of atomic data with reliable oscillator strengths and radiative transition probabilities. Available results and corresponding references are listed. A few cases are discussed briefly with a view to illustrate the progress made and the difficulties encountered.

1 Introduction

The IRON Project (IP) is an international collaboration whose aim is primarily to treat systematically the electron impact excitation of the ions of iron. Indeed, iron is a crucial element in many an astrophysical problem and there is a need for accurate atomic data related to it. However, the methods used in the IP may be applied quite naturally to many other processes and species. In particular, radiative transitions are considered every time there is a demand for accurate oscillator strengths or transition probabilities to complement collisional data and allow for firmer bases to models in astrophysics or controlled fusion, for example. An introduction to the IP goals and methods has been given by Hummer et al. (1993) in the first paper of the series "Atomic data from the IRON Project" appearing in *Astron. Astrophys.* and *Astron. Astrophys. Suppl. Ser.* There are at present well over forty papers in this series. More details on the IP and a complete list of the papers published may be found on the project web page hosted by K. Butler in Munich [**http://www.usm.uni-muenchen.de/people/arch4/iron-project.html**]. Progress is also reviewed regularly in major conferences (see, for example, Butler 1996; Pradhan 1999; Mendoza 1999).

The basic methods employed have been inherited from the OPACITY Project (OP, Seaton 1987; Seaton et al. 1992, 1994; Opacity Project Team 1995, 1996) and they have been adapted to the specific aims of the IP. Naturally, frequent improvements and extensions of the codes used figure prominently among the tasks of the IP team. The collisional formalism selected is based on the close-coupling approximation (Burke and Seaton 1971) as implemented in the RMATRX package of Burke et al. (1971) and

CP543, *Atomic and Molecular Data and Their Applications,* edited by K. A. Berrington and K. L. Bell

Table 1: IP papers on oscillator strengths and radiative transition probabilities. The column labelled IP contains the number in the IP series published by *Astron. Astrophys.* and *Astron. Astrophys. Suppl. Ser.* Crosses indicate the types of transition treated and PI stands for photoionisation.

IP	Reference	Ion(s)	E1	E2 & M1	PI
VII	Nahar (1995)	Fe II	x		
XVI	Bautista (1996)	Fe V	x		x
XVII	Nahar & Pradhan (1996)	Fe III	x	x	
XIX	Quinet et al. (1996)	Fe II		x	
XX	Bautista (1997)	Fe I	x		x
XXII	Galavís et al. (1997)	C- & O-like		x	
XXVI	Bautista & Pradhan (1997)	Fe IV	x		x
XXIX	Galavís et al. (1998)	B-like		x	
XXXIII	Mendoza et al. (1999)	C-like	x		
XXXV	Nahar & Pradhan (1999)	Fe XXIV & XXV	x		
XXXIX	Bautista (1999)	Ni II	x		x
XLIII	Nahar et al. (2000)	Fe V	x	x	
XLIV	Chen & Pradhan (2000)	Fe VI	x	x	

Berrington et al. (1974, 1978), with developments for the asymptotic region by Seaton (1985) and Berrington et al. (1987). A detailed description of this package is given by Berrington et al. (1995). The treatment of relativistic effects can be performed by including the mass, one-body Darwin and spin-orbit Breit-Pauli (BP) corrections in the Hamiltonian, as first introduced in the RMATRX code by Scott and Burke (1980) and Scott and Taylor (1982). Alternatively, an LS-coupling calculation can be extended by means of algebraic transformations such as these implemented in the code JAJOM (Saraph 1972, 1978).

The IP structure calculations are performed with two codes: CIV3 (Hibbert 1975, Hibbert et al. 1991, see also the invited paper by A. Hibbert in this volume) and SUPERSTRUCTURE (Eissner and Nussbaumer 1969, Eissner et al. 1974, Nussbaumer and Storey 1978, Eissner 1991). Both codes allow for configuration interaction (CI), relativistic effects through the BP formalism (one- and two-body interactions) and semi-empirical corrections to *ab initio* calculations. In the IP, structure computations are necessary for two reasons: 1. in collisional work, it is imperative to build reliable wave-functions which provide a realistic description of the tighly bound N-electron system usually called "target"; 2. for specific transitions such as E2 and M1 "forbidden" lines, only structure codes allow for the computation of the required data. The choice between CIV3 and SUPERSTRUCTURE depends on the case considered and, mainly, on the experience of the workers involved in a particular study. Hence, IP targets have been treated either with one or the other code. For purely structural work, it so happens that, so far, SUPERSTRUCTURE has been selected for all the IP computations, even though comparisons with results yielded by independent CIV3 findings remain one of the best ways to assess reliability. The IP papers which are relevant to the topic of this

ICAMDATA conference are listed in Table 1.

2 SUPERSTRUCTURE

Full details may be found in the references listed in the present introduction. A brief summary follows. The code builds a wavefunction Φ for a system with N electrons while solving the Schrödinger equation:

$$H^N \Phi = E^N \Phi \,, \tag{1}$$

where H^N is the Hamiltonian and E^N the energy of the system.

2.1 Non-relativistic problem

The wavefunction Φ_i for a given state i is expanded over a configuration basis ϕ_k and configuration interaction (CI) is taken into account:

$$\Phi_i(x_1...x_N) = \sum_k b_{ik}\phi_k(x_1...x_N), \tag{2}$$

where x_j stands for the ensemble of space and spin coordinates of the j-th electron. The b_{ik} coefficients reflect the mixing of the various configurations in the chosen physical model for each state and they are obtained by diagonalising the non-relativistic Hamiltonian H_{nr} of the N-electron system in the configuration representation. The configuration functions must be antisymmetric in order to obey the exclusion principle of Pauli. In SUPERSTRUCTURE, use is made of Slater determinants of mono-electronic spin-orbital functions.

2.2 Radial wavefunctions and potential

The problem consists in finding as good an approximation as possible for the radial functions $P_{nl}(r)$ which are solutions of

$$\left\{\frac{\mathrm{d}^2}{\mathrm{d}r^2} - \frac{l(l+1)}{r^2} + 2V(\lambda_{nl}, r) + \varepsilon_{nl}\right\} P_{nl}(r) = 0 \tag{3}$$

with proper limiting conditions.

In SUPERSTRUCTURE, the potential $V(\lambda_{nl}, r)$
either (a) is taken from the statistical model of Thomas–Fermi–Dirac with scaling parameters λ_{nl}:

$$V(\lambda_{nl}, r) = \frac{V_{TF}(Z, N, r/\lambda_{nl})}{r} \tag{4}$$

with the conditions

$$\lim V_{TF}(Z, N, r/\lambda_{nl}) = Z \tag{5}$$

when $r \to 0$ and

$$\lim V_{TF}(Z, N, r/\lambda_{nl}) = Z - (N-1) = z \tag{6}$$

when $r \to \infty$.

Z is the atomic number and z is the reduced charge.
or (b) is a scaled Coulomb potential:

$$V_C(\lambda_{nl}, r) = \frac{Z \exp(-Zr/2) + \xi_{\bar{n}l}(1 - \exp -(Zr/2))}{r} \tag{7}$$

In Practice, radial wavefunctions are optimised through an energy minimisation procedure:

$$\delta F = \delta \left\{ \sum_{i=1}^{\mathrm{NE}} g_i E_i(\lambda_{1s}, \lambda_{2s}, \ldots) \right\} = 0 \tag{8}$$

where E_i stands for the non-relativistic energy of state i. The user must decide on the energies to be minimised and on the details of the variation of λ parameters.

2.3 Relativistic corrections to the Hamiltonian

The BP Hamiltonian:

$$H_{\mathrm{BP}} = H_{\mathrm{nr}} + H_{\mathrm{rc}} \tag{9}$$

is implemented in the code. H_{rc} includes one-body and two-body relativistic corrections which will not be repeated here. See the paper by Eissner et al. (1974) for technical details. Of course, terms such as spin-orbit, mass, Darwin, spin-spin, etc. are present.

2.4 The Term Energy Corrections (TEC) technique

Very few *ab initio* calculations yield "exact" wavefunctions. Hence, the energies and the A-values obtained are in error by margins which depend on the difficulty of the case treated and/or the care taken in designing a realistic physical model. Once the best possible representation has been built, how can its remaining defects be attenuated? Semi-empirical corrections are the answer, even if they must be manipulated with caution. In SUPERSTRUCTURE, use is made of the TEC technique as implemented by Zeippen et al. (1977).

If **H** is the matrix Hamiltonian in the representation of functions Φ_i [see equation (2)],

$$\mathbf{H}\,\mathbf{B} = \mathbf{B}\,\mathbf{E} \tag{10}$$

where **B** is the matrix of the b_{ik} coefficients and **E** is the diagonal matrix of energies. In *LS*-coupling:

$$\mathbf{H}_{\mathrm{nr}}\,\mathbf{B}_{\mathrm{nr}} = \mathbf{B}_{\mathrm{nr}}\,\mathbf{E}_{\mathrm{nr}}^{[\mathrm{th}]}\,, \tag{11}$$

where $\mathbf{E}_{\mathrm{nr}}^{[\mathrm{th}]}$ is the diagonal matrix of theoretical (*"ab initio"*) *LS* energies. Equation (11) can be written as:

$$\mathbf{H}_{\mathrm{nr}} = \mathbf{B}_{\mathrm{nr}}\,\mathbf{E}_{\mathrm{nr}}^{[\mathrm{th}]}\,[\mathbf{B}_{\mathrm{nr}}]^{-1}\,. \tag{12}$$

If experimental energies are known for the case under consideration, semi-empirical corrections may be introduced:

$$E_{\mathrm{nr}}^{i\ [\mathrm{th}]} + \Delta E_{\mathrm{nr}}^{i}\,. \tag{13}$$

so as to reduce the error on the level energies of interest in the calculation.

Thus, an "effective" non-relativistic Hamiltonian is obtained:

$$\mathbf{H}'_{\rm nr} = \mathbf{B}_{\rm nr} \left[\mathbf{E}^{[\rm th]}_{\rm nr} + \Delta\mathbf{E}_{\rm nr}\right] [\mathbf{B}_{\rm nr}]^{-1}, \tag{14}$$

The final equation of the problem becomes:

$$[\mathbf{H}'_{\rm nr} + \mathbf{H}_{\rm rc}]\, \mathbf{B} = \mathbf{B}\, \mathbf{E}, \tag{15}$$

where $\mathbf{H}_{\rm rc}$ is the matrix containing the BP relativistic corrections to the Hamiltonian [see equation (9)].

The TECs change the CI mixing coefficients. Note that the code CIV3 includes a similar feature under the name of "fine-tuning" (Hibbert 1996). Experience has shown that this type of procedure usually helps with the accuracy of a calculation if the "non-tuned" model is sufficiently reliable already. In this context, it is hardly necessary to repeat that the measurements performed by spectroscopists towards reliable and extended sets of level energies are of crucial importance.

2.5 Relativistic corrections to the magnetic dipole operator

Extending the work of Drake (1971) on two-electron systems, Eissner and Zeippen (1981) implemented relativistic corrections to the magnetic dipole operator in the code SUPERSTRUCTURE for a general N-electron system. The corrected operator is

$$\mathbf{M1C} = \sum_{j=1}^{N} \mathbf{M}(j) + \sum_{i=1}^{N} \sum_{j<i} \mathbf{M}(ij), \tag{16}$$

where $\mathbf{M}(j)$ contains one-body operators including the first-order magnetic dipole operator

$$\mathbf{M} = \mathbf{L} + 2\mathbf{S} \tag{17}$$

plus corrections of order α^2, while $\mathbf{M}(ij)$ contains two-body operators also of order α^2 where α is the fine-structure constant. Technical details are rather involved and will not be repeated here.

3 The Breit-Pauli R-matrix (BPRM) formalism

Full details may be found in the references given in the present introduction. Recent developments are reported by Nahar and Pradhan (2000). A brief sketch of the basic points follows.

3.1 Theory

In the close coupling (CC) approximation, an $(N+1)$-electron system is divided into an N-tighly bound-electron "target" and an extra electron which may be bound or free, depending on the eigenvalues of the $(N+1)$-electron Hamiltonian being negative or positive, respectively. The expansion for the wavefunction $\Psi(E)$ corresponding to a total spin and angular symmetry $SL\pi$ or $J\pi$ of the $(N+1)$-electron system is taken to be

$$\Psi_E(e+ion) = A\sum_i \chi_i(ion)\theta_i + \sum_j c_j\Phi_j, \tag{18}$$

where A is the antisymmetrisation operator, χ_i is the target wavefunction for a specific term $S_iL_i\pi_i$ or level $J_i\pi_i$, θ_i is the wavefunction for the $(N+1)$-th electron in a channel labelled $S_iL_i(J_i)\pi_i\ k_i^2\ell_i(SL\pi)\ [J\pi]$ and k_i^2 is the incident kinetic energy. In the second sum, the Φ_j's are correlation wavefunctions for the $(N+1)$-electron system. They compensate for orthogonality conditions between the continuum and the bound orbitals and they represent additional short-range correlation.

The BPRM method yields the solutions of the relativistic CC equations using a Breit-Pauli Hamiltonian for the $(N+1)$-electron system. Note that here The BP Hamiltonian is limited to

$$H_{N+1}^{\rm BP} = H_{N+1} + H_{N+1}^{\rm mass} + H_{N+1}^{\rm Dar} + H_{N+1}^{\rm so}, \tag{19}$$

where H_{N+1} is the non-relativistic Hamiltonian,

$$H_{N+1} = \sum_{i=1}^{N+1}\left\{-\nabla_i^2 - \frac{2Z}{r_i} + \sum_{j>i}^{N+1}\frac{2}{r_{ij}}\right\}, \tag{20}$$

and the additional terms are only the one-body mass correction, the Darwin and the spin-orbit terms, respectively. The best known structure codes include more corrections in the BP formalism (see Section 2.3).

Two energy ranges are distinguished:

$$\begin{aligned} E &= k^2 > 0 \longrightarrow \text{continuum channel (scattering)} \\ E &= -\tfrac{z^2}{\nu^2} < 0 \longrightarrow \text{bound channel (state)}, \end{aligned} \tag{21}$$

where ν is the effective quantum number relative to the core level. Determination of the quantum defect ($\mu(\ell)$), defined as $\nu_i = n - \mu(\ell)$ where ν_i is relative to the core level $S_iL_i\pi_i$, is helpful in establishing the ℓ-value associated with a given channel (level).

Ψ_E is a CI-type wavefunction over a large number of electronic configurations depending on the target states included in expansion (18). It is clear that the final degree of accuracy reached in such collisional calculations is closely related to the quality of the target representation. Matrix elements may be calculated with these Ψ_E, and the E1, E2, M1 or other operators to obtain the corresponding transition probabilities. The present version of the BPRM codes only includes the E1 operator which allows for calculating dipole-allowed and intercombination transition probabilities.

3.2 Energy levels and identification

The identification of bound levels computed in intermediate coupling with the BPRM method is a rather complex task. These levels are associated with collision complexes of the $(N+1)$-electron system, initially identified by their total angular momenta and parity, $J\pi$. A scheme had to be developed to attach to all levels a complete set of spectroscopic information consisting of

$$C_t(\ S_t\ L_t)\ J_t\ \pi_t n\ell\ [K]\mathrm{s}\ \ J(SL)\ \pi, \tag{22}$$

and also to label each of them with a possible $SL\pi$ symmetry. C_t denotes the target configuration.

In the present scheme, most of the spectroscopic information related to a computed level is extracted from the few closed and bound channels that dominate the wavefunction of that level. A new code, PRCBPID, has been designed to carry out the identification, including an analysis of the quantum defects and of the angular momentum algebra of the dominant channels.

Two important problems are also tackled. The correspondence between the computed fine-structure levels and the standard LS-coupling designation is examined. Completeness checks for the set of all fine-structure components within all computed LS multiplets are carried out. A correspondence between the sets of $SL\pi$ and $J\pi$ in the same configuration can be established from the set of $SL\pi$ symmetries, formed from the target term, $S_tL_t\pi_t$, nl quantum numbers of the valence electron, and $J\pi$ of the fine-structure level belonging to the LS term. The full identification procedure is described in detail by Nahar and Pradhan (2000).

4 The IP radiative data

As shown in Table 1, the current set of IP radiative data concerns 12 ions of iron and some other cases. Depending on the particular ion considered, E1, E2 and/or M1 transitions have been treated as well as the photoionisation of Fe I, Fe IV, Fe V and Ni II. Of course, work is in progress on other ions.

It would be useless to repeat here the presentation of the data or the analyses reported in the respective IP papers. But it is perhaps relevant to comment on the reliability of the IP data. Assessing the accuracy of theoretical findings is not always easy. Recently, Hibbert (1996) and Wiese (1996) have stressed the need for theoreticians and experimentalists to self-rate their respective results. Where applicable, this procedure is certainly recommended. However, when dealing with enormous calculations yielding hundreds or thousands of transition rates in a complex ion of iron, for example, it is not possible fully to check the convergence of the computation in all cases.

Of course, the agreement with independent calculations or with experiment remains a major element in the rating process. For example, in their study of Fe II forbidden lines, Quinet et al. (1996) used the HFR code of Cowan (1981) as well as SUPERSTRUCTURE to allow for a statistical comparison between the two sets of data which resulted in a firmer rating of the IP results. Again, Galavís et al. (1997, 1998) or Mendoza et al. (1999) carried out detailed comparisons with other theoretical or experimental work. A few cases are discussed briefly below in order to illustrate the quality obtained but also the difficulties encountered.

Needless to say, critical compilations (see in particular the invited paper by J. Reader about the NIST effort in this volume), bibliographies (e.g. Biémont and Zeippen 1991, 1996) or databases (see, for example, the NIST service at **http://physics.nist.gov/PhysRefData/contents-atomic.html**) are extremely useful when assessing an atomic data set accuracy.

4.1 An extended set of radiative data for Fe V

In the most recently published IP paper, Nahar et al. (2000) propose an extensive set of dipole-allowed, intercombination, and forbidden transition probabilities for Fe V. The BPRM method (see Section 3) was used to calculate 1.46×10^6 oscillator strengths for the allowed and intercombination E1 transitions among 3,865 fine-structure levels dominated by configuration complexes with $n \leq 10$ and $l \leq 9$. Also, a CI calculation with SUPERSTRUCTURE (see Section 2) produced relativistic E2 and M1 A-values for transitions among 65 low-lying levels dominated by the $3d^4$ and $3d^3\ 4s$ configurations. Level identification was carried out extensively (see Section 3.2). The target Fe VI is represented by an eigenfunction expansion of 19 fine-structure levels of $3d^3$ and correlation configurations. Fe V bound levels are obtained with angular and spin symmetries $SL\pi$ and $J\pi$ of the (e + Fe VI) system such that S = 5,3,1, $L \leq 10$, $J \leq 8$ of even and odd parities. The completeness of the calculated dataset is verified in terms of all possible bound levels and transitions in correspondence with the LS terms. The fine-structure averaged relativistic values are compared with previous OP LS-coupling data and other results. The forbidden E2 and M1 transition rates are in good agreement with those computed by Garstang (1957) for the $3d^4$ transitions.

The uncertainties of the BPRM rates for the allowed transitions are expected to be within a few percent for the strong transitions, about 10-30% for the weak ones (see Nahar et al. 2000 for details). The agreement between the results obtained with the length and velocity forms of the E1 operator is generally good but significant dispersion is seen for gf-values smaller than 0.01. However, it should be noted that in the CC R-matrix calculations the length formulation is likely to be more accurate than the velocity formulation since the wavefunctions are better represented in the asymptotic region that dominates the contribution to the length form of the oscillator strength.

This study exemplifies the future potential of computational spectroscopy with the BPRM method complemented by related atomic structure calculations. The huge set of data presented is fairly complete and its statistical accuracy should be high enough to help considerably with the calculation of monochromatic opacities and with the analysis of Ultraviolet spectra from astrophysical and laboratory sources where non-local thermodynamic equilibrium (NLTE) atomic models with many excited levels are needed.

4.2 The $2s2p^3\ ^5S^o_2$ level in C-like ions

Radiative decay rates have been computed with SUPERSTRUCTURE for the $2s2p^3\ ^5S^o_2$-$2s^22p\ ^3P_2$, $^3P_1$ and $^1D_2$ intercombination transitions in more than twenty members of the carbon isoelectronic sequence (Mendoza et al. 1999). CI contributions and relativistic corrections were studied in detail. The spin-spin interaction was shown to be of considerable importance for low Z. In particular, the $^5S^o_2$ radiative lifetime and the $B = A(^5S^o_2 - ^3P_2)/A(^5S^o_2 - ^3P_1)$ branching ratio cannot be estimated accurately without it.

In his analysis of relativistic corrections, Jones (1970) shows that one-body and two-body (Breit) relativistic terms respectively scale as $\alpha^2(Z-S)^4$ and $\alpha^2(Z-S)^3$, where α is the fine-structure constant and S is a screening constant. Thus the Breit contribution decreases in importance as Z increases along the sequence.

In the present study, the spin-spin contribution increases $A(^5S^o_2 - ^3P_1)$ by only $\sim 5\%$ at low Z but the reductions in $A(^5S^o_2 - ^3P_2)$ and $A(^5S^o_2 - ^1D_2)$ are as large as 20%. Consequently, Mendoza et al. (1999) had to take the Breit interaction explicitly into account to treat the intercombination lines considered here.

The $^5S^o_2$ radiative lifetime τ and the branching ratio B are affected differently by the two-body relativistic corrections along the sequence. The exclusion of the spin–spin interaction only leads to comparatively small differences in τ: less than 11 % at the neutral end and less than 5% for $Z > 10$. Regarding B, the inclusion of the spin–spin contribution causes a large decrease at low Z: up to 20% or so. This is in agreement with the findings of Fleming and Brage (1997). By $Z = 20$, the spin-spin effect is reduced to less than 5%.

This case clearly illustrates that caution is required before neglecting two-body relativistic corrections in the Hamiltonian when treating "light" elements. The data proposed by Mendoza et al. (1999) are in satisfactory to good agreement with the theoretical findings of Brage et al. (1997) and Fleming and Brage (1997) who used CIV3 in a systematic way and with experimental studies by Träbert et al. (1998), Curry et al. (1997), Bridges et al. (1997), Musielok et al. (1996), Calamai and Johnson (1991) or Johnson et al. (1991). The overall accuracy rating is estimated to be 10% for strong transitions and 20% for weak ones. See also Section 4.4.2.

4.3 The [O III] 5007, 4959Å line-intensity ratio

For a long time, there has been a discrepancy between the best theoretical and observed line-intensity ratio of the [O III] lines at 5007, 4959Å. Until recently, the best calculated ratio was 2.88 (Nussbaumer and Storey 1981) or 2.89 (Galavís et al. 1997 for the IP). The value observed in very high signal-to-noise spectra of nebulae is somewhat higher: 3.03 ± 0.03 (Rosa 1985), 3.17 ± 0.04 (Iye et al. 1987), 3.00 ± 0.08 (Leisy and Dennefeld 1996), 3.10 ± 0.05 (Aller et al. 1996), 3.00 ± 0.013 (Mathis and Liu 1999). No improvement in the wavefunctions used to calculate the atomic data involved in this ratio ever brought any significant reduction of the discrepancy between physics and astronomy. Of course, the difference of some 4% does not look impressive but, as it appears to be well established, it generates some doubts regarding the quality of atomic physics calculations.

It can be shown that, in the non-relativistic limit, when all magnetic interactions become negligible, the ratio of the line strengths corresponding to the two transitions $^1D_2 - ^3P_2$ and $^1D_2 - ^3P_1$ is always 3, regardless of the configuration expansion used. Therefore, the more elaborate physical models only modify the E2 contribution to the radiative rates which is very small anyway. This explains the stability of the theoretical line ratio over the years and it shows that the use of the standard formulation for the magnetic dipole operator will always bring a line-intensity ratio close to the present value of 2.89.

This is reminiscent of another case where the first-order M1 operator was found to be inadequate, ie. the half-filled shells $2p^3$ and $3p^3$ where the spin-orbit interaction vanishes to first order and comparatively small relativistic effects become very important (see Zeippen 1980, Eissner and Zeippen 1981). Using the formalism described in Section 2.5 resulted in the resolution of a rather large discrepancy between an observed and a theoretical line-intensity ratio in [O II]. The ions in the N and P isoelectronic sequences were treated with the relativistically corrected M1 operator (Butler and Zeippen 1984, Zeippen 1987, Becker et al. 1989, Mendoza and Zeippen 1982).

Such effects are expected to be small for ions where the spin-orbit interaction does not vanish to first order, such as O III, but not small enough to be neglected when differences of a few percent are at issue. This is why Storey and Zeippen (2000) used the IP wavefunctions of Galavís et al. (1997) with the corrected M1 operator mentionned in Section 2.5. The outcome of this improved calculation is that the theoretical line-intensity ratio goes up to 2.98 which differs by 2% or less from the astronomical data deduced by Rosa (1985), Leisy and Dennefeld (1996) or Mathis and Liu (1999). This is an interesting example of the need for very sophisticated physics when attempting to resolve small but persistent discrepancies with reliable observations. In conclusion, there is now no reason to question the validity of the atomic wavefunctions in this particular case. The present recommended values for the relevant Einstein's coefficients may be found in Storey and Zeippen (2000).

4.4 Comparisons with recent experimental data

As already stated at the beginning of this Section, it is not always easy to assess the accuracy of atomic data calculations. One basic criterion is of course the agreement (or otherwise) of theoretical findings with experimental measurements where available. For lack of space, it would be impossible to review here all the comparisons carried out by IP members and reported in the various papers in the IP series to which the reader is referred. What follows is a brief discussion of comparisons with some very recent studies by Träbert and his colleagues using storage ring (SR) devices to obtain accurate experimental A-values and lifetimes. For detailed information on this important campaign of measurements, see the review by Träbert (2000).

4.4.1 B-like and F-like ions of Ti

In the non-relativistic limit, the M1 contribution to transition $2s^22p\ ^2P^o$ ($J' = 3/2 \rightarrow J = 1/2$) is not affected at all by CI effects. Its A-value depends only on the quantun numbers of the initial and final states and the line strength is equal to 1.333. Therefore, if experimental level energies are used together with this line strength, the "exact" value of the M1 rate is easily obtained. As the E2 contribution to this transition is negligible, there should be very good agreement between theory and experiment in this case. The result obtained by Träbert et al. (1999b) for Ti^{17+} (A = 1595±25 s^{-1}) compares very well with the data computed by Kaufman and Sugar (1986) or Verhey et al. (1987) ($A = 1590\ s^{-1}$) or the IP finding of Galavís et al. (1998) ($A = 1599\ s^{-1}$), for example.

Turning now to the transition $2s^22p^5\ ^2P^o$ ($J' = 3/2 \rightarrow J = 1/2$), the problem is very similar: here, the $2p$ orbital contains five electrons and one hole instead of one electron and five holes in the previous case. Thus, the M1 line strength is again equal to 1.333 in the non-relativistic limit. As the E2 contribution is also negligible, the agreement between theory and experiment is expected to be excellent. Unfortunately, the measurement of Träbert et al. (1999b) for Ti^{13+} ($A = 1949 \pm 38\ s^{-1}$) turns out to be some 3% higher than recent computations ($A = 1890\ s^{-1}$) by Kaufman and Sugar (1986) and the IP (Galavís et al. in progress). These IP theoretical values are obtained using the experimental level energies of Edlén (1981a and b) with the calculated line strength. Because this is a rather simple case, the small difference between theory and experiment is worth investigating. Note that the discrepancy is also present in Sc^{12+} (Träbert et al. 1999c).

4.4.2 C-like ions of O and F

The mean life for the $2s2p^3\ ^5S_2^o$ level in O^{2+} was measured by Johnson et al. (1991) using an ion trap device and more recently by Träbert et al. (2000). There is very good agreement between the two experiments ($\tau = 1.22 \pm 0.08$ and 1.25 ± 0.013 ms, respectively) and two recent calculations by Fleming and Brage (1997) using CIV3, and Mendoza et al. (IP, 1999) ($\tau = 1.21 \pm 0.06$ and 1.22 ms, respectively). Previous theoretical work lies further away from experiment and there is some scatter among the data published.

In the same isoelectronic sequence, Träbert et al. (2000) also measured the mean life for the $2s^22p^2\ ^1S_0$ level in O^{2+} and F^{3+}. In oxygen, there is satisfactory agreement between the experimental result ($\tau = 530 \pm 25$ ms) and seven independent computations by Garstang (1951, 1968), Nussbaumer (1971), Nussbaumer and Rusca (1979), Baluja and Doyle (1981), Froese Fischer and Saha (1985), Vilkas et al. (1996) and Galavís et al. (IP, 1997) who propose $\tau = 546$, 527, 517, 546, 519 and 559 ± 56 ms, respectively. In fluorine, the experimental lifetime is 304 ± 5 ms, while the "best" calculated values were proposed by Garstang (1951), Froese Fischer and Saha (1985), Vilkas et al. (1996) and Galavís et al. (IP, 1997): 312, 310, 299 and 312 ± 31 ms, respectively. Again, in this case, it appears that experiment and some theoretical findings agree satisfactorily.

4.4.3 O-like fluorine

The radiative lifetime for the $2s^22p^4\ ^1S_0$ level in F^+ was measured by Calamai et al. (2000): $\tau = 420 \pm 12$ ms. The theoretical result which comes closest to this finding is that calculated for the IP by Galavís et al. (1997): 435 ms. This is due to the closer agreement between the SR and the IP A-values for the $^1S_o \rightarrow {}^1D_2$ transition: 1.90 ± 0.13 and 1.83 s^{-1} respectively, as compared to other computations by Garstang (1951), Froese Fischer and Saha (1983) or Baluja and Zeippen (1988).

4.4.4 The doublet-quadruplet intersystem lines in C II and N III

Due to their astrophysical importance, the $2s^22p\ ^2P^o - 2s2p^2\ ^4P$ transitions in C^+ and N^{2+} have been the subject of quite a few theoretical studies and of three recent experimental efforts. There is the work of Fang et al. (1993a and 1993b) who used a radio frequency ion trap, of Träbert et al. (1999a) based on a SR device, and of Smith et al. (1999) employing an electrostatic ion trap. A more detailed discussion may be found in Träbert et al. (1999a). Suffice to say here that the wavefunctions built by Biémont et al. (1994) and used with slight modifications by Galavís et al. (1998) for the IP work allow for the best agreement with the experiments of Fang et al. (1993a and 1993b). It should be stressed that the optimisation procedure selected was not designed to yield such a result but to reproduce as realistically as possible a large number of level energies both in B-like carbon and nitrogen. Biémont et al. (1994) show that the energy separations and fine-structure splittings obtained were good in this respect. Even so, these authors concluded that more experimental work was needed on the doublet-quadruplet intersystem radiative rates in view of the remaining discrepancies between their work and the measurements of Fang et al. (1993a and 1993b).

More recently, the findings of Träbert et al. (1999a) and Smith et al. (1999) turned out to be lower than the first experimental data. As a result, this new work now tends to favour the computations of Froese Fischer (1994) based on the MCHF + core-polarisation method in the case of B-like carbon and the calculations of Bell et al. (1995) and Brage et al. (1995) in the case of B-like nitrogen. The author of the present review is busy reassessing the situation using the SUPERSTRUCTURE + TEC method.

4.5 The TIP/TOPbase facility

The OP data have been available to users for many years from the TOPbase service (Cunto et al. 1993) hosted by the CDS (telnet: cdsarc.u-strasbg.fr [130.79.128.5] [user-name: **topbase** and password: **Seaton+**]; **http://vizier.u-strasbg.fr/OP.html**) and HEASARC (**http://heasarc.gsfc.nasa.gov/topbase/**). This database is under development and extension. The TIPbase facility soon will be open offering easy access to the IP collisional and radiative data. More details may be found in the invited paper by C. Mendoza in this volume and in the poster contribution to this conference by Mendoza and Zeippen.

5 Conclusion

The methods of the IP have been summarised briefly. They are based on elaborate formalisms and are regularly developed and extended. Large sets of radiative data are already available. Their size and accuracy are such that they should be very useful in many applications. The work in progress and to be carried out remains considerable and will call for the time and expertise of every member of the IP team. Great attention is devoted to the dissemination of the data produced towards the community of atomic data users.

Acknowledgements. Friendly greetings are adressed to David Hummer who was our first coordinator and continues to help greatly with his critical reading of every IP paper before submission. The excellent team spirit within the project is gratefully acknowledged.

6 References

Aller, L.H., Hyung, S., and Feibelman, W.A., *Publ. Astr. Soc. Pac.* **108**, 488 (1996).
Baluja, K.L., and Doyle, J.G., *J. Phys. B* **14**, L11 (1981).
Baluja, K.L., and Zeippen, C.J., *J. Phys. B* **21**, 1455 (1988).
Bautista, M.A., *Astron. Astrophys. Suppl. Ser.* **119**, 105 (1996).
Bautista, M.A., *Astron. Astrophys. Suppl. Ser.* **122**, 167 (1997).
Bautista, M.A., *Astron. Astrophys. Suppl. Ser.* **137**, 529 (1999).
Bautista, M.A., and Pradhan, A.K., *Astron. Astrophys. Suppl. Ser.* **126**, 365 (1997).
Becker, S.R., Butler, K., and Zeippen, C.J., *Astron. Astrophys.* **221**, 375 (1989).

Bell, K.L., Hibbert, A., Stafford, R.P., and Brage, T., Mon. Not. R. Astr. Soc. **272**, 909 (1995).
Berrington, K.A., Burke, P.G., Butler, K., Seaton, M.J., Storey, P.J., Taylor, K.T., and Yu, Y., J. Phys. B **20**, 6379 (1987).
Berrington, K.A., Burke, P.G., Chang, J.J., Chivers, A.T., Robb, W.D., and Taylor, K.T., *Comput. Phys. Commun.* **8**, 149 (1974).
Berrington, K.A., Burke, P.G., Le Dourneuf, M., Robb, W.D., Taylor, K.T., and Vo Ky, L., *Comput. Phys. Commun.* **14**, 367 (1978).
Berrington, K.A., Eissner, W., and Norrington, P.H., *Comput. Phys. Commun.* **92**, 290 (1995).
Biémont, E., Delahaye, F., and Zeippen, C.J., *J. Phys. B* **27**, 5841 (1994).
Biémont, E., and Zeippen, C.J., *J. Physique IV* **C1**, 209 (1991).
Biémont, E., and Zeippen, C.J., *Phys. Scr.* **T65**, 192 (1996).
Brage, T., Froese Fischer, C., and Judge, P.G., Astrophys. J. **445**, 457 (1995).
Brage, T., Hibbert, A, and Leckrone, D.S., *Astrophys. J.* **478**, 423 (1997).
Bridges, J.M., Wiese, W.L., and Griesmann, U., quoted by Curry et al. (1997).
Burke, P.G., Hibbert, A., and Robb, W.D., *J. Phys. B* **4**, 153 (1971).
Burke, P.G., and Seaton, M.J., *Methods Comput. Phys.* **10**, 1 (1971).
Butler, K., *Phys. Scr.* **T65**, 63 (1996).
Butler, K., and Zeippen, C.J., *Astron. Astrophys.* **141**, 274 (1984).
Calamai, A.G., Gwinner, G., Tordoir, X., Träbert, E., and Wolf, A., *Phys. Rev. A* in press (2000).
Calamai, A.G., and Johnson, C.E., *Phys. Rev. A* **44**, 218 (1991).
Chen, G.X., and Pradhan, A.K., *Astron. Astrophys. Suppl. Ser.* in press (2000).
Cowan, R.D., *The Theory of Atomic Structure and Spectra*, Univ. California Press (Berkeley), (1981).
Cunto, W., Mendoza, C., Ochsenbein, F., and Zeippen, C.J., *Astron. Astrophys.* **275**, L5 (1993).
Curry, J.J., Gibson, N.D., and Lawler, J.E., *Astron. Astrophys.* **321**, 1021 (1997).
Drake, G.W.F., *Phys. Rev. A* **3**, 908 (1971).
Edlén, B., *Phys. Scr.* **22**, 593 (1981a).
Edlén, B., *Phys. Scr.* **23**, 1079 (1981b).
Eissner, W., *J. Phys. IV* **C1**, 3 (1991).
Eissner, W., Jones, M., and Nussbaumer, H., *Comput. Phys. Commun.* **8**, 270 (1974).
Eissner, W., and Nussbaumer, H., *J. Phys. B* **2**, 1028 (1969).
Eissner, W., and Zeippen, C.J., *J. Phys. B* **14**, 2125 (1981).
Fang, Z., Kwong, V.H.S., and Parkinson, W.H., *Astrophys. J.* **413**, L141 (1993a).
Fang, Z., Kwong, V.H.S., Wang, J., and Parkinson, W.H., *Phys. Rev. A* **48**, 1114 (1993b).
Fleming, J., and Brage, T., *J. Phys. B* **30**, 1385 (1997).
Froese Fischer, C., *Phys. Scr.* **49**, 323 (1994).
Froese Fischer, C., and Saha, H.P., *Phys. Rev. A* **28**, 3169 (1983).
Froese Fischer, C., and Saha, H.P., *Phys. Scr.* **32**, 181 (1985).
Galavís, M.E., Mendoza, C., and Zeippen, C.J., *Astron. Astrophys. Suppl. Ser.* **123**, 159 (1997).
Galavís, M.E., Mendoza, C., and Zeippen, C.J., *Astron. Astrophys. Suppl. Ser.* **131**,

499 (1998).
Garstang, R.H., *Mon. Not. R. Astr. Soc.* **111**, 115 (1951).
Garstang, R.H., *Mon. Not. R. Astr. Soc.* **117**, 393 (1957).
Garstang, R.H., in *Planetary Nebulae* (IAU Symposium 34) (D.E. Osterbrock and C.R. O'Dell, Eds.), Reidel (Dordrecht), p. 143 (1968).
Hibbert, A., *Comput. Phys. Commun.* **9**, 141 (1975).
Hibbert, A., *Phys. Scr.* **T65**, 104 (1996).
Hibbert, A., Glass, R., and Froese Fischer, C., *Comput. Phys. Commun.* **64**, 455 (1991).
Hummer, D.G., Berrington, K.A., Eissner, W., Pradhan, A.K., Saraph, H.E., and Tully, J.A., *Astron. Astrophys.* **279**, 298 (1993).
Iye, M., Ulrich, M.H., and Peimbert, M., *Astron. Astrophys.* **186**, 84 (1987).
Johnson, B.C., Smith, P.L., and Parkinson, W.H., quoted by Calamai and Johnson (1991).
Jones, M., *J. Phys. B* **3**, 1571 (1970).
Kaufman, V., and Sugar, J., *J. Phys. Chem. Ref. Data* **15**, 321 (1986).
Leisy, P., and Dennefeld, M., *Astron. Astrophys. Suppl. Ser.* **116**, 95 (1996).
Mathis, J.S., and Liu, X.W., *Astrophys. J.* **521**, 212 (1999).
Mendoza, C., *Comput. Phys. Commun.* **121-122**, 74 (1999).
Mendoza, C., and Zeippen, C.J., *Mon. Not. R. Astr. Soc.* **198**, 127 (1982).
Mendoza, C., Zeippen, C.J., and Storey, P.J., *Astron. Astrophys. Suppl. Ser.* **135**, 159 (1999).
Musielok, J., Bridges, J.M., Djurović, S., and Wiese, W.L., *Phys. Rev. A* **53**, 3122 (1996).
Nahar, S.N., *Astron. Astrophys.* **293**, 967 (1995).
Nahar, S.N., Delahaye, F., Pradhan, A.K., and Zeippen, C.J., *Astron. Astrophys. Suppl. Ser.* **144**, 141 (2000).
Nahar, S.N., and Pradhan, A.K., *Astron. Astrophys. Suppl. Ser.* **119**, 509 (1996).
Nahar, S.N., and Pradhan, A.K., *Astron. Astrophys. Suppl. Ser.* **135**, 347 (1999).
Nahar, S.N., and Pradhan, A.K., *Phys. Scr.* in press (2000).
Nussbaumer, H., *Astrophys. J.* **166**, 411 (1971).
Nussbaumer, H., and Rusca, C., *Astron. Astrophys.* **72**, 129 (1979).
Nussbaumer, H., and Storey, P.J., *Astron. Astrophys.* **64**, 139 (1978).
Nussbaumer, H., and Storey, P.J., *Astron. Astrophys.* **99**, 177 (1981).
The Opacity Project Team, *The Opacity Project - Volume 1* [ISBN 0 7503 0288 7], Institute of Physics Publishing (Bristol & Philadelphia) (1995).
The Opacity Project Team, *The Opacity Project - Volume 2* [ISBN 0 7503 0174 0], Institute of Physics Publishing (Bristol & Philadelphia) (1996).
Pradhan, A.K., *Phys. Scr.* **T83**, 69 (1999).
Quinet, P., Le Dourneuf, M., and Zeippen, C.J., *Astron. Astrophys. Suppl. Ser.* **120**, 361 (1996).
Rosa, M., *The Messenger* **39**, 15 (1985).
Saraph, H.E., *Comput. Phys. Commun.* **3**, 256 (1972).
Saraph, H.E., *Comput. Phys. Commun.* **15**, 247 (1978).
Scott, N.S., and Burke, P.G., *J. Phys. B* **12**, 4299 (1980).
Scott, N.S., and Taylor, K.T., *Comput. Phys. Commun.* **25**, 349 (1982).
Seaton, M.J., *J. Phys. B* **18**, 2111 (1985).

Seaton, M.J., *J. Phys. B* **20**, 6363 (1987).
Seaton, M.J., Yu, Y., Mihalas, D., and Pradhan, A.K., *Mon. Not. R. Astr. Soc.* **266**, 805 (1994).
Seaton, M.J., Zeippen, C.J., Tully, J.A., Pradhan, A.K., Mendoza, C., Hibbert, A., and Berrington, K.A., *Rev. Mex. Astron. Astrof.* **23**, 19 (1992).
Smith, S.J., Chutjian, A., and Greenwood, J.B., *Phys. Rev. A* **60**, 3569 (1999).
Storey, P.J., and Zeippen, C.J., Mon. Not. R. Astr. Soc., **312**, 813 (2000).
Träbert, E., *Phys. Scr.* **61**, 257 (2000).
Träbert, E., Calamai, A.G., Gillaspy, J.D., Gwinner, G., Tordoir, X., and Wolf, A., *Phys. Rev. A* submitted (2000).
Träbert, E., Gwinner, G., Knystautas, E.J., Tordoir, X., and Wolf, A., *J. Phys. B* **32**, L491 (1999a).
Träbert, E., Gwinner, G., Wolf, A., Tordoir, X., and Calamai, A.G., *Phys. Lett. A* **264**, 311 (1999b).
Träbert, E., Wolf, A., Linkemann, J., and Tordoir, X., *J. Phys. B* **32**, 537 (1999c).
Träbert, E., Wolf, A., Pinnington, E.H., Linkemann, J., Knystautas, E.J., Curtis, A., Bhattacharya, N., and Berry, H.G., *Phys. Rev. A* **58**, 4449 (1998).
Verhey, T.R., Das, B.P., and Perger, W.F., *J. Phys. B* **20**, 3639 (1987).
Vilkas, M.J., Martinson, I., Merkelis, G., Gaigalas, G., and Kisielius, R., *Phys. Scr.* **54**, 281 (1996).
Wiese, W.L., *Phys. Scr.* **T65**, 188 (1996).
Zeippen, C.J., *J. Phys. B* **13**, L485 (1980).
Zeippen, C.J., *Astron. Astrophys.* **173**, 410 (1987).
Zeippen, C.J., Seaton, M.J., and Morton, D.C., *Mon. Not. R. Astr. Soc.* **181**, 527 (1977).

Panel Discussions

Atomic and Molecular Databases in Japan and Asia

Izumi Murakami and Takako Kato

Data and Planning Center, National Institute for Fusion Science, Oroshi-cho, Toki-shi, Gifu 509-5292, Japan

Abstract. Atomic and molecular data activities and databases in Japan and Asia are introduced. In China IAPCM has a database on windows95/NT. In Korea KAERI has A+M databases with original engines. In Japan JAERI shows evaluated data as graphs on WWW and NIFS offers numerical A+M databases on WWW.

INTRODUCTION

Several institutes in Asian countries have taken the initiative in activies for atomic and molecular (A+M) data and databases. As their own research activities and also as members of the international Data Center Network organized by the IAEA A+M Unit, data activies, such as data production, compilation, evaluation, and database development, have been carried out since 1970s. Information exchanges and collaborations between the data centers have also been carried out.

In the panel discussion I introduced the atomic and molecular data activities at National Institute for Fusion Science (NIFS) and Japan Atomic Energy Research Institute (JAERI) as Japanese representatives; and Institute of Applied Physics and Computational Mathematics (IAPCM) as the Chinese representative. Korea Atomic Energy Researh Institute (KAERI) in Korea has A+M data activities and their databases. Dr. Rhee will introduce their databases in this volume.

DATA ACTIVITIES AT IAPCM

Institute of Applied Physics and Computational Mathematics in Beijin, China, has atomic and molecular databases working on Windows98/NT. The databases have retrievable sub systems which can search or calculate data and output results in tables, figures, or ascii files. The databases are as follows: **SPEC**, database of experimental and theoretical spectral data such as wavelengths, transition probabilities, and absorption oscillator strengths; **POTL**, database of ionization potential energies of atoms and ions compiled originally at ORNL; **LEVL**, database of energy

CP543, *Atomic and Molecular Data and Their Applications*, edited by K. A. Berrington and K. L. Bell

Dissociative Collisions

$H + H_2 \rightarrow H + (H^+ + H^+) + 2e$
$H + H_2 \rightarrow H^+ + (H^+ + H^+) + 3e$
$H + H_2 \rightarrow H^- + (H^+ + H^+) + e$
$H + H_2 \rightarrow H + (H + H^+) + e$, via $1s\sigma_g$ State
$H + H_2 \rightarrow H + (H + H^+) + e$, via $2p\sigma_u$ State
$H + H_2 \rightarrow H^+ + (H + H^+) + 2e$, via $1s\sigma_g$ State
$H + H_2 \rightarrow H^+ + (H + H^+) + 2e$, via $2p\sigma_u$ State
$H + H_2 \rightarrow H^- + (H^+ + H)$, via $1s\sigma_g$ State
$H + H_2 \rightarrow H^- + (H^+ + H)$, via $2p\sigma_g$ State
$H^+ + H_2 \rightarrow H^+ + H^+ + H^+ + 2e$
$He^+ + H_2 \rightarrow He^+ + H^+ + H^+ + 2e$
$He^{2+} + H_2 \rightarrow He^{2+} + (H^+ + H) + e$, via $1s\sigma_g$ State of H_2^+ Ion
$He^{2+} + H_2 \rightarrow He^{2+} + (H^+ + H) + e$, via $2p\sigma_u$ State of H_2^+ Ion
$He^{2+} + H_2 \rightarrow He^+ + (H^+ + H)$, via $1s\sigma_g$ State of H_2^+ Ion
$He^{2+} + H_2 \rightarrow He^+ + (H^+ + H^+) + e$
$He^{2+} + H_2 \rightarrow He^{2+} + (H^+ + H^+) + 2e$
$He^{2+} + H_2 \rightarrow He + (H^+ + H^+)$
$H_2^+ + H_2 \rightarrow H^+$ (Projectile)
$H_2^+ + H \rightarrow H^+$ (Projectile)
$H_2^+ + H_2 \rightarrow H$ (Projectile)
...

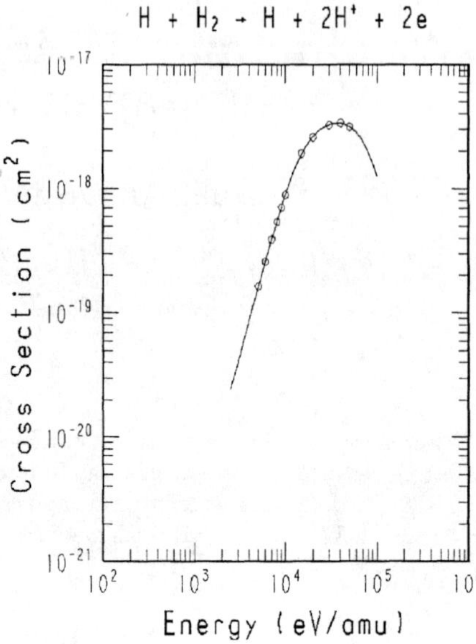

FIGURE 1. An example of process lists in JEAMDL (left). Evaluated cross section of H and H_2 collision from JEAMDL (right).

levels of atoms and ions; **EXCT** and EION, cross sections and rate coefficients in analytical formula for excitation and ionization of ions by electron impact; **BIBL**, bibliographic database; and **OPCT**, database for opacity.

DATA ACTIVITIES AT JAERI

JAERI has worked on A+M data evaluation for fusion plasmas in collaboration with NIST, ORNL, and others. **JEAMDL** is the database of evaluated data given as analytically fitted curves for collisional ionization, molecular dissociation, photon emission, and so on. The list of publications and figures in the publications are seen at URL=http://wwwndc.tokai.jaeri.go.jp/JEAMDL/. There we find lists of A+M processes for each publication and figures are linked to the corresponding process. Figure 1 shows an example of the list of processes and a graph in JEAMDL [1].

DATA ACTIVITIES AT NIFS

NIFS has worked on A+M databases as well as A+M data evaluation for plasma and fusion research. The NIFS database is accessible through WWW and open to registered users free of charge. The registration form can be submitted from the database home page at URL=http://dbshino.nifs.ac.jp/ and a user ID and password will be sent by email later. Since September 1997 the database has started on WWW and number of users has increased. As of 1th March 2000, we have 353 registered users, about half of which are international (Fig.2). Access and retrieve counts to the numerical databases exceed 250 and 800 per month, respectively.

We have four A+M and PWI numerical databases: **AMDIS**, cross sections of ionization, excitation, recombination, and dissociation by electron impact; **CHART**, cross sections of ionization and charge transfer by ion-atom collision; **SPUTY**, sputtering yields for monatomic solid (physical sputtering); and **BACKS**, back scattering coefficients for energy and angle from solid surface. These databases have original numerical data as well as bibliographic information. Numerical data can be shown as a data table or a graph with an empirical curve. Figure 3 shows examples of graphic output from AMDIS ionization database for ionization cross section of carbon atom and ions, and from AMDIS excitation database for excitation cross section for O^{2+} ion.

Database for recombination was added to AMDIS in 1999. We extend the AMDIS database for recombination to also have rate coefficients in order to include experimental data of recombination rate coefficients, such as data from merged beam experiments. Figure 4 shows an example of graphic output from AMDIS recom-

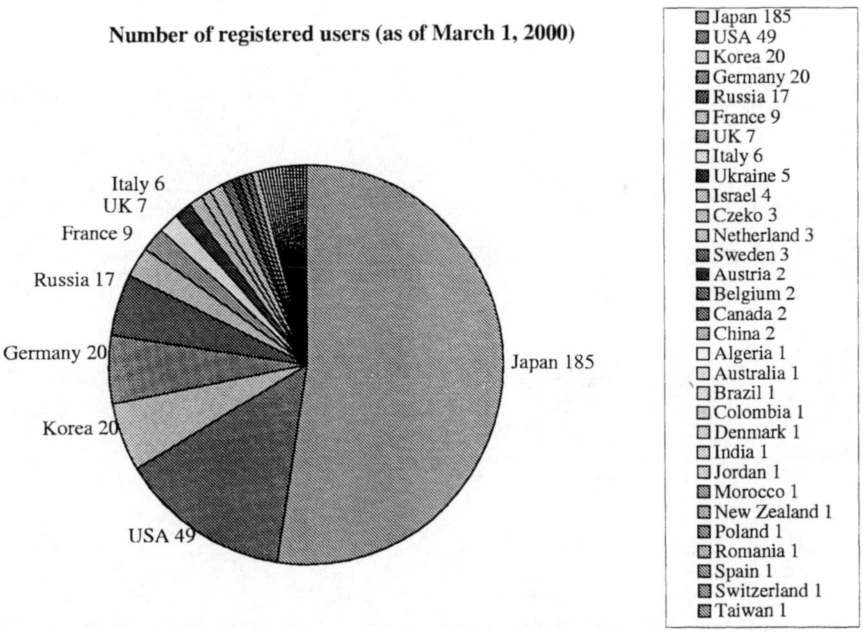

FIGURE 2. Distribution of registered users for the NIFS database as of March 1,2000.

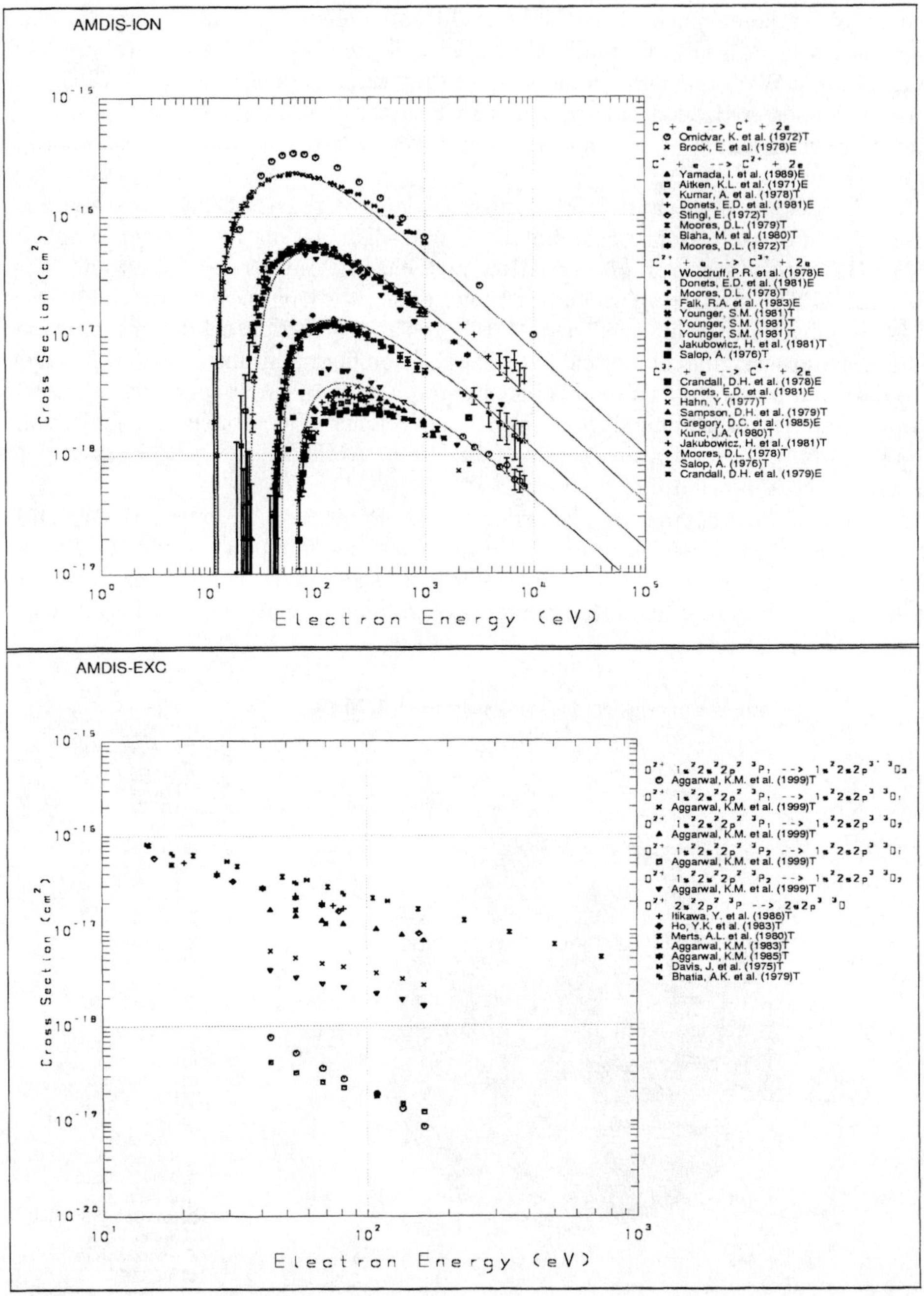

FIGURE 3. Example of AMDIS Ionization output (upper panel) for ionization cross section of carbon atom and ions with the Lotz empirical formulae (solid lines). Lower panel is an example of AMDIS excitation output for excitation cross section of O^{2+} ion.

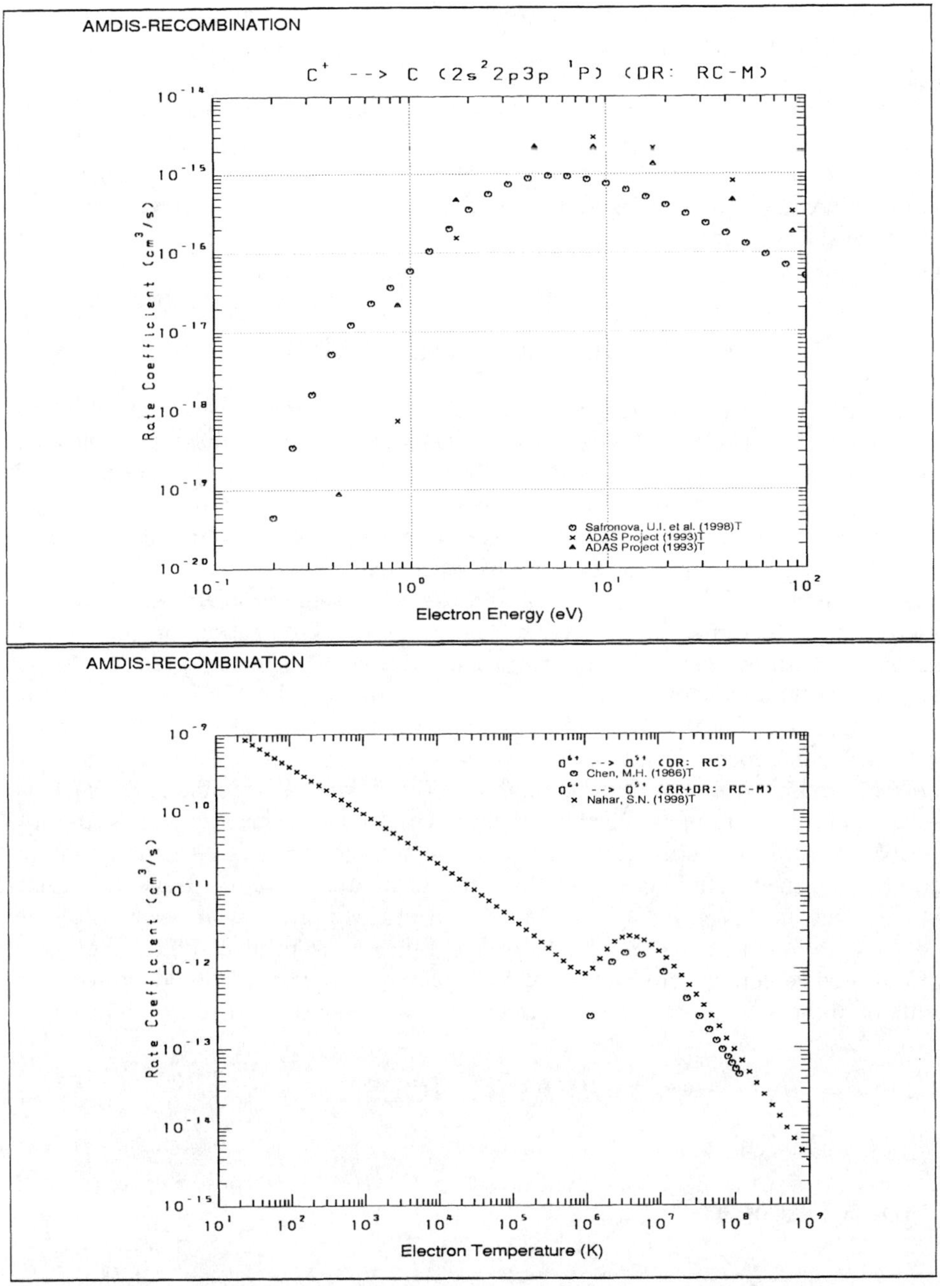

FIGURE 4. Examples of AMDIS Recombination output. Dielectronic recombination rate coefficients from C^{+} ion to an excited state of C atom with a Maxwellian electron velocity distribution (upper panel) and dielectronic and radiative recombination rate coefficients from O^{6+} ion to O^{5+} ion with a Maxwellian electron velocity distribution (lower panel).

bination database for dielectronic recombination rate coefficients from the ground state of C^+ ion to an excited state of C atom. We are collecting data for recombination.

We have worked on data evaluation for A+M data relevant for plasma and fusion research as domestic and international collaborations since 1970s. Evaluated data have been publised as IPP-AM series (-1989) and NIFS-DATA series (1990-). Our databases are useful for storing and comparing data for evaluations. Recent projects include evaluating cross sections and rate coefficients for hydrocarbon [2] and Li ions.

CONCLUDING REMARKS

Feedback from the database users and their cooperation is very important to improve databases. Comments and data submission to our database are welcome. We are especially grateful for data submission and information on numerical data.

To have better A+M data environments for users, we might need to consider about having mirror sites of the databases for faster access through WWW, centralized database of evaluated data for easy use, and a data archive center.

Establishing WWW mirror sites of databases has some problems such as data base software (commercial or free software), or synchronization of versions of databases. Improvement of network technologies could help for faster access to databases without mirror sites.

Centralization of evaluated databases for easy use might be needed. But we need to discuss which form of databases is easier to use for users.

A data archive center would be good for storing data of large file sizes, such as excitation cross sections by electron impact with many resonances, which are not suitable to store in existing databases because of large data size. Also it would be good to keep such data in electronic form for future use. Not all authors keep their data in electronic form for a long time. Data in electronic files are easily lost. It is quite laborious to make an electronic file of data from printed tables or graphs. A data archive center could avoid such problems. Properly it should protect the rights of authors and references to data sets should be attached to the data sets.

REFERENCES

1. Ito R., Tanabe T., Shirai T., and Phaneuf R.A., *JAERI-DATA/Code 96-024* (1996).
2. Janev R., Wang J.G., Kato T., and Kato M., *ICAMDATA2000* poster paper (No.4 of Tuesday) (2000).

Spectroscopic Measurements and AMO Data Centre in KAERI

Yongjoo Rhee, S. K. Kim, H. M. Park and Jongmin Lee

Laboratory for Quantum Optics
Korea Atomic Energy Research Institute
P.O.Box 105, Yusong, Taejon, 305-600, Korea

Abstract. An introduction to the AMO (Atomic, Molecular, and Optical) database system in KAERI is given with some experimental aspects related to the AMO data production. Data sources, constructing concepts and current status of the database are described. An example of measured data which have been compiled internally is given.

INTRODUCTION

Since the complete establishment of the TCP/IP protocol in the 1980's the internet technologies have been developed very fast and almost all the computers in the world connected to internet are using HTTP technology, commonly known as world-wide web (WWW). The information flow in the AMO research areas has also been affected by this electronic breakthrough and *cyber*-libraries based on the www technology are now emerging in the world. However, there must be data sources or producers somewhere for the database to be updated. It is understood that the balance between the data production and data compilation are very important in the database management. In Korea, the necessity of AMO data in the scientific community caused a natural notion of AMO database and KAERI established a database system, with a main application to the high precision trace analysis for the nuclear safety, as well as to a fusion science and AMO studies.

AMO DATA PRODUCTION

For the production of high resolusion atomic parameters, doppler-free spectroscopy experiments have been performed with a single mode diode laser system. Fig. 1 shows the schematic diagram of spectroscopy experimental setup. The wealth of spectral lines of Sm (Samarium) has attracted many attentions and spectroscopry parameters such as autoionization levels, isotope shift and hyperfine structures have been investigated repeatedly.[1] [2]

In our experiments, atomic vapor of Sm metal was produced in a long metal tube vapor cell which generated an isothermal vapor of uniform density.[3] Ar gas was used as a buffer gas to prevent the Sm vapor from coating the internal

CP543, *Atomic and Molecular Data and Their Applications*, edited by K. A. Berrington and K. L. Bell

surface of windows. The temperature of the vapor is about 700 °C producing atomic vapor of density of 3×10^{12} cm^{-3}. After the laser beam was split into a strong pump beam and a weak probe beam, the two divided beams were aligned to propagate through the Sm vapor in an opposite direction to each other. The intensity of pump beam was modulated by a mechanical chopper to generate a modulated signal to a lock-in amplifier, with a frequency of 3.8 kHz. The probe beam, after interaction with atoms, was measured with a photo-diodes and analyzed with a digital lock-in amplifier combined with the transmission signal from a Fabry-Perot interferometer. By scanning the diode laser wavelength over the range of 635 - 690 nm, we could obtain 20 Doppler-free spectra of transition lines from $4f^6 6s^2\ ^7F_J$ ($J = 0$ to $J = 6$) to $4f^6 6s6p\ ^9D^o$, $^9F^o$, and $^7G^o$ levels with a resolution better than 8MHz. The specific mass shift, field shift, and changes in the mean square nuclear charge radii were evaluated by King plot analysis.

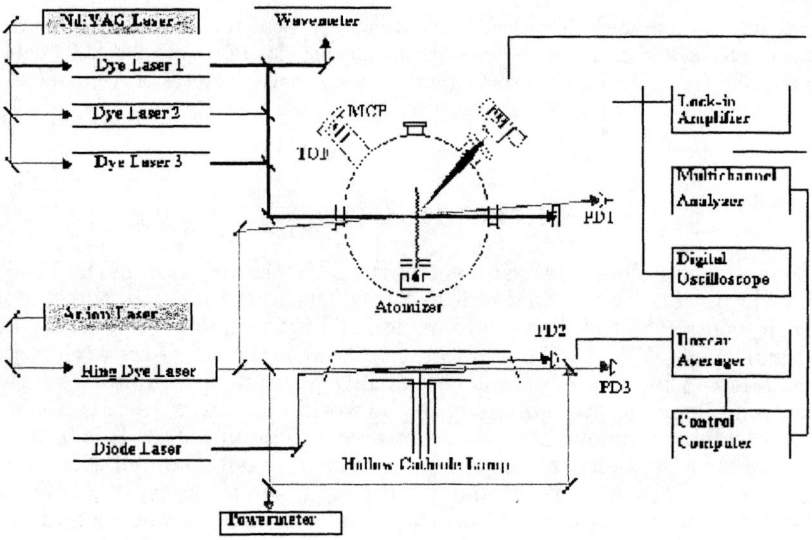

Fig. 1 Experimental setup for high resolution spectroscopy

DATA COMPILATION

AMODS (Atomic, Molecular, and Optical Database System) accessible through http://amods.kaeri.re.kr (Fig. 2) has been established since March 1997 after about one year of internal preparation in KAERI. AMODS is first launched as a data source for the spectroscopy community of Korean scientists to support people who need AMO data and it is still growing to be copious. AMODS is now comprised of 8 separate sub-databases; i.e. atomic spectral lines (ASL), transition probabilities (TP), atomic energy levels (AEL), atomic transition lines (ATL), ALADDIN ONLINE execution (ALAD), fundamental constants (CONST), electron impact excitation cross sections (IMPACT), and internal database (ID), among which the last one (ID) is for internal use only and can be accessed only by some specific IP addresses.

In the establishment of AMODS, the basic principles of construction are to make it 1) easily and directly accessible for numerical data with minimum number of clicks, 2) a preferred system by experts who are fairly familiar with the meaning of search results to minimize the redundant helps, 3) a DB with minimum decorative graphics to maintain the best speed performance, and 4) secure against curious hackers.

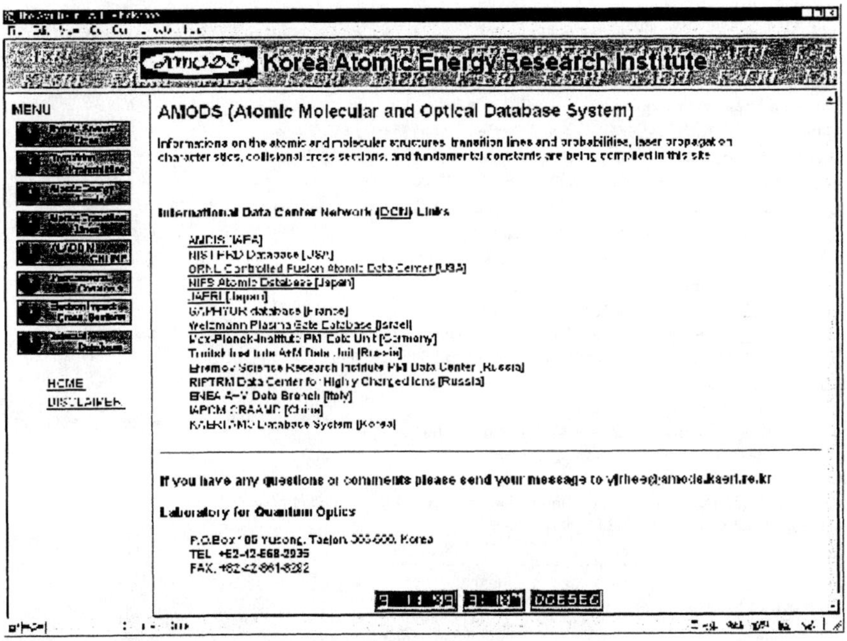

Fig.2 AMODS homepage

The original archived numerical data of ASL, TP, AEL and ATL have been obtained from the CDS database of Strasbourg University in France, followed by stratification and structural arrangement accoding to atomic elements in a periodic table style so that searchable elements can be easily chosen by users. The ASL database contains spectral information on most of the atomic elements in the periodic table and is most frequently accessed by users. It can give the first step spectroscopic information for AMO studies.

Fortran source codes and related data files of ALADDIN database have been acquired from A+M Data Unit of IAEA. Two of the original fortran files have been slightly modified in the I/O routines so that they can be executed from a web browser communicating with CGI programs of web server to obtain the calculating parameters from a user. Numerical search results can be output as an ascii format or graphic format. Fig 3 shows the initial menu screen of the ALADDIN DB. Graphical output data of ALADDIN ONLINE excution is generated on-the-fly by the gnuplot graphics.

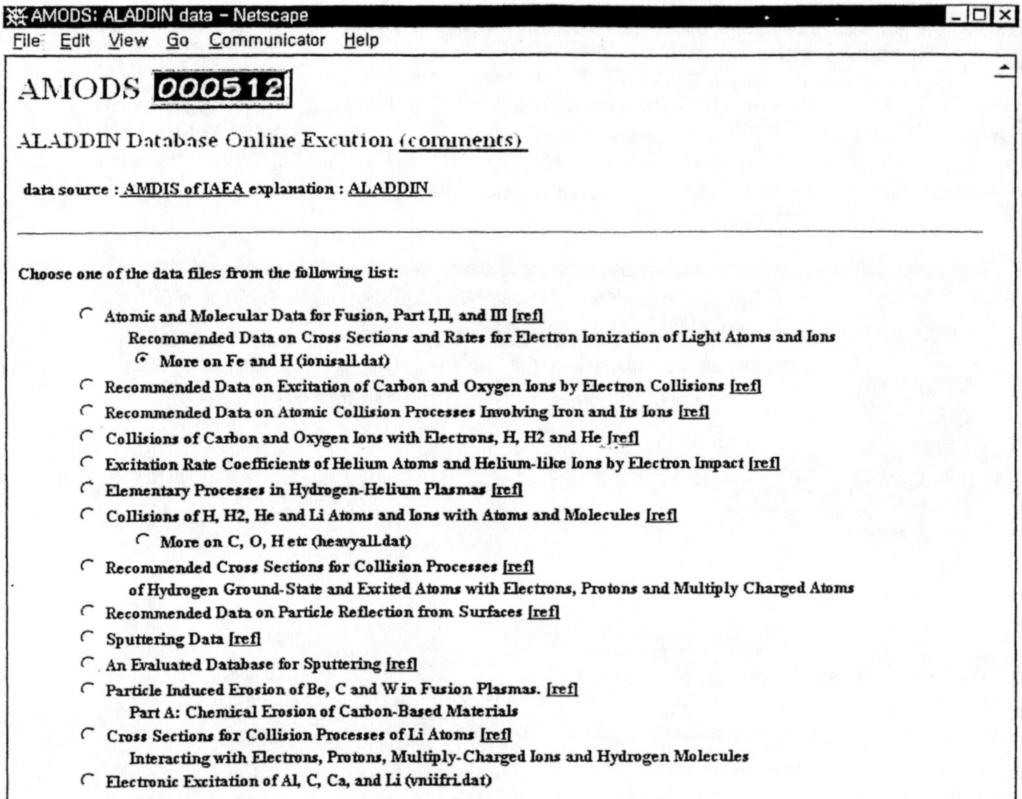

Fig.3 ALADDIN ONLINE screen and input data files

Some fundamental constants recommended by CODATA86 have been compiled in Acrobat pdf format for easy reference of contants with high degree of resolution. This DB will be updated when the new CODATA values become available.

Electron impact excitation cross-sections of molecules are being compiled in IMPACT. Cross-sectional data based on cubic interpolation of the oxygen molecule obtained in the crossed-beam experiments has been compiled[4] and still more data of other molecules are being worked on. The search results can be seen as a graphic format.

The internal database is access-controlled both by perl script and by the access-control file of the web server since it contains some of our raw experimental results which have not been published yet and some copy-right sensitive information. A perl script makes the menu of the home page look different according to the connecting web browser's IP address and the internal database can not be accessed by the users outside of specific IP address ranges. Fig 4 is a compiled data of isotope shift of Sm I measured with diode laser system by Doppler-free spectroscopy in our lab.

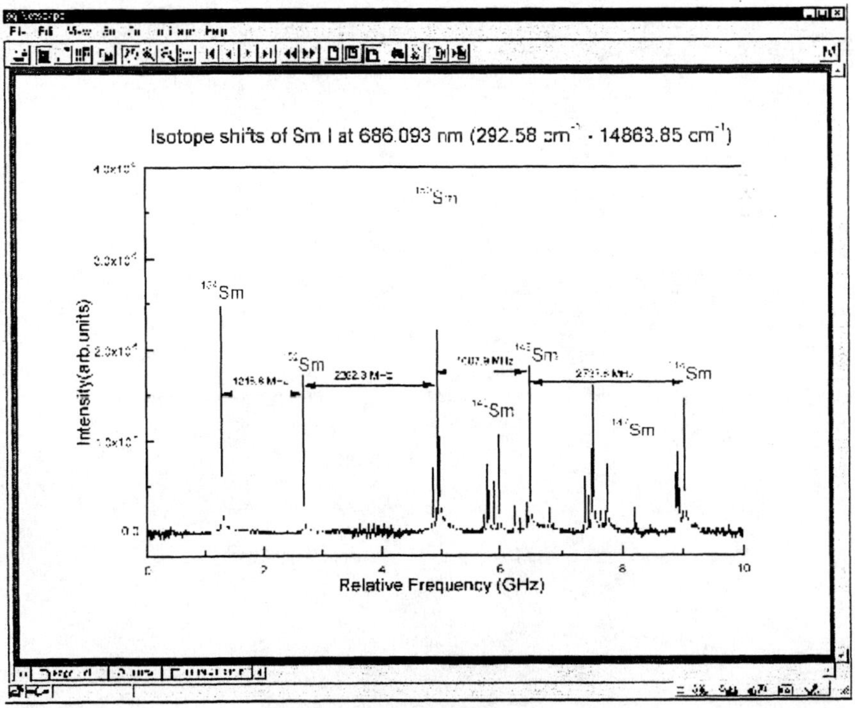

Fig.4 Isotope shift of Sm I

CONCLUSIONS

Through AMODS, we have tried to make international collaboration on the AMO databases, resulting in being a mirror site of NIST PRD database. NIFS in Japan and AMDIS at IAEA are also good collaborators with KAERI in the improvement of AMO database systems.

AMODS is continuously evolving toward the completion of a database of numerical data on atoms, molecules and quantum optics as well as toward the highest availability in the internet environment.

REFERENCES

[1] Park, H.M., Kim, H. C.,Yi, J. H., Han, J. M., Rhee, Y. J., and Lee, J, *J. Kor. Phy. Soc.*, **33**, No 3, 288, (1998), Park, H. M., Kim, H. C., Yi, J. H., Han, J. M., and Lee, J., *J. Kor. Phy. Soc.*, **30**, No 2, 453 (1997)

[2] Martin, W. G., Zalubas, R., and Hagan, L., *Atomic Energy levels - The rare earth elements*, Natl. Stand. Ref. Data Ser. **60** (1978)

[3] Park, H. M., Lee, M., Jung, E. C., Yi, J. H., Rhee, Y. J., and Lee, J., *J. Opt. Soc. Am. B*, **16**, No 7, 1169 (1999)

[4] (a) Shyn, T. W. and Sharp, W. E., *Phys. Rev. A*, **26**, No 3, 1369 (1982), (b) Shyn, T. W. and Sweeney, C. J., *Phys. Rev. A*, **48**, No 2, 1214 (1993), (c) Shyn, T. W. and Sweeney, C. J., *Phys. Rev. A*, **47**, No 2, 1006 (1993). (d) Shyn, T. W., Sweeney, C. J., Grafe, A., and Sharp, W. E., *Phys. Rev. A*, **50**, No 6, 4794 (1994), (e) Shyn, T. w., Sweeney, C. J., and Grafe, A., *Phys. Rev. A*, **49**, No 5, 3680 (1994), (f) Shyn, T. W. and Sharp, W. E., *Phys. Rev. A*, **43**, No 5, 2300 (1991)

The NIST Atomic Spectra Database

Wolfgang L. Wiese

National Institute of Standards and Technology, Gaithersburg, MD 20899–8420 USA

Abstract. A summary of the contents and features of the new NIST Atomic Spectra Database is presented. All material in this large database has been obtained from critical data assessments done over the last 30 years. Only reference data, i.e., data fulfilling certain quality requirements, are included. For each spectral line and energy level, *one* numerical value––the "best" available value––is tabulated.

The U. S. National Institute of Standards and Technology (NIST) has issued databases on atomic spectroscopy on the World Wide Web (WWW) since the spring of 1995. The first version (1.0) was relatively incomplete, with numerous gaps between covered elements and spectra. Several smaller improvements were undertaken in the following years (versions 1.1 to 1.3). But in the meantime, work also started on a much larger database with a completely new search engine and WWW interface. This new, greatly expanded version 2.0 was released in March 1999, it is called the NIST Atomic Spectra Database (ASD) [1] and is now the principal NIST spectroscopic database. It contains critically evaluated compiled reference data for 900 spectra and includes at least some data on all natural elements as well as some man–made elements, from Hydrogen (atomic number Z = 1) through Einsteinium (Z = 99). Specifically, it comprises 70000 energy levels, 91000 wavelengths of lines ranging from about 1Å to 2000μm, and 45000 transition probabilities (oscillator strengths). Wavelengths for prominent observed transitions are included for all 99 elements. Classified lines with transition probabilities are included for most spectra of H through Ni (Z = 1–28). For heavier elements, wavelengths and transition probabilities are given for prominent lines of the neutral atoms and the first four ions. Rather comprehensive lists of the wavelengths of observed lines are included for all spectra of Mg, Al, S, Sc, plus Be I, O II and Ne I. Energy level data are presented for most spectra of Hydrogen through Krypton (Z=1–36), plus the first five spectra of the lanthanides (Z=57–71).

The Atomic Spectra Database has been set up specifically to handle queries on energy levels and transitions (lines), and it offers a large range of options and selections to be specified by the user. Thus, for queries on spectral lines, the transitions may be arranged by wavelength or in multiplet order, they may be listed in Angström units or nanometers, and given in air or vacuum. The transition strength may be displayed by any combination of oscillator strength (or f–value), log gf, transition probability and line strength, and an estimated

CP543, *Atomic and Molecular Data and Their Applications,* edited by K. A. Berrington and K. L. Bell
2000 American Institute of Physics 1-56396-971-8

ASD DATA — INFORMATION — LINES LEVELS List of SPECTRA GROUND STATES & IONIZATION ENERGIES Bibliography Help

NIST Atomic Spectra Database Lines Data (Wavelength Ordered)
Wavelength=2900 Å, ± 100 Mg I
16 Lines of Data Found (page 1 of 2)

Wavelength Air (Å)	Rel. Int.	A_{ki} ($10^8\ s^{-1}$)	Acc.	E_i (cm^{-1})		E_k (cm^{-1})	Configurations	Terms	J_i-J_k	Type	TP Refs.	Line Refs.
2 809.756	3			47 957.045	-	83 536.84	$3s3d - 3p3d$	$^3D - {^3D^o}$	3-3			R1
2 811.048	2*bl*			47 957.027	-	83 520.47	$3s3d - 3p3d$	$^3D - {^3D^o}$	2-2			R1
2 811.780	1			47 957.058	-	83 511.25	$3s3d - 3p3d$	$^3D - {^3D^o}$	1-1			R1
2 846.7167	12	1.5e-01	D-	21 850.405	-	56 968.271	$3s3p - 3s5d$	$^3P^o - {^3D}$	0-1		1s	R1
2 848.344		1.1e-01	D-	21 870.464	-	56 968.271	$3s3p - 3s5d$	$^3P^o - {^3D}$	1-1		1s	
2 848.3456	14	2.1e-01	D-	21 870.464	-	56 968.248	$3s3p - 3s5d$	$^3P^o - {^3D}$	1-2		1s	R1
2 851.652		7.6e-03	E	21 911.178	-	56 968.271	$3s3p - 3s5d$	$^3P^o - {^3D}$	2-1		1s	
2 851.654		6.8e-02	D-	21 911.178	-	56 968.248	$3s3p - 3s5d$	$^3P^o - {^3D}$	2-2		1s	
2 851.6562	16	2.7e-01	D-	21 911.178	-	56 968.218	$3s3p - 3s5d$	$^3P^o - {^3D}$	2-3		1s	R1
2 852.1261	50*g*	4.95e+00	B	0.000	-	35 051.264	$3s^2 - 3s3p$	$^1S - {^1P^o}$	0-1		3,4,5	R1
2 902.9255	2			21 870.464	-	56 308.381	$3s3p - 3s5d$	$^3P^o - {^1D}$	1-2			R1
2 906.3617	4			21 911.178	-	56 308.381	$3s3p - 3s5d$	$^3P^o - {^1D}$	2-2			R1
2 915.453	3*d*			46 403.065	-	80 693.01	$3s3d - 3p3d$	$^1D - {^1D^o}$	2-2			R1
2 936.741	10	1.7e-02	D-	21 850.405	-	55 891.80	$3s3p - 3s6s$	$^3P^o - {^3S}$	0-1		1s	R1
2 938.473	12	5.2e-02	D-	21 870.464	-	55 891.80	$3s3p - 3s6s$	$^3P^o - {^3S}$	1-1		1s	R1

uncertainty is presented. Energy level information may be either included or suppressed with the line data. Of course, energy level data may also be displayed separately. ASD includes a comprehensive bibliography of the original literature sources used for the data tables, and it also includes an extensive help file with a compendium on atomic spectroscopy. Some sample outputs are shown in Figures 1, 2, and 3, which give data displays in formats similar to those in the NIST reference data books on energy levels [2], spectra [3] and transition probabilities [4].

FIGURE 1. Sample ASD output of transition data from a single spectrum, in wavelength order. A detailed explanation of the symbols, code letters, etc. is given in the Help page of ASD [1].

ASD | DATA: LINES, LEVELS | INFORMATION: List of SPECTRA, GROUND STATES & IONIZATION ENERGIES, Bibliography, Help

NIST Atomic Spectra Database Levels Data
Mg I   267 Lines of Data Found (page 1 of 35)

Configuration	Term	*J*	Level (cm^{-1})	Leading percentages			
$3s^2$	^1S	0	0.000	93	7	$3p^2$	^1S
$3s3p$	$^3P°$	0	*21 850.405*	98	1	$3p3d$	$^3P°$
		1	*21 870.464*	98	1	$3p3d$	$^3P°$
		2	*21 911.178*	98	1	$3p3d$	$^3P°$
$3s3p$	$^1P°$	1	*35 051.264*	94	6	$3p3d$	$^1P°$
$3s4s$	^3S	1	41 197.403	98	2	$3d^2$	^3S
$3s4s$	^1S	0	43 503.333	96	2	$3d^2$	^1S
$3s3d$	^1D	2	46 403.065	77	22	$3p^2$	^1D
$3s4p$	$^3P°$	0	*47 841.119*	99	1	$3p3d$	$^3P°$
		1	*47 844.414*	99	1	$3p3d$	$^3P°$
		2	*47 851.162*	99	1	$3p3d$	$^3P°$
$3s3d$	^3D	2	47 957.027	98	1	$3p4f$	^3D
		3	47 957.045	98	1	$3p4f$	^3D
		1	47 957.058	98	1	$3p4f$	^3D

FIGURE 2. Sample ASD output of energy levels for a specified spectrum. A detailed explanation of the symbols, code letters, etc. is given in the Help page of ASD [1].

ASD — DATA: LINES · LEVELS — INFORMATION: List of SPECTRA · GROUND STATES & IONIZATION ENERGIES · Bibliography · Help

NIST Atomic Spectra Database Lines Data (Wavelength Ordered)
Wavelength=2850 Å, ± 2
62 Lines of Data Found (page 1 of 4)

Spec.	Wavelength Air (Å)	Rel. Int.	A_{ki} (10^8 s^{-1})	Acc.	E_i (cm^{-1})		E_k (cm^{-1})	g_i-g_k	Type	TP Refs.	Line Refs.
W I	2 848.02	810									1
Th I	2 848.084	100									434
Fe II	2 848.106	15	7.0e-01	D	44 784.761	-	79 885.493	6-6		7	56
Sc IV	2 848.159	7			448 607.8	-	483 708.0	7-9			S1
Zr II	2 848.19	210									1
Mo II	2 848.23	1700									1
Os I	2 848.25	420									1
La IV	2 848.30	50000*w*									79
Fe II	2 848.320	15	1.1e+00	D-	45 079.879	-	80 177.975	6-4		3n	56
Mg I	2 848.344		1.1e-01	D-	21 870.464	-	56 968.271	3-3		ls	
Mg I	2 848.3456	14	2.1e-01	D-	21 870.464	-	56 968.248	3-5		ls	R1
Er II	2 848.37	270									1
Eu III	2 848.44	10*c*									312
Yb II	2 848.44	100									1
Zr I	2 848.52	350									1

FIGURE 3. Sample ASD output of transition data, from all spectra of all elements, wavelength ordered over a specified range. A detailed explanation of the symbols, code letters, etc. is given in the Help page of ASD [1].

It should be emphasized that the NIST Atomic Spectra Database is a tabulation of critically evaluated reference data. For each spectral line and its strength and for each energy level, the database contains only one "best" value, which is determined from the critical assessment of all available literature data. This best value may be either from a single source or may be the average of several sources. Furthermore, the restriction to "reference" data means that only data fulfilling certain quality requirements are included. This is especially important for transition probabilities (or oscillator strengths) which are in many cases still rather uncertain. Only data with uncertainties estimated not to exceed $\pm$ 50% are normally included.

The data evaluation process is necessarily quite time consuming and involves a detailed analysis of all pertinent literature sources. An elaborate scheme for the critical assessment has been established and is consistently being followed. A key ingredient are technique–specific critical factors that need to be considered for each numerical determination of oscillator strengths. Reference [5] contains a detailed discussion of these factors.

The currently active compilers (part–time or full–time) of the NIST Atomic Spectra Database are: W. C. Martin, A. Musgrove, E. Saloman, C. Sansonetti, and J. Sugar, (Energy Levels and Wavelengths); J. R. Fuhr, D. E. Kelleher, L. Podobedova, W. L. Wiese (Transition Probabilities). The database was developed by D. E. Kelleher, P. Mohr and K. Olsen.

REFERENCES

1. URL address: http://physics.nist.gov/asd.
2. See for example, Sugar, J., and Corliss, C. H., "Atomic Energy Levels for the Iron–Period Elements: Potassium through Nickel," *J. Phys. Chem. Ref. Data* **14**, Suppl. 2, 1985.
3. See for example, Reader, J., Corliss, C. H., Wiese, W. L., and Martin, G. A., "Wavelengths and Transition Probabilities for Atoms and Atomic Ions," *Natl. Stand. Ref. Data Ser., Natl. Bur. Stand.* U.S. **68**, 1980.
4. See for example, Wiese, W. L., Fuhr, J. R., and Deters, T. M., "Atomic Transition Probabilities of Carbon, Nitrogen and Oxygen," *J. Phys. Chem. Ref. Data*, Monograph 7, 1996.
5. Wiese, W. L., and Kelleher, D. E., "The Critical Assessment of Atomic Transition Probabilities," in "*Atomic and Molecular Data and Their Applications* (ICAMDATA 97), edited by P. J. Mohr and W. L. Wiese, AIP Conference Proceedings 434, 1998, pp. 105–118.

The Atomic Data and Analysis Structure

H P Summers and M G O'Mullane

Department of Physics and Applied Physics, University of Strathclyde, 107 Rottenrow, Glasgow G4 0NG, UK

I INTRODUCTION

The Atomic Data and Analysis Structure (ADAS) Project is a shared activity of a world-wide consortium of fusion and astrophysical laboratories directed at developing and maintaining a common approach to analysing and modelling the radiating properties of plasmas. The members of the ADAS project enjoy an identical set of atomic modelling and analysis codes and very large fundamental and derived databases of atomic data collectively called ADAS.

ADAS had its origins at the JET Joint Undertaking experiment of the European fusion programme in 1984. It was decided to develop a substantial capability for modelling and analysing radiating properties of the plasma in support of the powerful spectroscopic equipment to be installed at the experiment. Strategic guidelines were established including identifying a level of sophistication in atomic modelling sufficient for the long term future of the fusion programme, central maintenance of fundamental and derived atomic data, precise and optimised interfacing to experimental diagnostic analysis and quality control of theoretical plasma modelling data from the more testing environment of a large experimental spectroscopy division. In 1988, ADAS was centralized and unified and, within the guidelines, the Atomic Data and Analysis Structure began to evolve into a professionally coded interactive computational support system linking to virtually all spectroscopic diagnostic and plasma modelling activities at JET.

In the early nineties, there was a request from a number of European fusion laboratories and solar astrophysics laboratories that ADAS should be converted into a UNIX based system with IDL as its graphical interface for their use. A pilot study was carried out to demonstrate feasibility funded by Rutherford Appleton Laboratory. In 1994, the ADAS Project was set up with JET's support, funded by subscription of participant laboratories, to oversee the conversion from JET-ADAS to IDL-ADAS by a professional computer consultant firm over a period of two years from 1994-96.

CP543, *Atomic and Molecular Data and Their Applications*, edited by K. A. Berrington and K. L. Bell
© 2000 American Institute of Physics 1-56396-971-8/00/$17.00

TABLE 1. Current institutions that are members of the ADAS Project. The institutional contact is also the represenative on the ADAS Steering Committee.

Institution	Contact	Discipline
JET Joint Undertaking, UK	Dr. K-D Zastrow	Fusion
University of Strathclyde, UK	Prof. H. P. Summers	Astrophysics & Fusion
IPP Garching, Germany	Prof. K. H. Behringer	Fusion
IPP Juelich, Germany	Dr. P. Mertens	Fusion
IPF University of Stuttgart, Germany	Dr. K. Hirsch	Fusion
MPAE, Katlenburg-Lindau, Germany	Dr. D. Innes	Astrophysics
Rutherford Appleton, UK	Dr. J. Lang	Astrophysics
CEA Caderache, France	Dr. R. Guirlet	Fusion
ENEA Frascati, Italy	Dr. R. de Angelis	Fusion
IAP, Technical University, Vienna, Austria	Prof. H-P. Winter	Fusion
CCFM Varennes, Canada	Dr. E. Haddad	Fusion
CFADC, Oak Ridge, USA	Dr. D. Schultz	Fusion & Astrophysics
KTH, Stockholm, Sweden	Prof. E. Rachlew-Kallne	Fusion
General Atomics, San Diego, USA	Dr. T Evans	Fusion
OACT/IACT, University of Catania, Italy	Dr. A. Lanzafame	Astrophysics
JAERI, Naka, Japan	Dr. H. Kubo	Fusion
KAERI, Yusong, Korea	Dr. Y. Rhee	Fusion
NIFS, Toki, Japan	Dr. I. Murakami	Fusion

Membership of the ADAS Project has now increased from the original five to eighteen (see table 1). ADAS is maintained and developed centrally under the management of the University of Strathclyde and the guidance of the Project Steering Committee, on which all ADAS members are represented. The Steering Committee appraises the progress of the project annually, approves new members and makes adjustments to the scientific objectives of the project. Programmers maintain IDL-ADAS on participant laboratory workstations over the internet. New releases, updates and corrections are distributed and installed approximately every six months at the participant laboratories. The ADAS Project holds a workshop annually for sharing of experience in the use of ADAS and for co-ordinating scientific development of ADAS amongst the participants. By this method, a unique facility for the ADAS consortium has been created beyond the scope of any single laboratory.

The Atomic Data and Analysis Structure (ADAS) is an interconnected package of computer codes and data collections comprising three components - an interactive system, a library of key subroutines, and a large database of fundamental and derived atomic data. The interactive part provides immediate display of important fundamental and derived quantities used in analysis together with a substantial capability for preparation of derived data. It also allows exploration of parameter dependencies and diagnostic prediction of atomic population and plasma models. The second part is non-interactive but provides a set of subroutines which can be accessed from the user's own codes to draw in necessary data from the derived ADAS database. The database spans most types of data required for fusion and astrophysical application.

TABLE 2. ADAS series

Series	Content
ADAS1	Entry and validation of fundamental atomic data
ADAS2	Excited state populations of ions in a plasma
ADAS3	Charge exchange related emission
ADAS4	Recombination, ionisation and radiated power
ADAS5	General interrogation programs
ADAS6	Data analysis programs
ADAS7	Creating and Using Dielectronic Data
ADAS8	Structure and Excitation Calculations

The primary ADAS literature comprises a comprehensive user guide, manuals on special topic areas, bulletins which accompany each ADAS release, tutorials and data status documents. This documentation is accessible on the world-wide-web at *http://patiala.phys.strath.ac.uk/adas* and is subject to periodic updating. The ADAS manual is currently at version 2.2 [6] and the most recent bulletin was that of 18 March 2000.

II ADAS CODES

The various ADAS interactive codes are essentially of two types, namely those which are interrogative on the ADAS database (both fundamental and derived parts) and those which execute atomic modelling calculations. ADAS as a whole currently comprises ~70 codes, organised into eight code series (see table 2).

ADAS series 1 is concerned with fundamental, that is essentially individual reaction data. This is to be distinguished from the composite effects of many processes which are the subject of most of the rest of ADAS. It is also to be noted that ADAS series 1 does not include *ab intio* calculation of fundamental data. Rather it allows preparation and assessment of such data from other sources for inclusion in the ADAS database. Unit conversions, data display, comparison with, and adjustment of, simple expected functional forms for the data, Maxwell averaging, control of asymptotic behaviour and interpolation onto user preferred or ADAS standard temperature ranges are the main objectives.

ADAS series 2 is concerned with evaluating excited populations of specific ions in a plasma environment and then their radiation emission. There are five such codes of varying purpose and sophistication (including population solution with an escape factor approach to radiation self-absorption. These codes rely on availability of a reaction rate data collection (called a specific ion file) for the ion in the ADAS database. Series 2 has also provision the supplementation of a raw specific ion file with the extra data on state selective recombination and ionisation necessary for population modelling of plasmas in dynamic states of ionisation. More subtle codes allow manipulation of the resolution level of the populations (e.g. from LSJ to LS or vice versa), variation of the temperatures at which a specific ion file is tabulated and conduct sensitivity analysis of populations to uncertainties of the fundamental data in the specific ion file.

ADAS series 3 is concerned with those situations in a plasma when charge transfer from neutral hydrogen (or its isotopes) or neutral helium in beams is a primary mechanism. Analysis of charge exchange driven emission of hydrogen-like impurities in plasma, emission measure deduction and effective emission coefficient calculation are available in four routines designed for both detailed display and preparation of tabular scans over parameters. Two further routines perform a similar role but for the calculation of beam emission and beam stopping respectively with hydrogen and helium beams.

ADAS series 4 is concerned with the ground and metastable populations of ions in a plasma and particularly their preparation and calculation for dynamical plasma models. It therefore operates with generalised collisional dielectronic recombination and ionisation coefficients, associated power loss coefficients and metastable fractions. It also mediates a transfer from an iso-electronic sequence storage of data which is the preferred choice for atomic data preparation and manipulation to an iso-nuclear sequence storage preferred for applied studies in dynamical plasma models. Programs allow computation of equilibrium or time dependent ionisation balance by drawing on the iso-nuclear data. The scope of series 4 is quite large extending into short wavelength filters modifying observed radiative power, astrophysical contribution function generation and parameterisation of ionisation and recombination.

ADAS series 5 provides a set of programs for interrogating data collections in the fundamental and derived database. The data classes addressed are those which have been found of particular importance for reduction of calibrated observed data.

ADAS series 6 is concerned with analysis of spectral data. This includes full implementation of the method of differential emission measure analysis. Then there are spectral fitting codes - most simply basic Gaussian spectral line fitting but with full statistical analysis of errors. However the main thrust of series 6 is towards the fitting of special spectral features embedded in an unconnected background spectrum. Present codes handle Zeeman/Paschen-Back multiplet features and dielectronic satellite line features.

ADAS series 7 is dedicated to the calculation of fundamental dielectronic recombination, ionisation and excitation/auto-ionisation data including their organisation for applications. The series also includes special population modelling code which are similar to those of series 2 codes but extended to treat the doubly excited populations and associated emissions occurring because of dielectronic recombination.

ADAS series 8 is dedicated to providing ADAS with a capability for the calculation of fundamental atomic energy level data together with basic radiative and collisional data. It includes post-processing codes for automatic generation of specific ion files for subsequent excited population modelling.

ADAS provides very large Fortran and IDL subroutine libraries which are available for use in a user's own codes and analysis programs. The IDL subroutines support the user's interface to interactive ADAS and graphical presentation of results. Modular basic atomic calculation routines, utilities and access routines to the ADAS data classes are mostly provided in Fortran. The libraries now include ~700 subroutines.

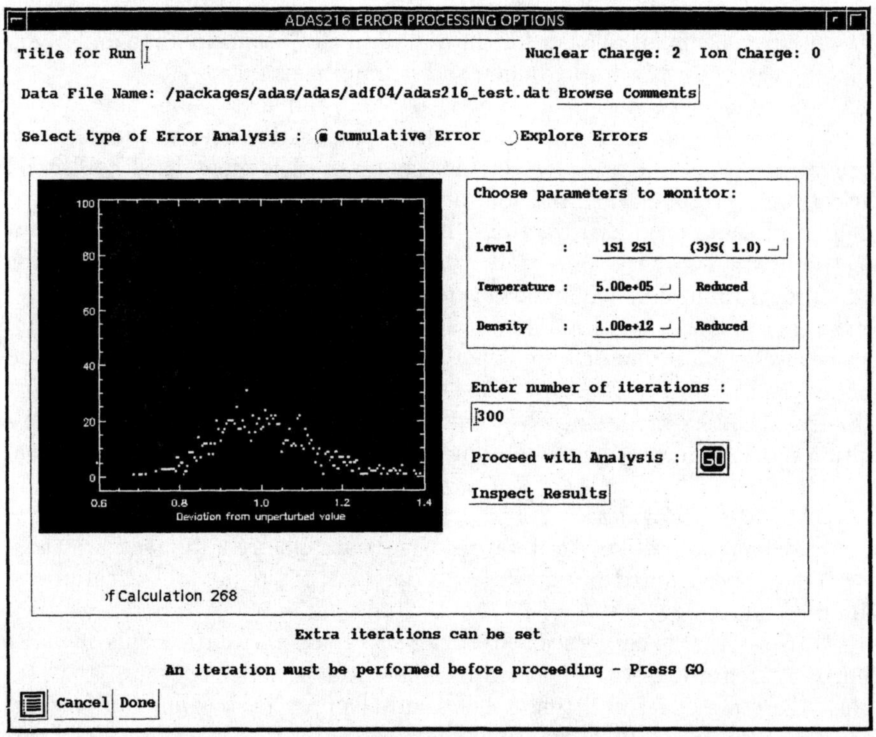

FIGURE 1. The processing screen for the sensitivity analysis code ADAS216. The display shows the build-up of a cumulative statistical error for the $1s2s^3S$ population of He^0 at electron temperature $5.0 \times 10^5 K$ and electron density 1.0×10^{12} cm^{-3}. Random samples are drawn from the individual errors (assumed Gaussian) of each reaction cross-section and the display is of the sample count Vs fractional deviation from the unperturbed population.

A Recent code developments

New substantive codes are currently introduced at the rate of 3-4 per annum. On the astrophysical side, attention has been stimulated by the SOHO satellite and the associated spectral analysis of the solar atmosphere [3]. Maximum likelihood spectral line fitting was introduced, which included detailed statistical analysis, tuned to data flow from SOHO spectrometers (CDS and SUMER). Also, the need to address opacity in the lower solar atmosphere [4] (a situation paralleled in low temperature technical plasmas and some fusion divertor plasmas) led to introduction of an escape factor model and extension of the series 2 population codes [2].

Generalised collisional-radiative population modelling and coefficient production for light elements has been a major theme for both fusion and astrophysics. In this context, the issue of the sensitivity of derived coefficients to uncertainty in fundamental reaction data has led to the introduction of a sensitivity analysis code. It is expected to be the first of a series of such codes. Figure 1 shows a screen shot of this code in action developing a statistical error in an effective emission coefficient. The spectral fitting initiative has been pursued further, but in a context of the fitting of special features - that is connected groups of spectral lines present in a spectrum from which physical fit parameters can be extracted. Zeeman/Paschen-Back and dielectronic satellite lines have so far been addressed. In the fusion context, the capability to model beam stopping and beam emission (of both hydrogen and helium) has been greatly extended with new population and extraction codes introduced in series 3 [1]. A major development has been the decision to introduce a fundamental data calculation capability into ADAS. This is reflected in the creation of the new code series 7 and series 8 for dielectronic recombination and its exploitation and general atomic structure and its exploitation respectively. These are continuing developments.

III ADAS DATA

There are 31 distinct data types in the ADAS databases occupying ~0.7Gbyte (see table 3). These divide broadly in to fundamental data types and derived data types. The latter, which corresponds to more than 70% of the database, are closely tuned to experiment analysis and the needs of plasma models. Many ADAS codes select a fundamental data type initially and then after the substantive calculations create derived coefficient data for direct inclusion in the database. Each data set type has its layout and content precisely described. These prescriptions are called ADAS data formats or *adf* numbers and must be rigidly adhered to in preparation of personal data for use by the ADAS package. The formats apply both to fundamental and derived data and the current types are given in Table 3.

TABLE 3. ADAS data formats.

Series	Content
adf01	bundle-n and bundle-nl charge exchange cross-sections
adf02	ion impact cross-sections with named participant
adf03	direct resolved radiative recombination coefficients
adf04	resolved specific ion data collections
adf05	general z excitation data collections
adf06	general z recombination/ionisation data collections
adf07	direct resolved electron impact ionisation coefficients
adf08	direct resolved radiative recombination coefficients
adf09	direct resolved dielectronic recombination coefficients
adf10	iso-electronic master files
adf11	iso-nuclear master files
adf12	charge exchange effective emission coefficients
adf13	ionisation per photon coefficients
adf14	thermal charge exchange coefficients
adf15	photon emissivity coefficients
adf16	generalised contribution functions
adf17	condensed projection coefficients
adf18	cross-referencing data
adf19	zero-density radiative power data
adf20	G(Te) functions
adf21	effective beam stopping coefficients
adf22	effective beam emission coefficients
adf23	state selective electron impact ionisation coefficients
adf24	state selective charge transfer cross-sections
adf25	driver data-sets for ADAS204 calculation
adf26	bundle-n populations of excited states in beams
adf27	driver data-sets for ADAS701 calculations
adf28	driver data-sets for ADAS702 post-processing
adf29	driver data sets for ADAS707 calculation
adf30	driver data-sets for ADAS708 post-processing
adf31	feature archives

A Recent data developments

The increase in the database in the last few years has been very substantial. Only a few items can be highlighted here. A main addition has been improved generalised collisional radiative data for light elements which span a number of ADAS data formats (adf10, adf11, adf13, adf15, adf17). In the astrophysical domain, production of one- and two-dimensional contribution functions, G(Te), G(Te,Ne), has become routine and extensive (adf20, adf16). For beam penetrated fusion plasma, large tabulations of stopping coefficients and beam emission coefficients for both hydrogen and helium beams have been generated together with useful population and projected population data (adf21, adf22, adf26, adf17). Large extensions to the charge exchange cross-section and derived emission coefficient data have gone in parallel with the beam developments and for divertor spectroscopy [5] (adf01, adf12, adf24). A continuing action has been the extension of the specific ion file collection, especially automatic preparation of these, to act as a basis for refinement in light element and heavy element generalised collisional-radiative modelling (adf04).

IV ADAS SPECIAL PROJECTS

Following planning discussions at ADAS annual workshops, an annual work plan is prepared. Small teams are established to implement parts of the work plan which draw together special interests and best expertise from the ADAS consortium and beyond. Teams meet as required for work periods of two or three weeks, with travel costs funded by ADAS, to see through key steps of the projects. ADAS is also linked to some much larger projects which are separately organised and resourced. These various projects are designed to keep ADAS at the leading scientific edge of atomic physics and spectroscopy of plasmas and to ensure that useful codes and fundamental and derived data for experiment modelling and analysis enter the ADAS system for the benefit of all ADAS members. Information is disseminated by special bulletins. Current projects include:

The GCR Project (D. Brooks - Porto Observatory, R. Dux - IPP Garching, H. Jentschke - Stuttgart, J. Lang - Rutherford Appleton, M. O'Mullane & H. P. Summers - Strathclyde/JET) - Calculation of generalised collisional-radiative coefficients, including effective ionisation, recombination, cross-coupling and emissivity coefficients, for light elements up to neon, to the highest precision, over all density and temperature ranges relevant to fusion and astrophysics.

The Helium beam Project (R. Hoekstra - KVI Groningen, H. Summers & M. O'Mullane - Strathclyde/JET, M. Brix - JET/IPP Juelich, M. Poschek & S. Menhart - TUW Vienna) - Calculation of beam stopping and beam emission coefficients for slow and fast helium beams in fusion plasmas. Singlet and triplet metastable resolution and collisional-radiative projection matrix methods are being used. Plasma analysis using the beams as unrelaxed edge probes and quasi-relaxed deep probes is being implemented.

The Heavy species Project (H. Summers, M. O'Mullane & S. Loch - Strathclyde/JET, P. de France - Louvain la Neuve, G. Senn - Innsbruck, I. Murakami & R. M. More -NIFS, T. Nakano - JAERI, Y. Rhee - KAERI) - Calculation of generalised collisional-radiative coefficients, including effective ionisation, recombination, cross-coupling and emissivity coefficients, for heavy elements in intermediate coupling to the highest precision, over all density and temperature ranges relevant to fusion materials studies and astrophysics.

The Rmax Project (N. R. Badnell - Strathclyde, K. Berrington - Sheffield) Calculation of inner-shell photo-excitation and electron impact excitation of

highly ionised species relevant to soft X-ray spectroscopy of cosmological plasmas. Most recent radiation damped R-matrix methods are being used.
The DR Project (N. R. Badnell - Strathclyde, D. Verner - Kentucky, M. O'Mullane - JET, M. Pindzola - Auburn, D. Griffin - Rollins, T. Gorcyzca) - Calculation of intermediate coupling state selective dielectronic coefficients to the highest quality in a J-resolved metastable picture for ions of elements up to zinc. The work is aimed at deep space astrophysics and modelling of heavy complex species in fusion.

REFERENCES

1. H Anderson, M G von Hellermann, R Hoekstra, L D Horton, A C Howman, R W T Konig, R Martin, R E Olson and H P Summers, Plasma Phys. Control. Fusion 42,??? (2000)
2. K H Behringer, Max-Planck-Institutt-fuer-Plasmaphysik report IPP10/5 (1997)
3. D H Brooks, G A Fischbacher, A Fludra, R A Harrison, D E Innes, E Landi, M Landini, J Lang, A C Lanzafame, S D Loch, R W P McWhirter, H P Summers and W T Thompson, Astron. Astrophys. 347, 277 (2000)
4. D H Brooks, G A Fischbacher, A Fludra, R A Harrison, D E Innes, E Landi, M Landini, J Lang, A C Lanzafame, S D Loch, R W P McWhirter and H P Summers , Astron. Astrophys. 357, 697 (2000)
5. C F Maggi, L D Horton and H P Summers, Plasma Phys. Control. Fusion 42,669 (2000)
6. H P Summers, "The ADAS User Manual", version 2.2 (2000), *http : //patiala.phys.strath.ac.uk/adas/*

TOPbase/TIPbase

C. Mendoza

Physics Center, Instituto Venezolano de Investigaciones Científicas (IVIC), PO Box 21827, Caracas 1020A, Venezuela

Abstract. Current developments that are being carried out within the context of the atomic database service TOPbase/TIPbase are briefly summarized.

INTRODUCTION

TOPbase/TIPbase is an atomic database service that has been available on the Internet since 1993 from both the Centre de Données Astronomiques de Strasbourg (CDS) [1], France, and HEASARC [2], NASA Goddard Space Flight Center, USA. Although the current WWW user interface is browser based, direct access to the original command based interface is still being offered at the CDS through `telnet` at `cdsarc.u-strasbg.fr` (IP=130.79.128.5) with account id= `topbase` and PW = `Seaton+`. TOPbase [3,4] contains term energies, f-values and photoionization cross sections for astrophysically abundant ions (Z =1–14; Z = 16; Z = 18; Z = 20 and Z = 26). They have been computed in LS coupling in the Opacity Project [5]. This database has averaged over 100 calls a month since it became operational. TIPbase (under development) will offer data from the current IRON Project [6], namely level energies, radiative transition probabilities, electron impact excitation collision strengths and rates for fine structure transitions of ions of astrophysical interest, with emphasis on the iron group. In the present paper we briefly describe the current developments that are being carried out to maintain and extend this service.

ATOMIC DATA

The atomic data contained in TOPbase/TIPbase have been calculated by an international group of experts from Canada, France, Germany, UK, USA and Venezuela with more than 15-year experience in large collaborative computational projects. State-of-the-art numerical methods are used such as those implemented in the atomic structure codes CIV3 [7] and SUPERSTRUCTURE [8] and collisional approaches based on the R-matrix method [9–11]. The emphasis is both on accuracy and completeness thus leading to series of papers where such issues have

CP543, *Atomic and Molecular Data and Their Applications*, edited by K. A. Berrington and K. L. Bell

been extensively discussed, namely the series "Atomic Data for Opacity Calculations (I–XXIII)" in the *Journal of Physics B* and the on-going series "Atomic Data from the IRON Project (I–XLIV)" in *Astronomy & Astrophysics Supplement Series.* A measure of the quality of the data produced by the OP and IP is given, for instance, by the large number of f-values that have been selected for a NIST reference compilation [12].

CURRENT DEVELOPMENTS

TOPbase

TOPbase is at present undergoing a major revision of both the data and software in order to release a new version (2.0). Improvements can be summarized as follows:

1. New data for ions with low cosmic abundance, namely P, Cl and K, and for Fe I and Fe II are to be included in the database. This will enhance considerably its completeness being particularly useful for both isoelectronic and isonuclear studies.
2. Data for selected ionic species are being updated, in particular the lowly ionized Fe ions (Fe III–VI) and ions in the He, O and Si isoelectronic sequences.
3. The high-energy tails of the photoionization cross sections for both the ground and excited states are being recalculated using a Distorted Wave code (see [13]). This will be an improvement from the existing extrapolated $1/E^3$ tails that do not display the ionization edges and can be inaccurate.
4. A large number of f-values for Fe VIII–Fe XIII computed with the atomic structure code SUPERSTRUCTURE, referred to as the PLUS data [14], will be included in the database. At present they are available from [15].
5. TOPbase will incorporate versatile facilities for computing and manipulating the opacities computed in the Opacity Project [16–18]. Data for selected mixtures can be retrieved at present from [15].
6. The database will also include radiative acceleration data calculated by Seaton [19] which have been recently used to estimate the diffusion of iron-group elements [20]. The data are current available from [15].
7. The web user interface is to include graphic processing at the client end through the implementation of Java applets.
8. An attempt will be made in the summer of 2000 to interconnect TOPbase with the GAPHYOR atomic database [21] in order to implement inter-database searches.

TIPbase

TIPbase will become operational by the end of 2000. The first version will give access to energy levels, radiative transition probabilities and electron excitation rates related to fine structure IR transitions.

REFERENCES

1. `http://cdsweb.u-strasbg.fr/OP.html`
2. `http://heasarc.gsfc.nasa.gov/topbase`
3. Cunto W., and Mendoza C., *Rev. Mex. Astron. Astrofis.* **23**, 107 (1992).
4. Cunto W., Mendoza C., Ochsenbein F., and Zeippen C.J., *Astron. Astrophys.* **275**, L5 (1993).
5. The Opacity Project Team, *The Opacity Project*, **Vol.1**, Bristol: IOPP, 1995.
6. Hummer D.G., Berrington K.A., Eissner W., et al., *Astron. Astrophys.* **279**, 298 (1993).
7. Hibbert A., *Comput. Phys. Commun.* **9**, 141 (1975).
8. Eissner W., Jones M., and Nussbaumer H., *Comput. Phys. Commun.* **8**, 270 (1974).
9. Burke P.G., Hibbert A., and Robb W.D., *J. Phys. B* **4**, 153 (1971).
10. Burke P.G., and Robb W.D., *Adv. Atom. Molec. Phys.* **11**, 143 (1975).
11. Burke P.G., in *Physics of Electronic and Atomic Collisions*, edited by S. Datz, Amsterdam: North-Holland, 1982, p. 447.
12. Wiese W.L., Fuhr J.R., and Deters T.M., *Atomic Transition Probabilities of Carbon, Nitrogen, and Oxygen - A Critical Data Compilation*, *J. Phys. Chem. Ref. Data* Monograph 7 (1996).
13. Berrington K., Quigley L., and Zhang H.L., *J. Phys. B* **30**, 5409 (1997).
14. Lynnas-Gray A.E., Seaton M.J., and Storey, P.J., *J. Phys. B* **28**, 2817 (1997).
15. `ftp://cdsarc.u-strasbg.fr/pub/cats/VI/89A`
16. Seaton M.J., Yan Y., Mihalas D., and Pradhan A.K., *Mon. Not. R. Astron. Soc.* **266**, 805 (1994).
17. Seaton M.J., *Mon. Not. R. Astron. Soc.* **265**, L25 (1993).
18. Seaton M.J., *Mon. Not. R. Astron. Soc.* **279**, 95 (1996).
19. Seaton M.J., *Mon. Not. R. Astron. Soc.* **289**, 700 (1997).
20. Seaton M.J., *Mon. Not. R. Astron. Soc.* **307**, 1008 (1999).
21. `http://gaphyor.lpgp.u-psud.fr`

Who is Where in Atomic and Molecular Data

J.L. Delcroix, M. Fitaire, D.Humbert, C. Leprince
Centre de Données GAPHYOR, Laboratoire de Physique des Gaz et des Plasmas
Université Paris-Sud Orsay
E-mail 1: gaphyor@lpgp.u-psud.fr E-mail 2: delcroix@supelec.fr

1. Introduction

The basic idea which led to the convening of the present series of conferences was the following: the Community of Atomic and Molecular Data (AMD) has difficulties to survive, because it is scattered all around the world into a large number of small institutions, and moreover there is no national or international Agency in charge of coordinating and funding it (with the relative exception of NIST and IAEA) The idea was then to gather representative people of the three groups participating in the Community:

- AMD Producers (experimentalists and theoreticians)
- AMD Databases Producers
- AMD Users in Universities, National Laboratories and Industrial Companies

It was thought that following these meetings the overall efficiency and possibly the funding of AMD system could be improved. In view of this philosophy we present here a geographical and sociological analysis of the AMD system. This is based on statistics taken from the GAPHYOR Database. One may think that the results are somewhat biased by the policy decisions in GAPHYOR. This is true in some way as we will see, but we believe that, in spite of this, the present analysis is significant indeed. To analyse the time evolution of this photography, we have divided our analysis into three periods of time:

- "Antique times" (before 1980)
- "Middle Age" (1980-1989)
- "Modern times" (after 1990)

The elemental unit in the statistics is ONE GAPHYOR RECORD. Note that it is not a bibliographic reference; it is a data extracted from a paper, and a given paper may generate one, several, and sometimes a large number of records. Let us describe quickly the structure of a GAPHYOR record (for more details please look at the GAPHYOR web site "http:/gaphyor.lpgp.u-psud.fr/"). Every GAPHYOR record includes three kinds of descriptors:

- Physicochemical factual descriptors: section (SE), initial reactants MOL1, MOL2, MOL3, process (PR), final products MOL4, MOL5, MOL6.
- Information and numerical descriptors: INFO, energies of reactants (ENVAL), numerical data (VAL), and remarks (REM).
- Bibliographical descriptors: publication (JO), year (AN), volume (VOL), page (PA), nation (NA), state (ST), and city (CI).

CP543, *Atomic and Molecular Data and Their Applications,* edited by K. A. Berrington and K. L. Bell

The geographical descriptors NA, ST, CI refer to the location of the first author laboratory. Note that it is somewhat biased in favour of large countries because smaller nations authors are often co-workers and hence not the first author of a given paper. They are then not taken into account in the statistics.

2. AMD Producers

Figure 1 below represents the geographical distribution of AMD Producers for the three periods of time defined above, and the overall total. Figure 2 gives the details for European Union. The nation (NA) of a record is taken as the one where lays the first author laboratory.

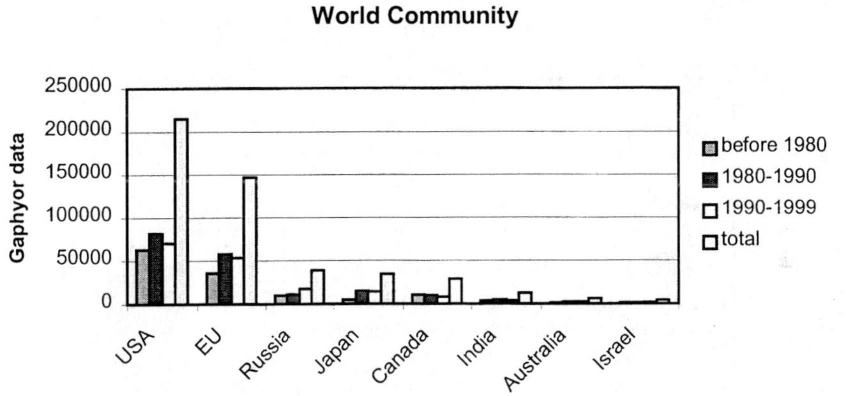

Figure 1: world geographical distribution of AMD producers

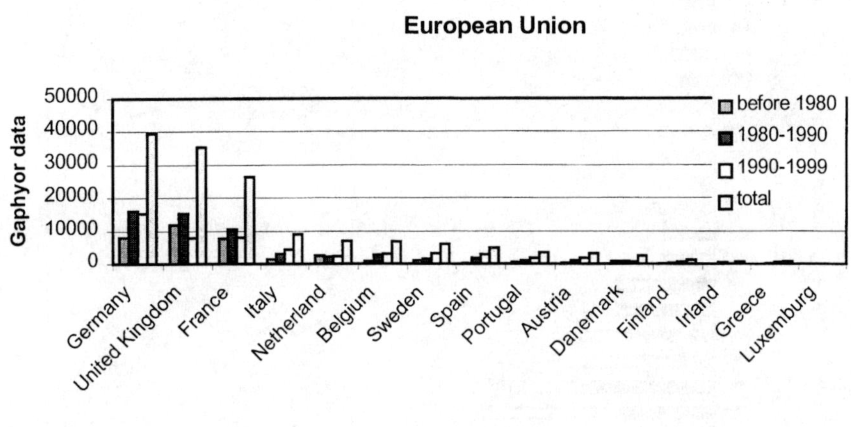

Figure 2: European Union geographical distribution of AMD producers

3. AMD Publishers

AMD are published in four kinds of media:

- Regular Journals (JOUR)
- Laboratory Reports (REPT), including Thesis
- Books (BOOK)
- Conferences (CONF)

Figure 3 shows the distribution of AMD publication into the four categories above. One sees that journals are largely dominant. But this is not the real distribution. It is somewhat biased by the political choice of the GAPHYOR team to concentrate on regular Journals. Figure 4 gives then the statistics of publication inside the main media (journals).

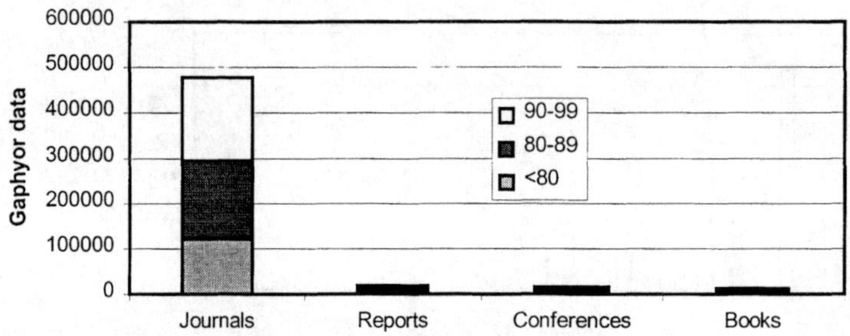

Figure 3: publication distribution into the four categories of media

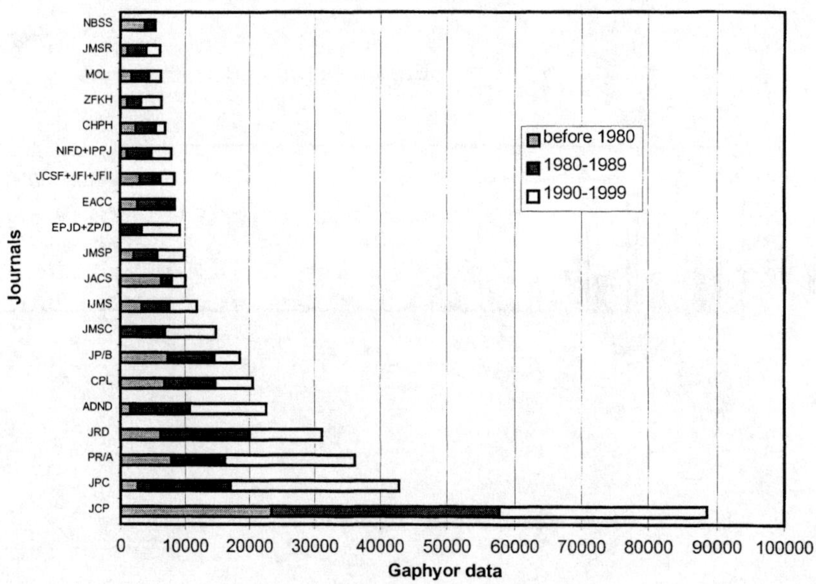

Figure 4: publication distribution into various journals

Many AMD are produced in a given country and published in another one. To analyse this situation we introduce a descriptor NJ (Nation of the Journal). Figure 5 gives then a statistics of records by NJ, i.e. the geographical distribution of AMD publication between nations.

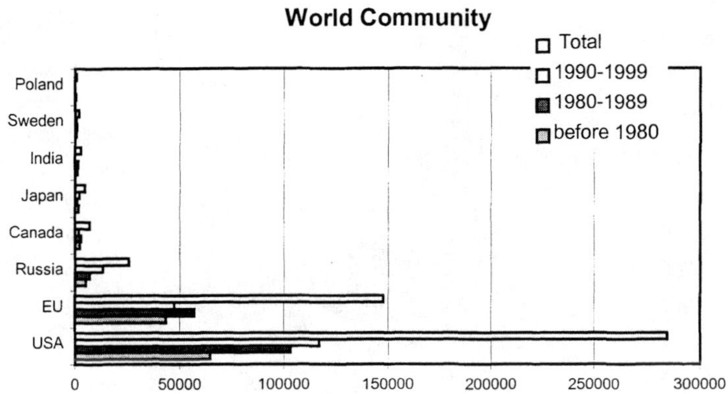

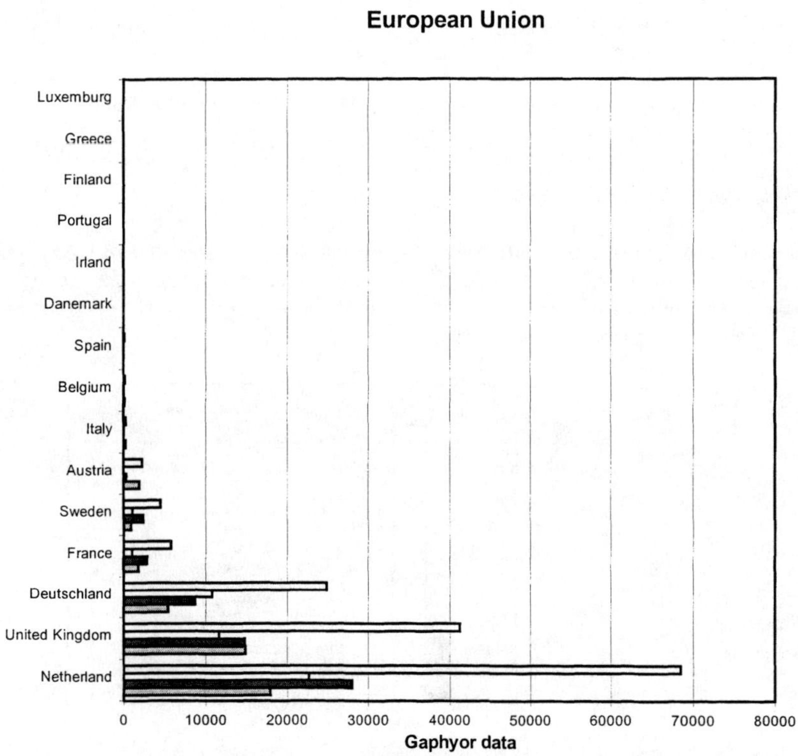

Figure 5: geographical distribution of publishers

Finally figure 6 exhibits the geographical correlations between AMD production and publication.

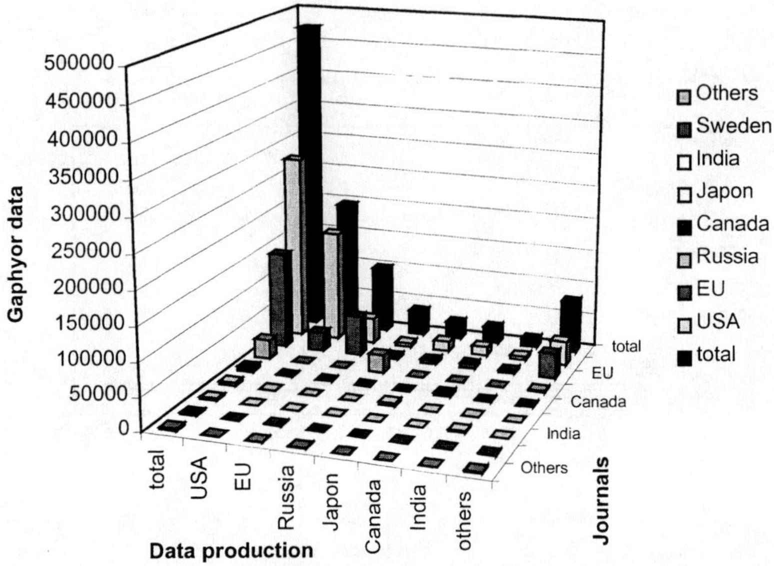

Figure 6: geographical correlations between AMD production and publication

4. AMD Users (GAPHYOR customers)

Let us now analyse the users population. For this we use again the statistics of GAPHYOR users. Our unit is now ONE GAPHYOR REQUEST, which may collect one, several or sometimes a large number of Data Sets. Figure 7 and Figure 8 give again the geographical distribution of users in the world and by nations

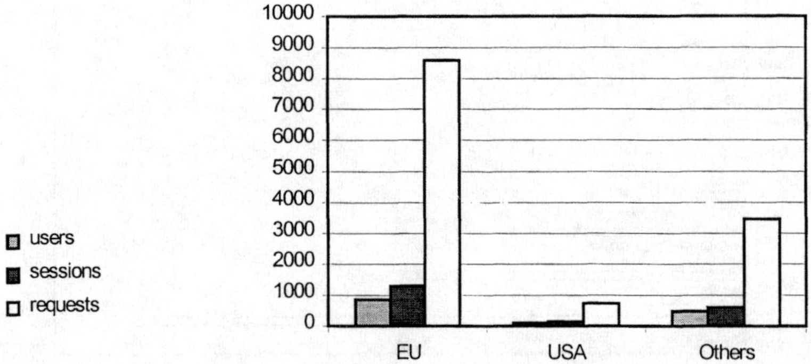

Figure 7: geographical distribution of users

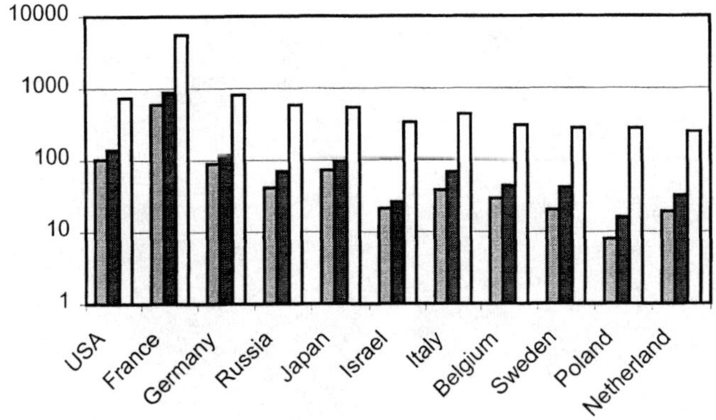

Figure 8: geographical distribution of users by nations

It is finally interesting to analyse the users publication into five categories of organizations:

- EDU: Academic organizations
- GOV:National laboratories and organizations
- COM: Industrial or other companies
- NET: Mainly individuals connected anonymously through a NET company
- XX: Various or unknown

Figure 9 shows the distribution of users between these five categories.

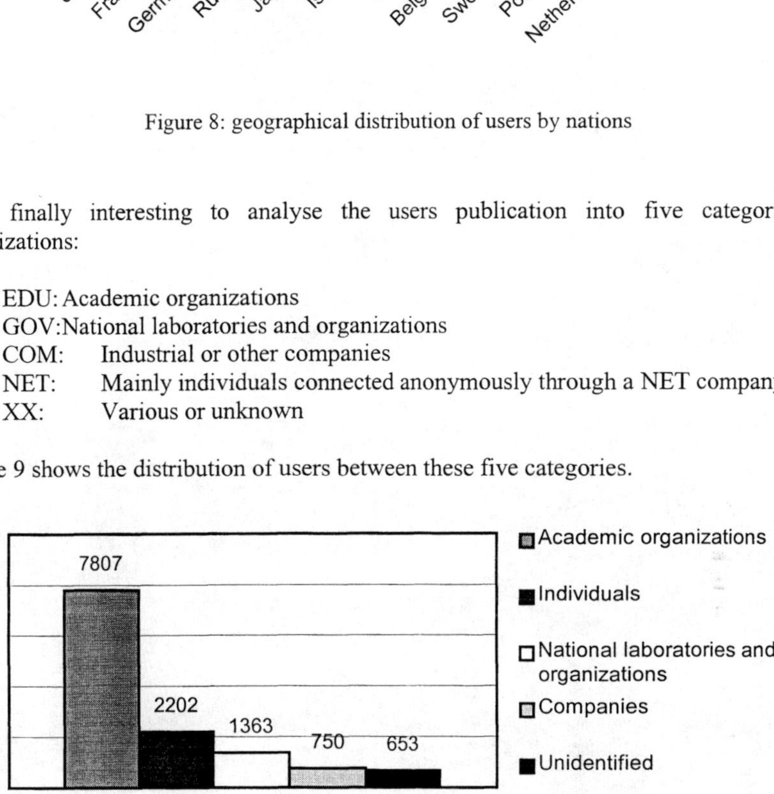

Figure 9: distribution of AMD users by category

Author Index

S

T

U

W

Z